EINSTEIN IN KAFKALAND

BY THE SAME AUTHOR

When I Grow Up:
The Lost Autobiographies
of Six Yiddish Teenagers

The Three Escapes of Hannah Arendt:
A Tyranny of Truth

Kvetch as Kvetch Can: Jewish Cartoons

EINSTEIN IN KAFKALAND

How Albert Fell Down the Rabbit Hole and Came Up with the Universe

KEN KRIMSTEIN

BLOOMSBURY PUBLISHING

NEW YORK · LONDON · OXFORD · NEW DELHI · SYDNEY

BLOOMSBURY PUBLISHING
BLOOMSBURY PUBLISHING INC.
1385 Broadway, New York, NY 10018, USA

BLOOMSBURY, BLOOMSBURY PUBLISHING, and the Diana logo
are trademarks of Bloomsbury Publishing Plc

First published in the United States 2024

Albert Einstein (1879-1955) German-Swiss mathematician, Relativity Einstein in 1905,
photo by Photo12/Universal Images Group via Getty Images;
Franz Kafka photo by Fine Art Images/Heritage Images/Getty Images

ISBN: HB: 978-1-63557-953-6; eBook: 978-1-63557-954-3

Library of Congress Cataloging-in-Publication Data is available

2 4 6 8 10 9 7 5 3 1

Typeset by Katya Mezhibovskaya
Printed and bound in Italy by L.E.G.O. SpA

To find out more about our authors and books visit www.bloomsbury.com
and sign up for our newsletters.

Bloomsbury books may be purchased for business or promotional use.
For information on bulk purchases please contact Macmillan Corporate
and Premium Sales Department at specialmarkets@macmillan.com.

For Alex,
my Prague

CONTENTS

AUTHOR'S NOTE: This Much Is True   ix

OVERTURE   1

I: Down the Rabbit Hole   9

II: The Pool of Tears   27

III: Meet Max Abraham   37

IV: The People are So Happy Here   45

V: Berta Fanta's Mad Tea Party   59

VI: Heeeeeeere's Mileva!   67

VII: Berta Fanta's Mad Tea Afterparty   77

VIII: Let's Bend Light   95

IX: The Crime of the Century   105

X: Albert & Paul's Lost Weekend   125

XI: The Duel of the Pens   146

XII: A Knock at the Door   157

XII: wtf   169

XIV: Einstein's Unified Theory of It All   187

CODA   197

ACKNOWLEDGMENTS   203

FURTHER READING, LISTENING, VIEWING   204

NOTES   210

WHAT THE HELL WAS GOING ON? A TIMELINE   212

AUTHOR'S NOTE: This Much Is True

The world changed, the universe changed, and you changed as a result of the events that follow.

This much is also true.

All the actual characters depicted were alive and in Prague during the fifteen months I describe, and all the actual events follow the sequence as recorded in letters, diaries, news stories, speeches, and personal testimonies.

As for the non-actual characters and events, the ones perhaps reminiscent of a certain "down the rabbit hole" fantasy of the 1860s, well, who can say?

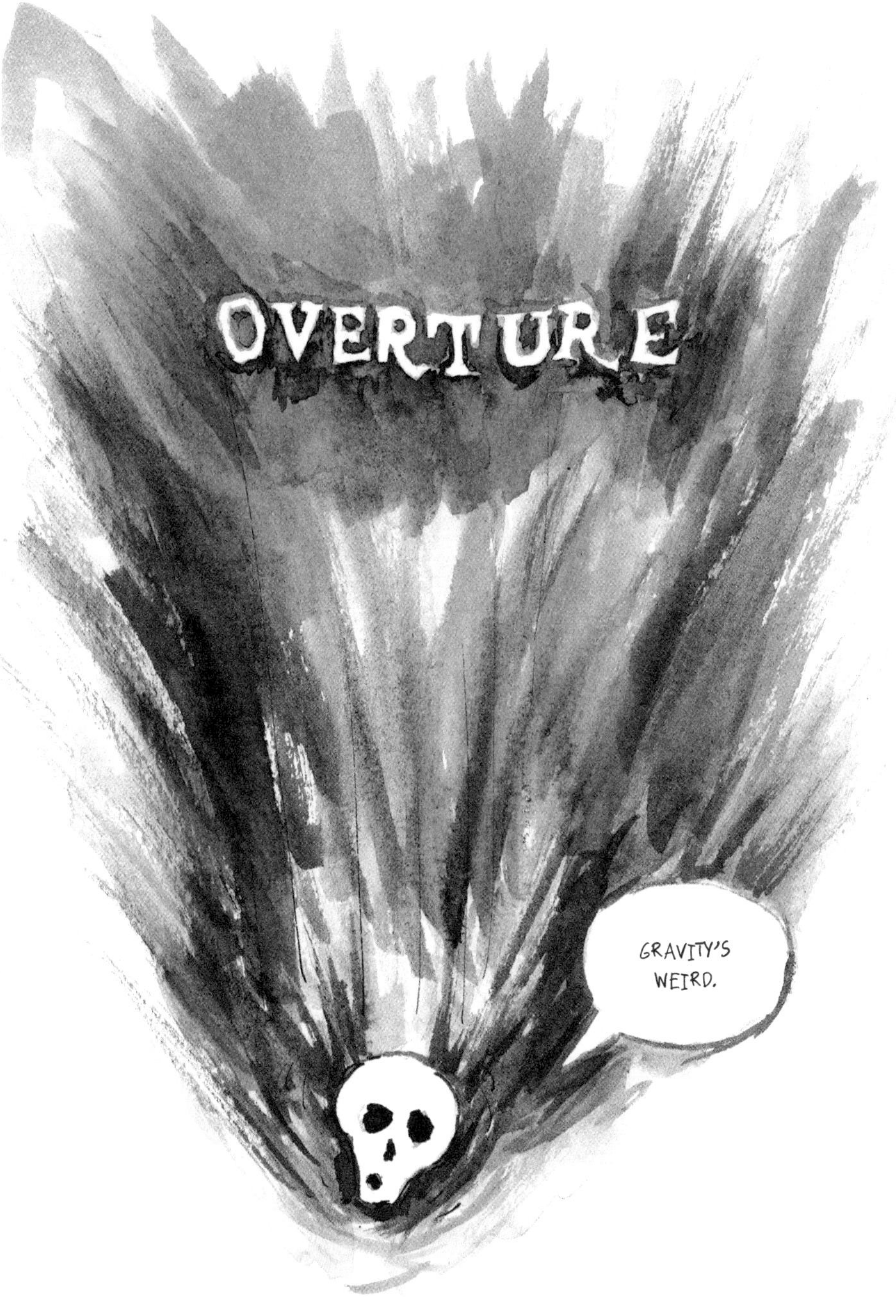
OVERTURE
GRAVITY'S WEIRD.

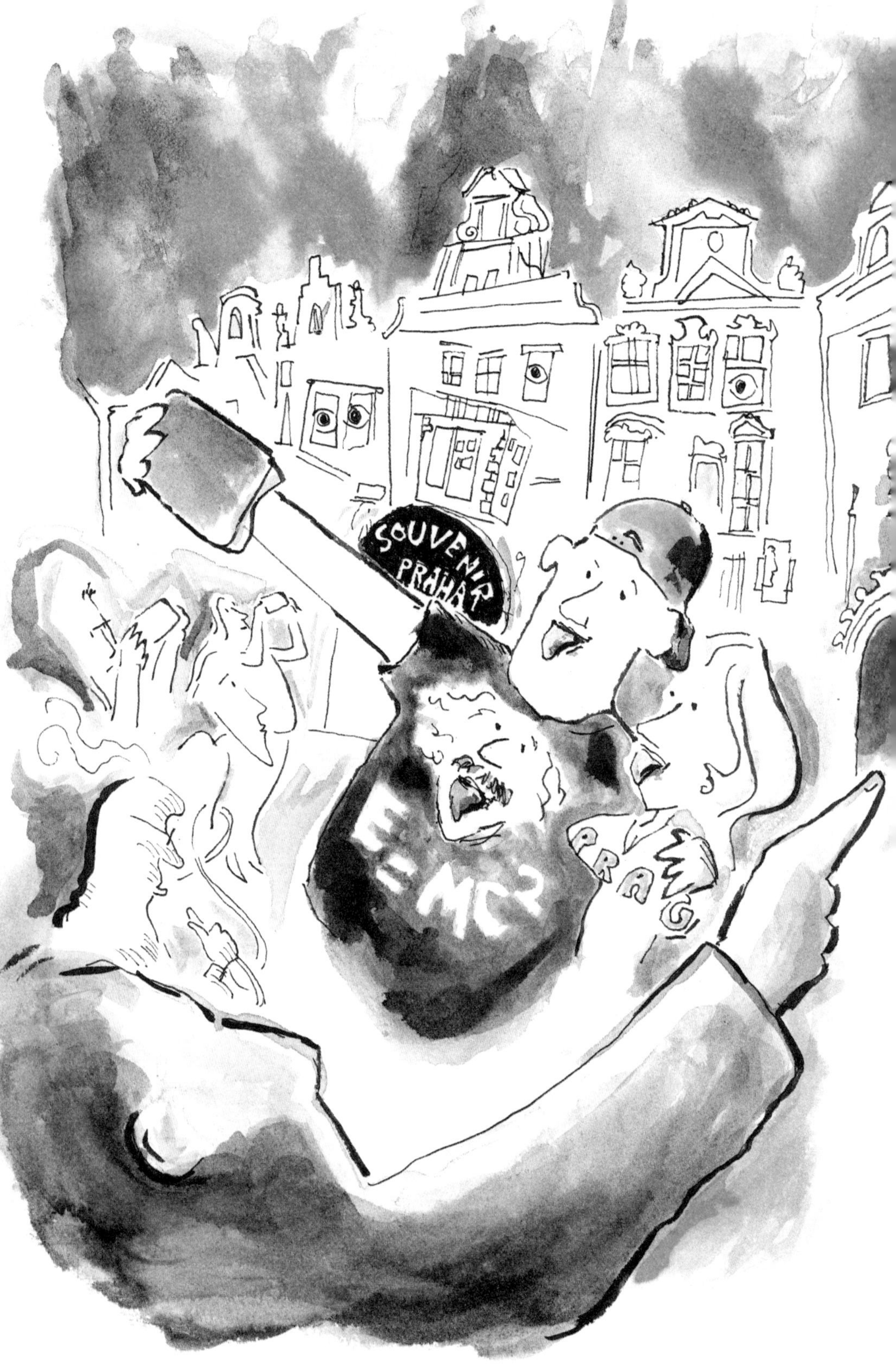
SOUVENIR PRAHA
E=MC2
PRAG

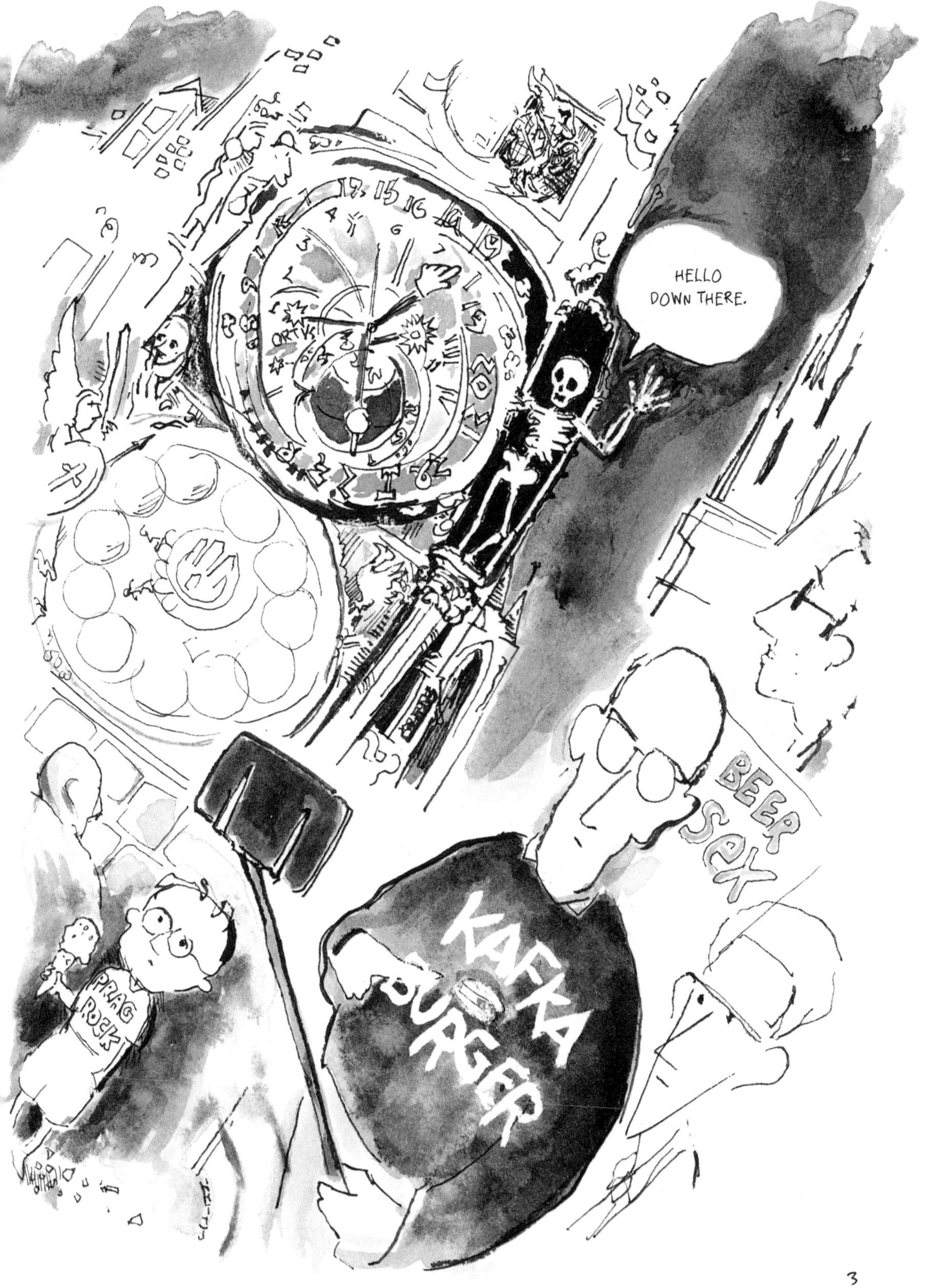
HELLO DOWN THERE.
ORTVS
BEER
SEX
KAFKA BURGER
PRAG ROCK

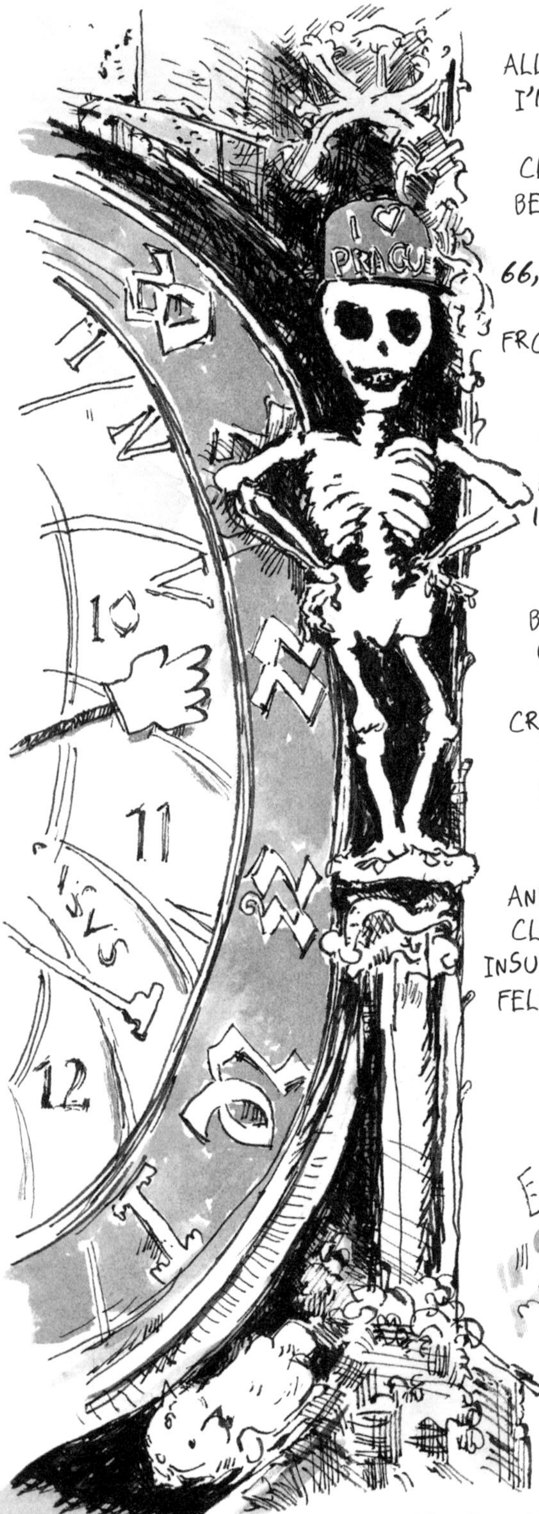
I ♥
PRAGUE

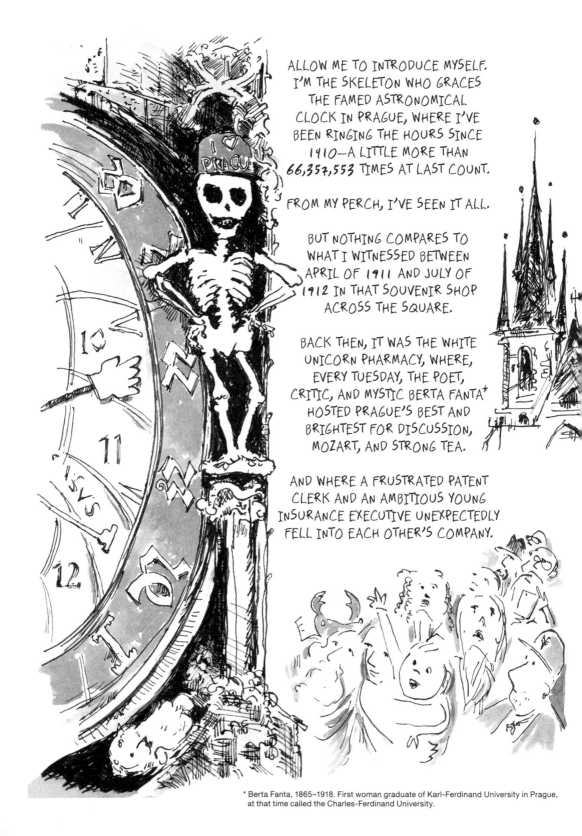

ALLOW ME TO INTRODUCE MYSELF. I'M THE SKELETON WHO GRACES THE FAMED ASTRONOMICAL CLOCK IN PRAGUE, WHERE I'VE BEEN RINGING THE HOURS SINCE 1410—A LITTLE MORE THAN 66,357,553 TIMES AT LAST COUNT.

FROM MY PERCH, I'VE SEEN IT ALL.

BUT NOTHING COMPARES TO WHAT I WITNESSED BETWEEN APRIL OF 1911 AND JULY OF 1912 IN THAT SOUVENIR SHOP ACROSS THE SQUARE.

BACK THEN, IT WAS THE WHITE UNICORN PHARMACY, WHERE, EVERY TUESDAY, THE POET, CRITIC, AND MYSTIC BERTA FANTA* HOSTED PRAGUE'S BEST AND BRIGHTEST FOR DISCUSSION, MOZART, AND STRONG TEA.

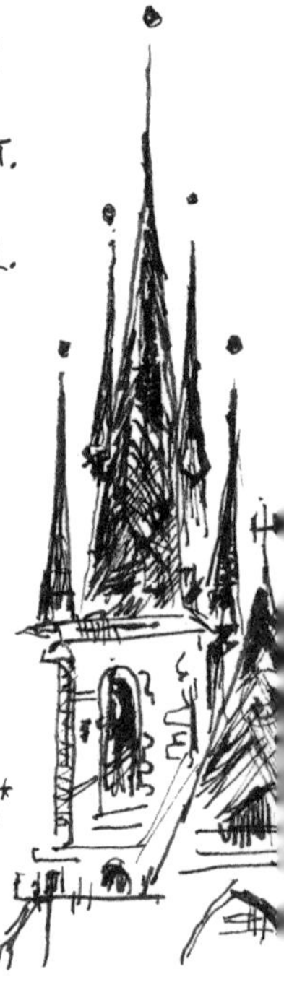

AND WHERE A FRUSTRATED PATENT CLERK AND AN AMBITIOUS YOUNG INSURANCE EXECUTIVE UNEXPECTEDLY FELL INTO EACH OTHER'S COMPANY.

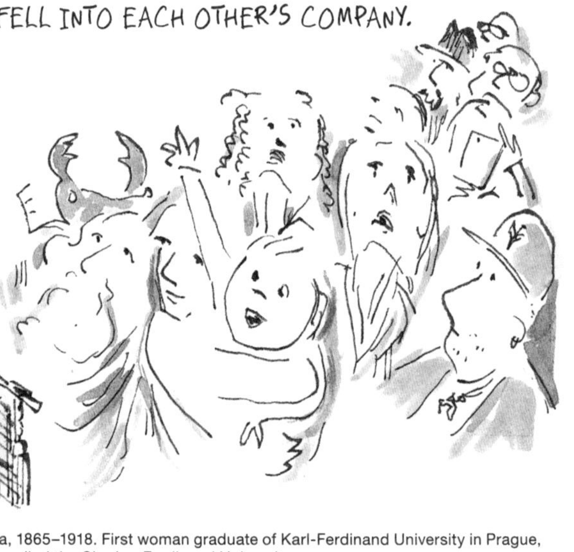

* Berta Fanta, 1865–1918. First woman graduate of Karl-Ferdinand University in Prague, at that time called the Charles-Ferdinand University.

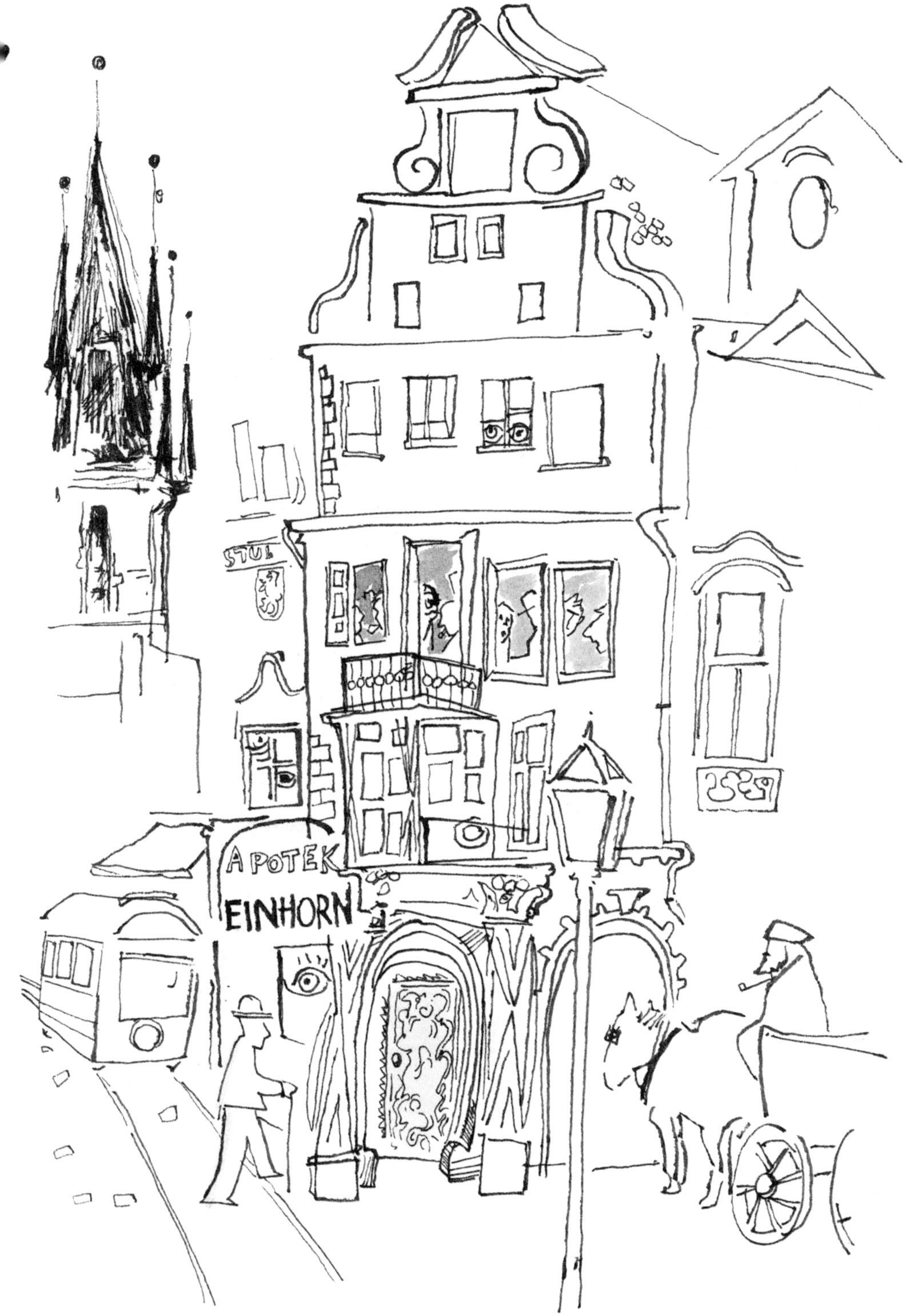
STUL
APOTEK
EINHORN

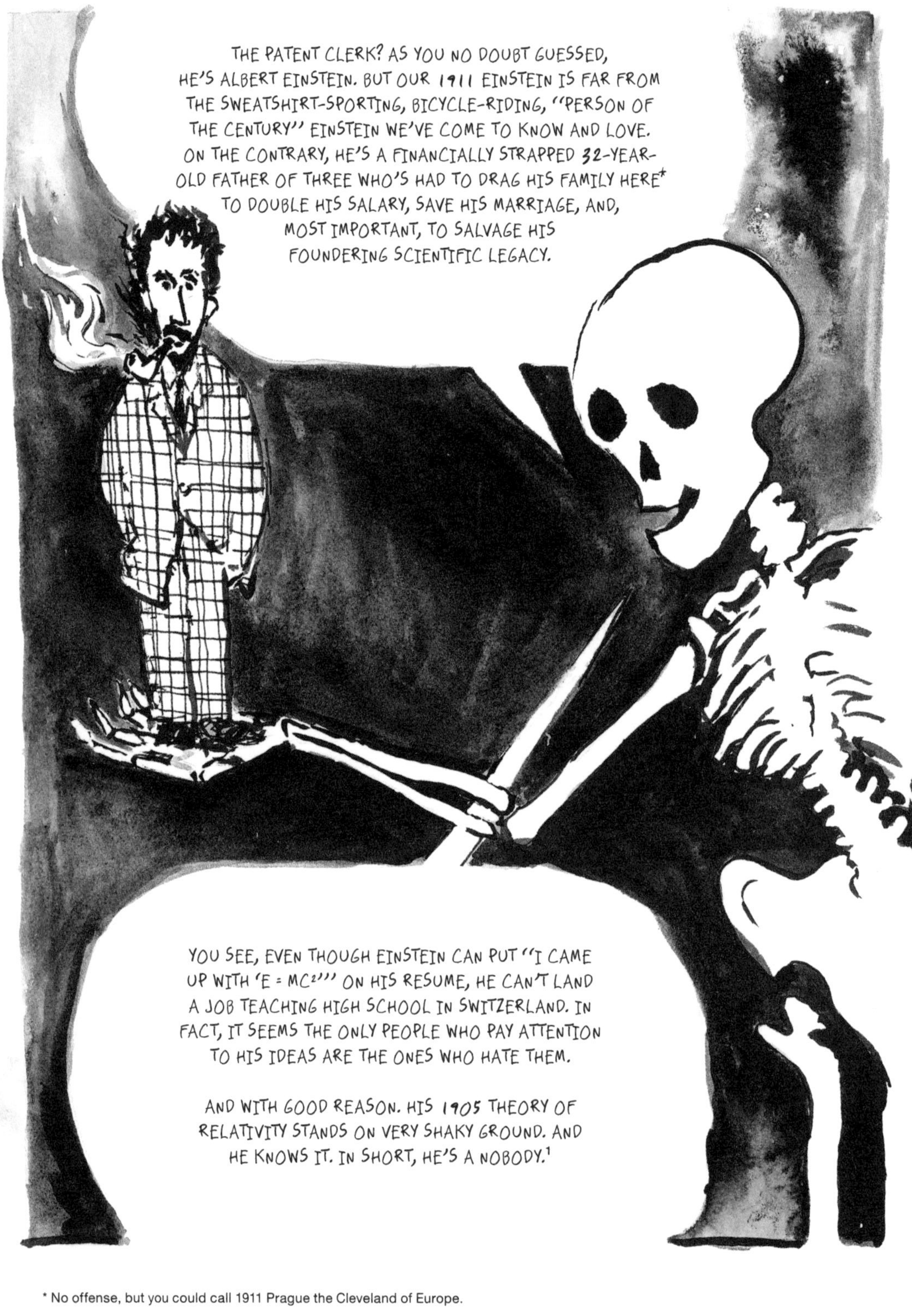
THE PATENT CLERK? AS YOU NO DOUBT GUESSED, HE'S ALBERT EINSTEIN. BUT OUR 1911 EINSTEIN IS FAR FROM THE SWEATSHIRT-SPORTING, BICYCLE-RIDING, "PERSON OF THE CENTURY" EINSTEIN WE'VE COME TO KNOW AND LOVE. ON THE CONTRARY, HE'S A FINANCIALLY STRAPPED 32-YEAR-OLD FATHER OF THREE WHO'S HAD TO DRAG HIS FAMILY HERE* TO DOUBLE HIS SALARY, SAVE HIS MARRIAGE, AND, MOST IMPORTANT, TO SALVAGE HIS FOUNDERING SCIENTIFIC LEGACY.
YOU SEE, EVEN THOUGH EINSTEIN CAN PUT "I CAME UP WITH '$E = MC^2$'" ON HIS RESUME, HE CAN'T LAND A JOB TEACHING HIGH SCHOOL IN SWITZERLAND. IN FACT, IT SEEMS THE ONLY PEOPLE WHO PAY ATTENTION TO HIS IDEAS ARE THE ONES WHO HATE THEM.
AND WITH GOOD REASON. HIS 1905 THEORY OF RELATIVITY STANDS ON VERY SHAKY GROUND. AND HE KNOWS IT. IN SHORT, HE'S A NOBODY.[1]

* No offense, but you could call 1911 Prague the Cleveland of Europe.

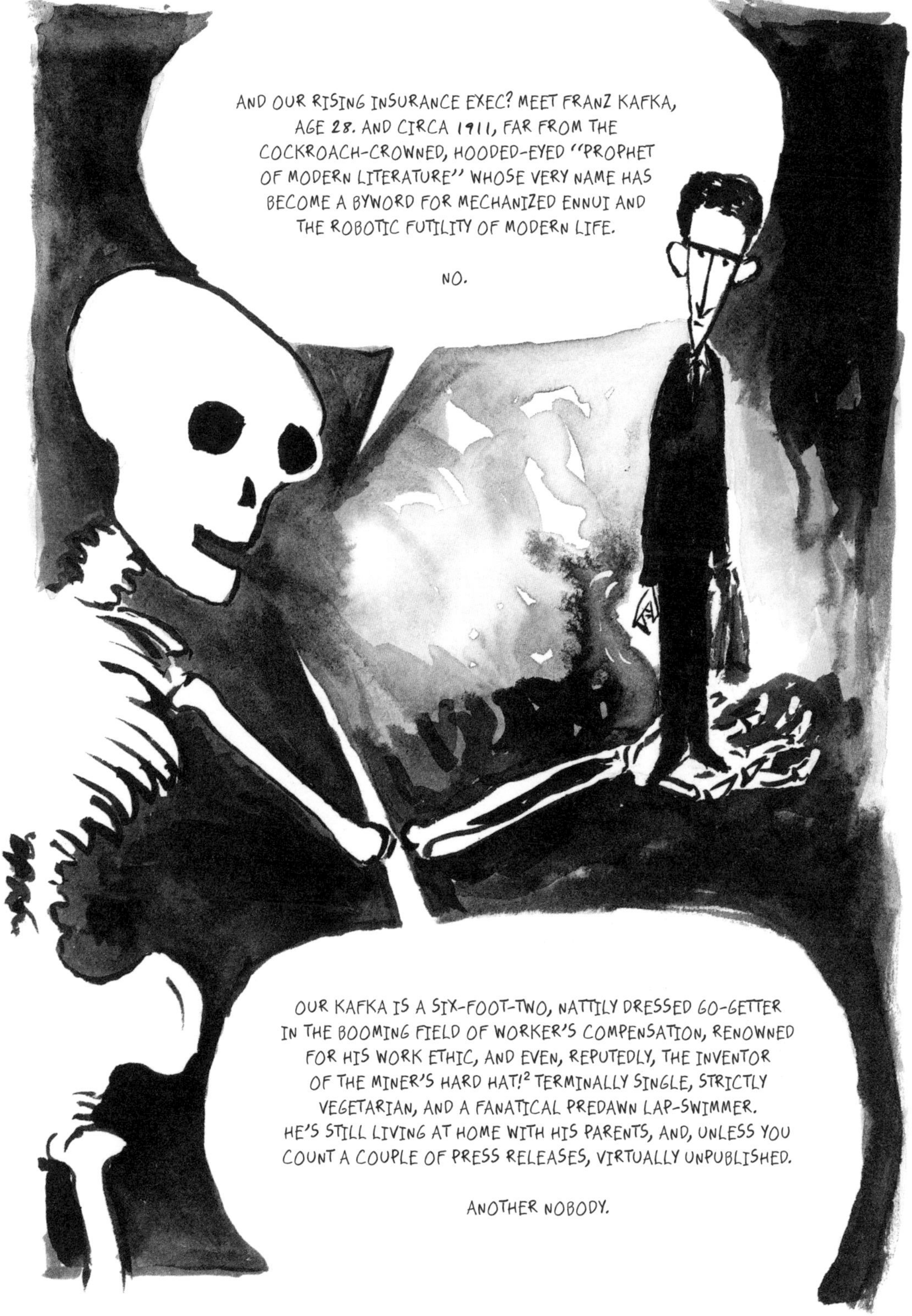
AND OUR RISING INSURANCE EXEC? MEET FRANZ KAFKA, AGE 28. AND CIRCA 1911, FAR FROM THE COCKROACH-CROWNED, HOODED-EYED "PROPHET OF MODERN LITERATURE" WHOSE VERY NAME HAS BECOME A BYWORD FOR MECHANIZED ENNUI AND THE ROBOTIC FUTILITY OF MODERN LIFE.
NO.
OUR KAFKA IS A SIX-FOOT-TWO, NATTILY DRESSED GO-GETTER IN THE BOOMING FIELD OF WORKER'S COMPENSATION, RENOWNED FOR HIS WORK ETHIC, AND EVEN, REPUTEDLY, THE INVENTOR OF THE MINER'S HARD HAT![2] TERMINALLY SINGLE, STRICTLY VEGETARIAN, AND A FANATICAL PREDAWN LAP-SWIMMER. HE'S STILL LIVING AT HOME WITH HIS PARENTS, AND, UNLESS YOU COUNT A COUPLE OF PRESS RELEASES, VIRTUALLY UNPUBLISHED.
ANOTHER NOBODY.

NEVERTHELESS, BY THE TIME EINSTEIN'S TRAIN PULLS OUT OF PRAGUE FIFTEEN MONTHS LATER, THE PHYSICIST WILL HAVE UNCOVERED THE KEY TO WHAT HE CALLED "SOLVING GRAVITY"—NOT ONLY RESCUING HIS LEGACY, BUT GIVING BIRTH TO WHAT'S BEEN CALLED EVERYTHING FROM "THE MOST PERFECT INTELLECTUAL ACHIEVEMENT OF MODERN PHYSICS"[3] TO "AMONG THE MOST BEAUTIFUL AND SIGNIFICANT ACHIEVEMENTS OF HUMAN UNDERSTANDING."[4]

AND KAFKA?

BY THE END OF 1912, HE'S PRODUCED HIS STORY "THE JUDGMENT," THE MASTERPIECE THAT CRACKED THE CODE OF THE MODERN WRITTEN WORD, LAUNCHING A BODY OF WORK THAT PHILIP ROTH SAID STANDS "AS A MONUMENT TO THE POWER OF LITERATURE TO TRANSCEND TIME AND PLACE, AND TO REVEAL THE HIDDEN DEPTHS OF HUMAN EXPERIENCE."

NOBODY KNOWS QUITE HOW THEY DID IT. OR WHY.

BUT I HAPPEN TO HAVE ASSEMBLED SOME PRETTY COMPELLING CLUES. AS I'VE SAID, I'VE SEEN IT ALL.

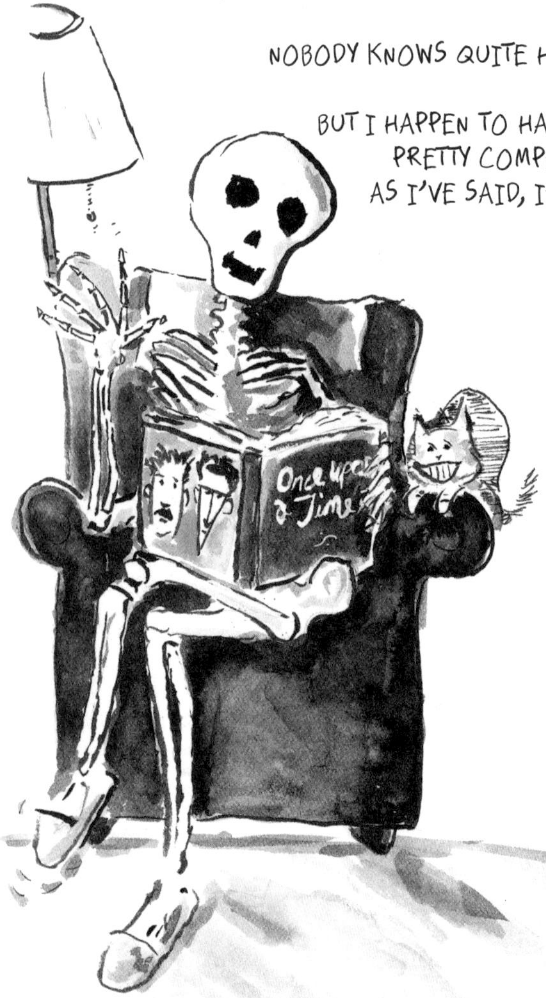

SO PULL UP A CHAIR, AND LET ME TELL YOU HOW IT ALL WENT DOWN.

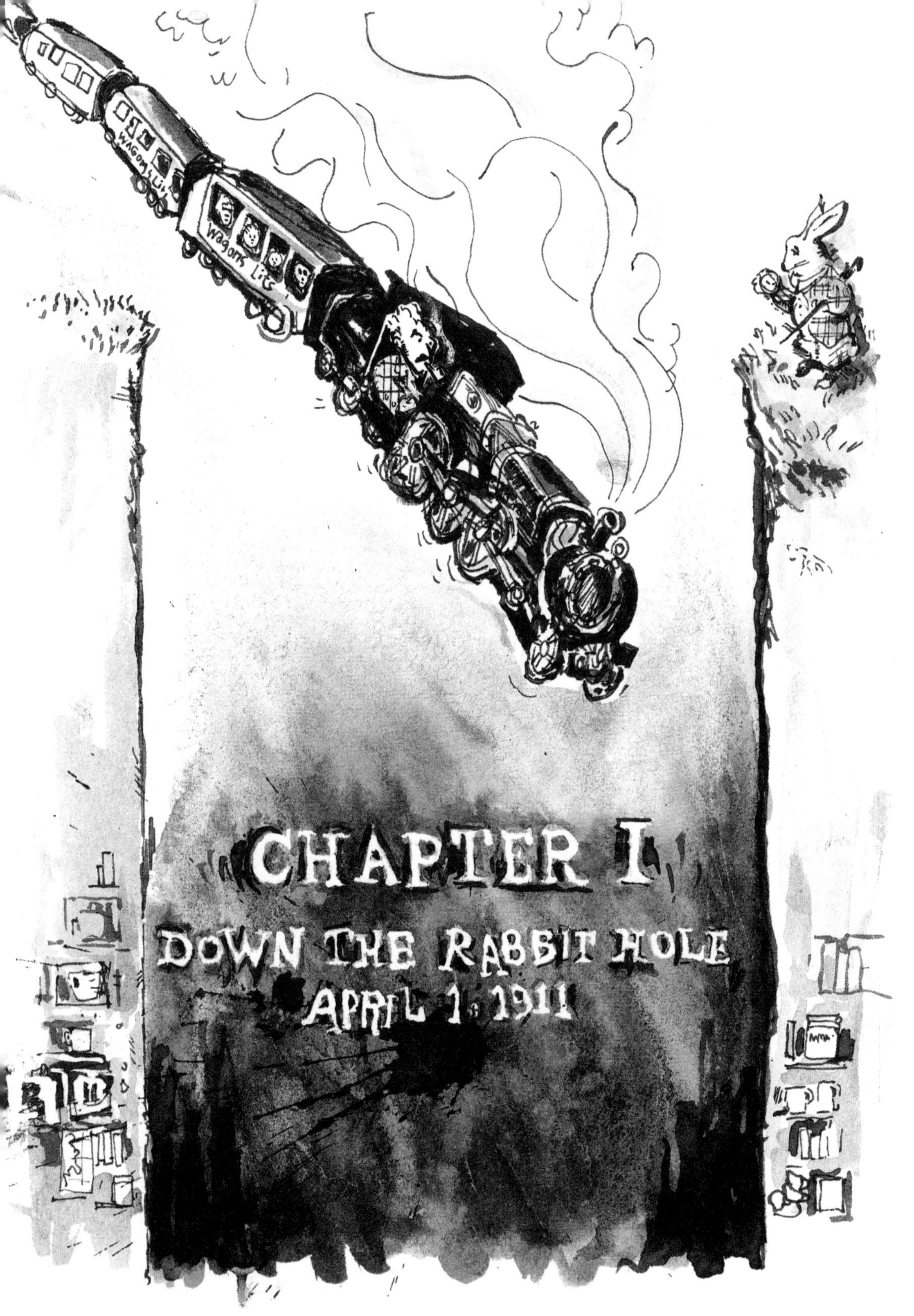
Wagons Lits
Wagons Lits
CHAPTER I
DOWN THE RABBIT HOLE
APRIL 1, 1911

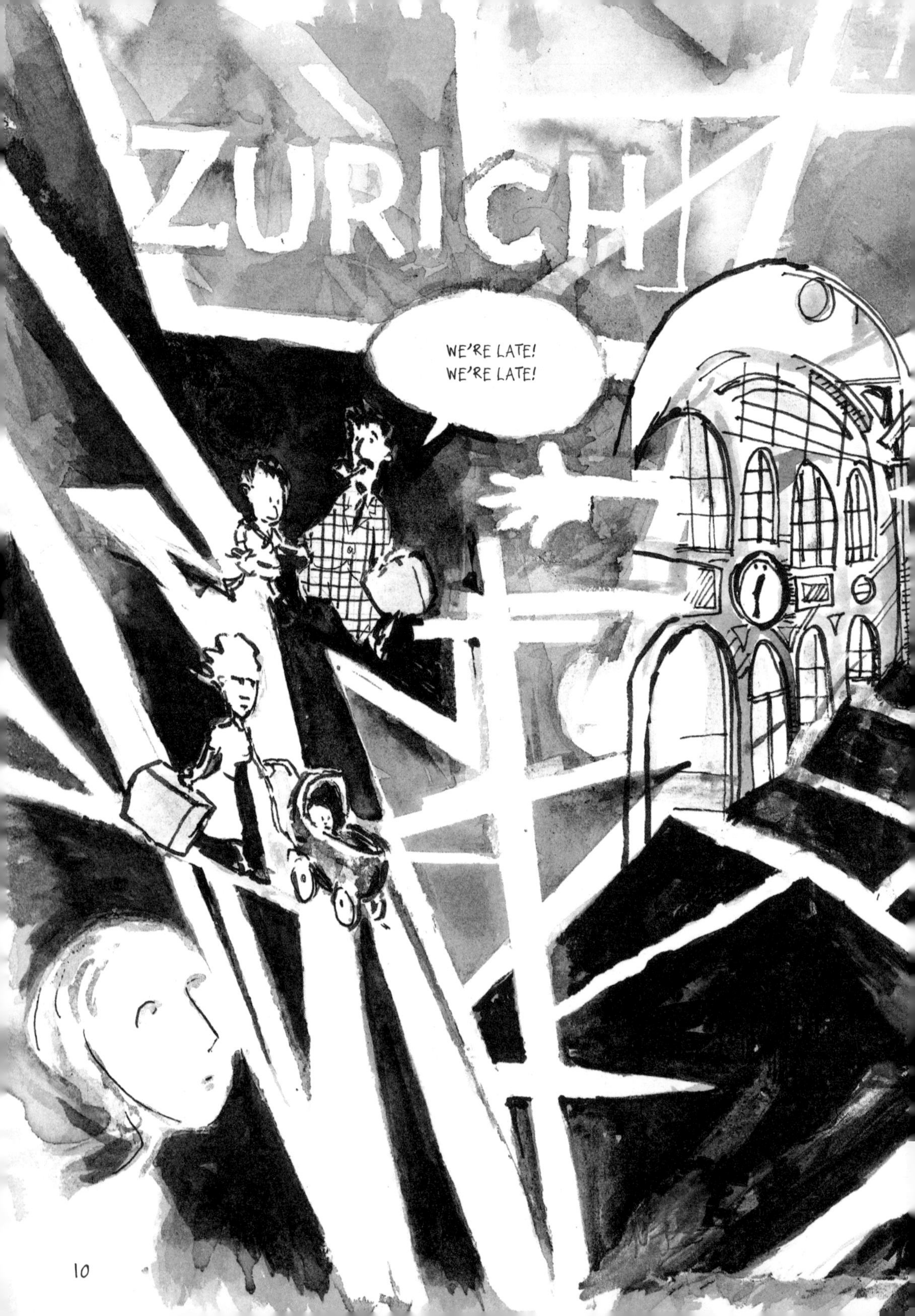
ZURICH
WE'RE LATE!
WE'RE LATE!

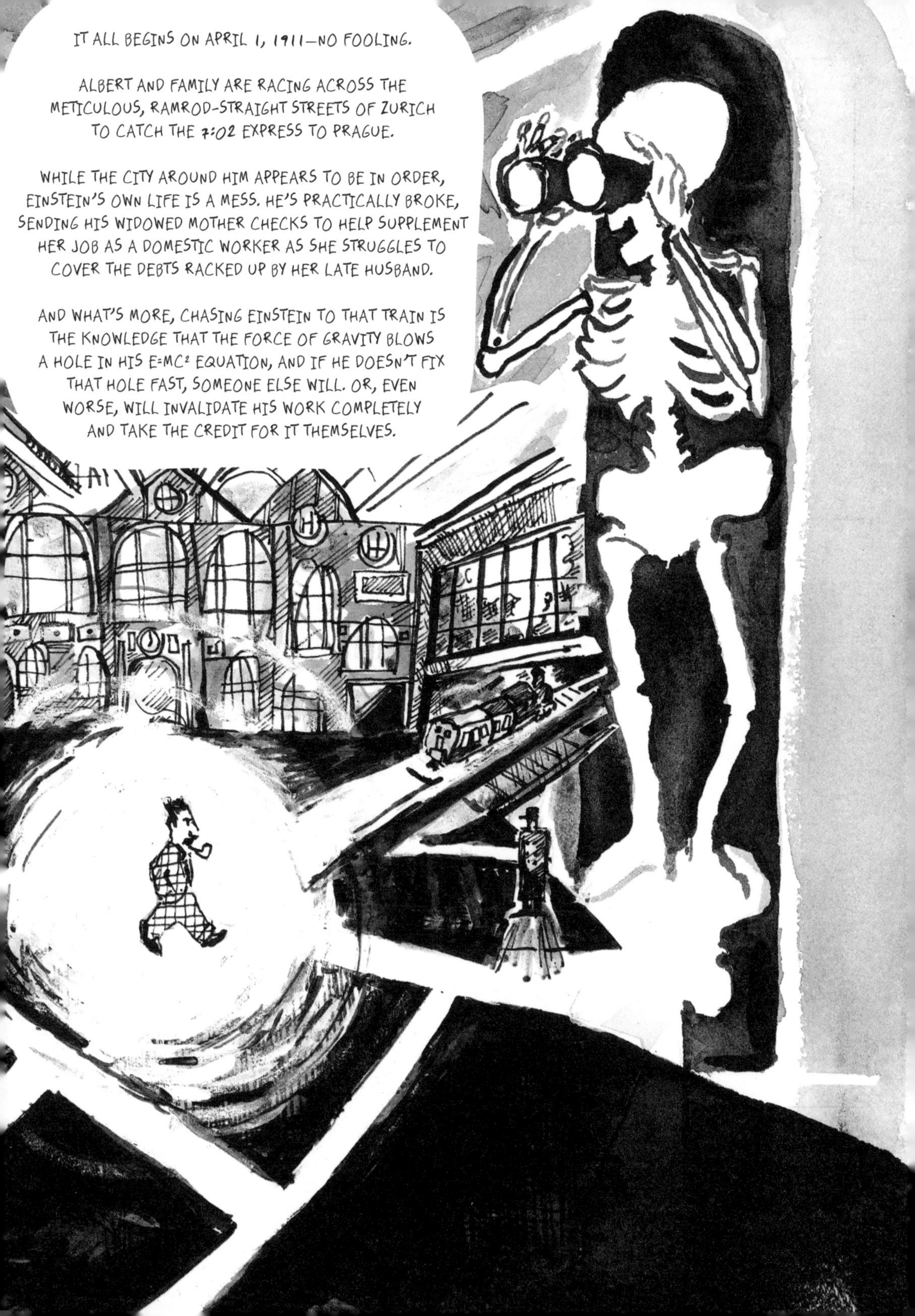
IT ALL BEGINS ON APRIL 1, 1911—NO FOOLING.
ALBERT AND FAMILY ARE RACING ACROSS THE METICULOUS, RAMROD-STRAIGHT STREETS OF ZURICH TO CATCH THE 7:02 EXPRESS TO PRAGUE.
WHILE THE CITY AROUND HIM APPEARS TO BE IN ORDER, EINSTEIN'S OWN LIFE IS A MESS. HE'S PRACTICALLY BROKE, SENDING HIS WIDOWED MOTHER CHECKS TO HELP SUPPLEMENT HER JOB AS A DOMESTIC WORKER AS SHE STRUGGLES TO COVER THE DEBTS RACKED UP BY HER LATE HUSBAND.
AND WHAT'S MORE, CHASING EINSTEIN TO THAT TRAIN IS THE KNOWLEDGE THAT THE FORCE OF GRAVITY BLOWS A HOLE IN HIS E=MC² EQUATION, AND IF HE DOESN'T FIX THAT HOLE FAST, SOMEONE ELSE WILL. OR, EVEN WORSE, WILL INVALIDATE HIS WORK COMPLETELY AND TAKE THE CREDIT FOR IT THEMSELVES.

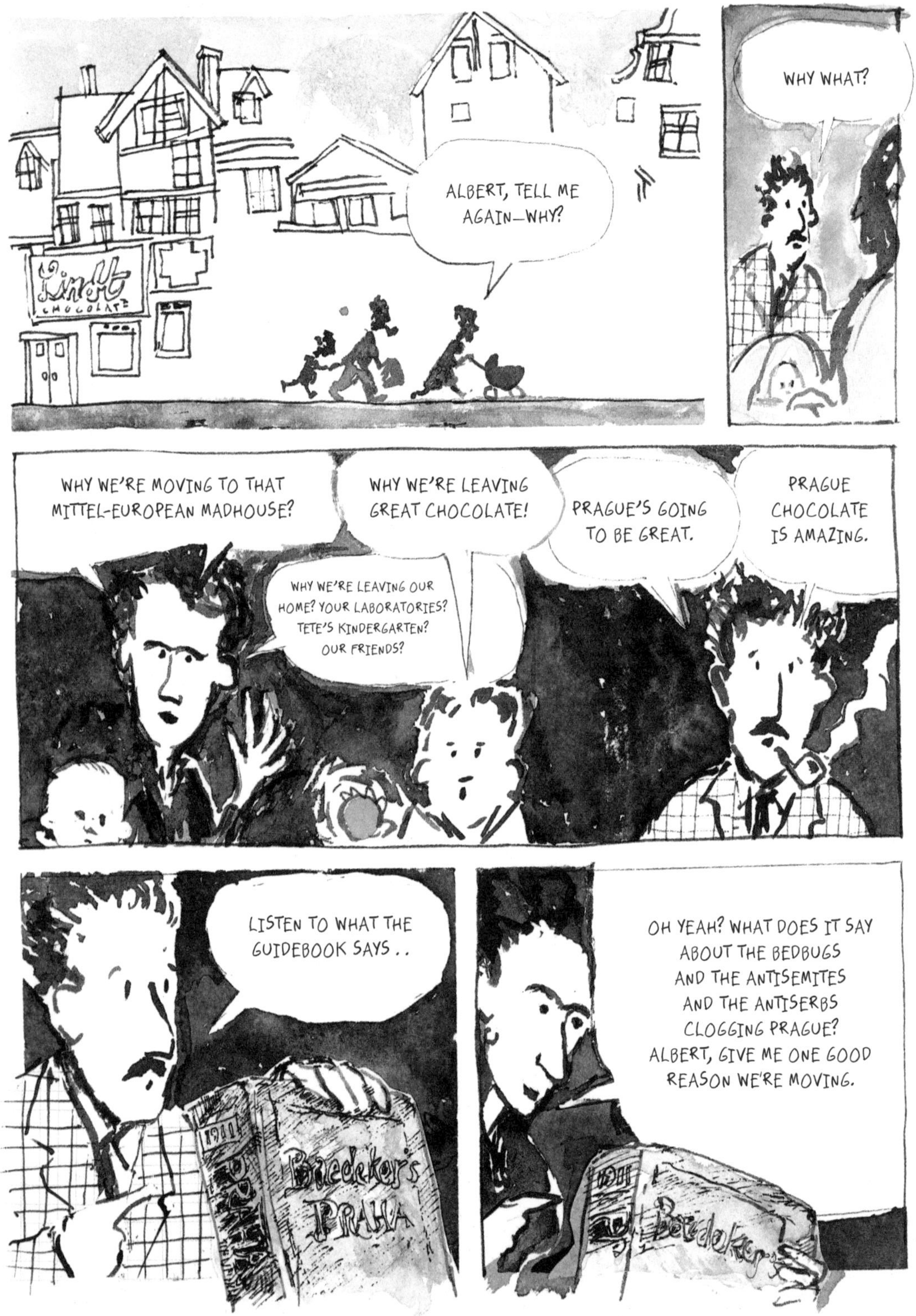
Lindt CHOCOLATE
ALBERT, TELL ME AGAIN—WHY?
WHY WHAT?
WHY WE'RE MOVING TO THAT MITTEL-EUROPEAN MADHOUSE?
WHY WE'RE LEAVING OUR HOME? YOUR LABORATORIES? TETE'S KINDERGARTEN? OUR FRIENDS?
WHY WE'RE LEAVING GREAT CHOCOLATE!
PRAGUE'S GOING TO BE GREAT.
PRAGUE CHOCOLATE IS AMAZING.
LISTEN TO WHAT THE GUIDEBOOK SAYS . .
1911
Baedeker's PRAHA
OH YEAH? WHAT DOES IT SAY ABOUT THE BEDBUGS AND THE ANTISEMITES AND THE ANTISERBS CLOGGING PRAGUE? ALBERT, GIVE ME ONE GOOD REASON WE'RE MOVING.
Baedeker's

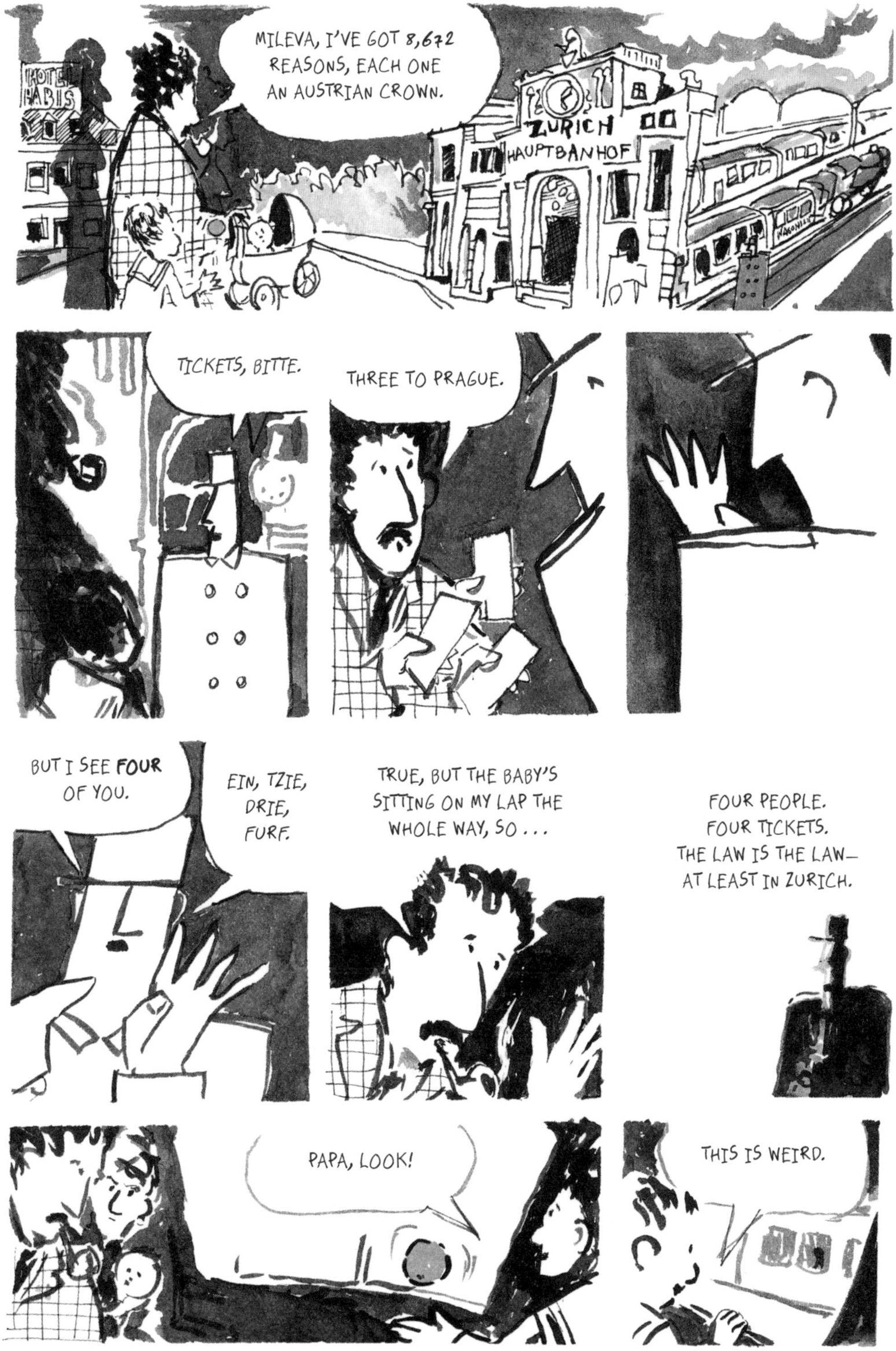
HOTEL HABIS
MILEVA, I'VE GOT 8,672 REASONS, EACH ONE AN AUSTRIAN CROWN.
ZURICH HAUPTBANHOF
TICKETS, BITTE.
THREE TO PRAGUE.
BUT I SEE **FOUR** OF YOU.
EIN, TZIE, DRIE, FURF.
TRUE, BUT THE BABY'S SITTING ON MY LAP THE WHOLE WAY, SO . . .
FOUR PEOPLE. FOUR TICKETS. THE LAW IS THE LAW— AT LEAST IN ZURICH.
PAPA, LOOK!
THIS IS WEIRD.

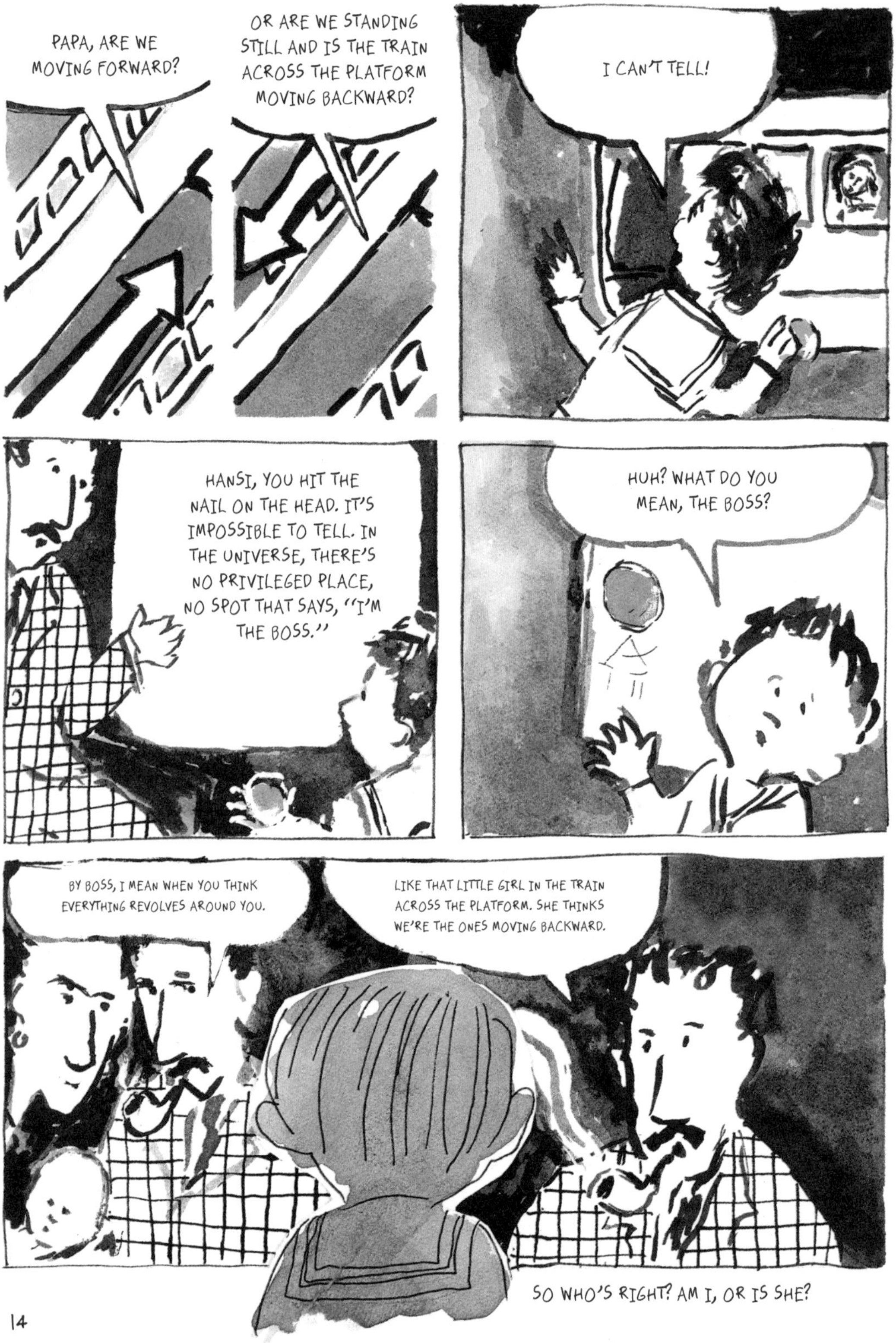
PAPA, ARE WE MOVING FORWARD?
OR ARE WE STANDING STILL AND IS THE TRAIN ACROSS THE PLATFORM MOVING BACKWARD?
I CAN'T TELL!
HANSI, YOU HIT THE NAIL ON THE HEAD. IT'S IMPOSSIBLE TO TELL. IN THE UNIVERSE, THERE'S NO PRIVILEGED PLACE, NO SPOT THAT SAYS, "I'M THE BOSS."
HUH? WHAT DO YOU MEAN, THE BOSS?
BY BOSS, I MEAN WHEN YOU THINK EVERYTHING REVOLVES AROUND YOU.
LIKE THAT LITTLE GIRL IN THE TRAIN ACROSS THE PLATFORM. SHE THINKS WE'RE THE ONES MOVING BACKWARD.
SO WHO'S RIGHT? AM I, OR IS SHE?

SO WHO'S THE BOSS?
BOTH OF YOU. IT'S ALL RELATIVE!
100 KM LATER...
HEY PAPA, ANOTHER QUESTION— EVERY TIME I THROW MY BALL UP,
IT LANDS RIGHT BACK IN MY HAND.
BUT WHEN IT'S NOT ATTACHED TO ME, SHOULDN'T IT FLY BACKWARD?
SHOULDN'T IT FLY BACK AND SMACK ME IN THE FACE?
HOW DOES THE BALL **KNOW** TO FALL BACK INTO MY HAND?

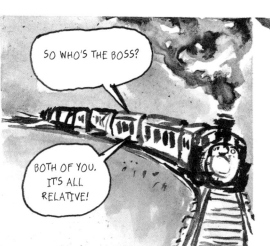

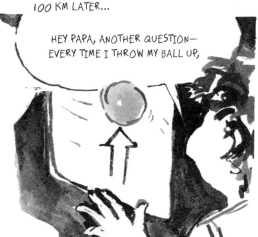

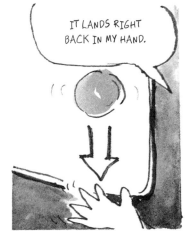

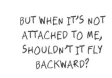

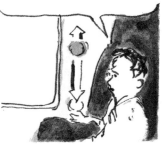

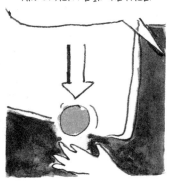

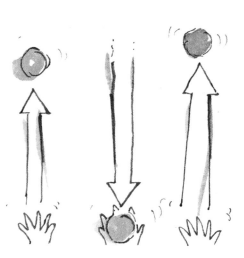

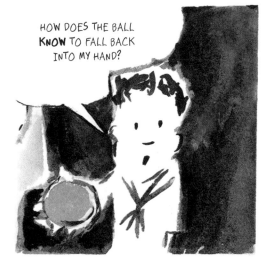

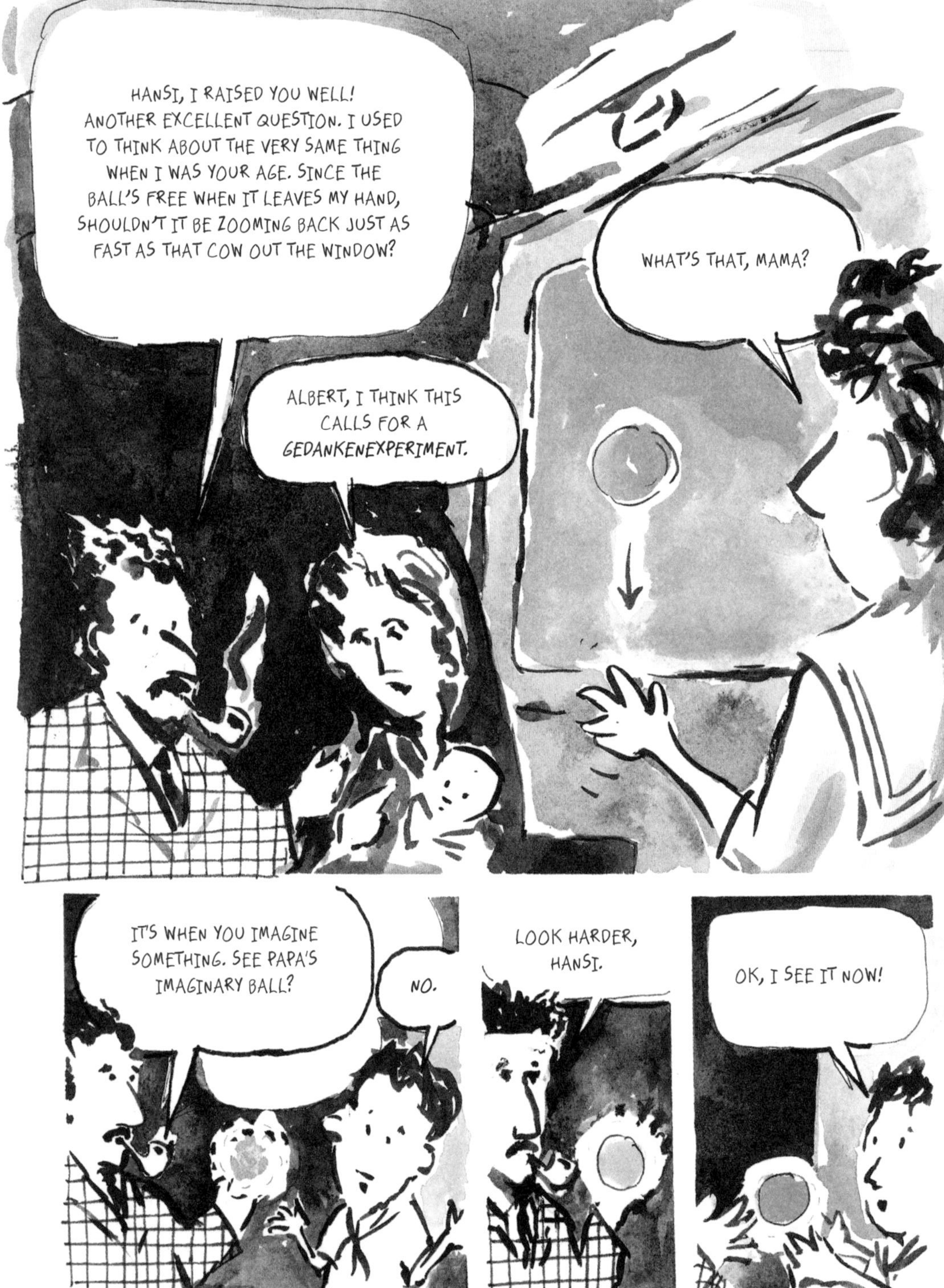

HANSI, I RAISED YOU WELL! ANOTHER EXCELLENT QUESTION. I USED TO THINK ABOUT THE VERY SAME THING WHEN I WAS YOUR AGE. SINCE THE BALL'S FREE WHEN IT LEAVES MY HAND, SHOULDN'T IT BE ZOOMING BACK JUST AS FAST AS THAT COW OUT THE WINDOW?
WHAT'S THAT, MAMA?
ALBERT, I THINK THIS CALLS FOR A *GEDANKENEXPERIMENT.*
IT'S WHEN YOU IMAGINE SOMETHING. SEE PAPA'S IMAGINARY BALL?
NO.
LOOK HARDER, HANSI.
OK, I SEE IT NOW!

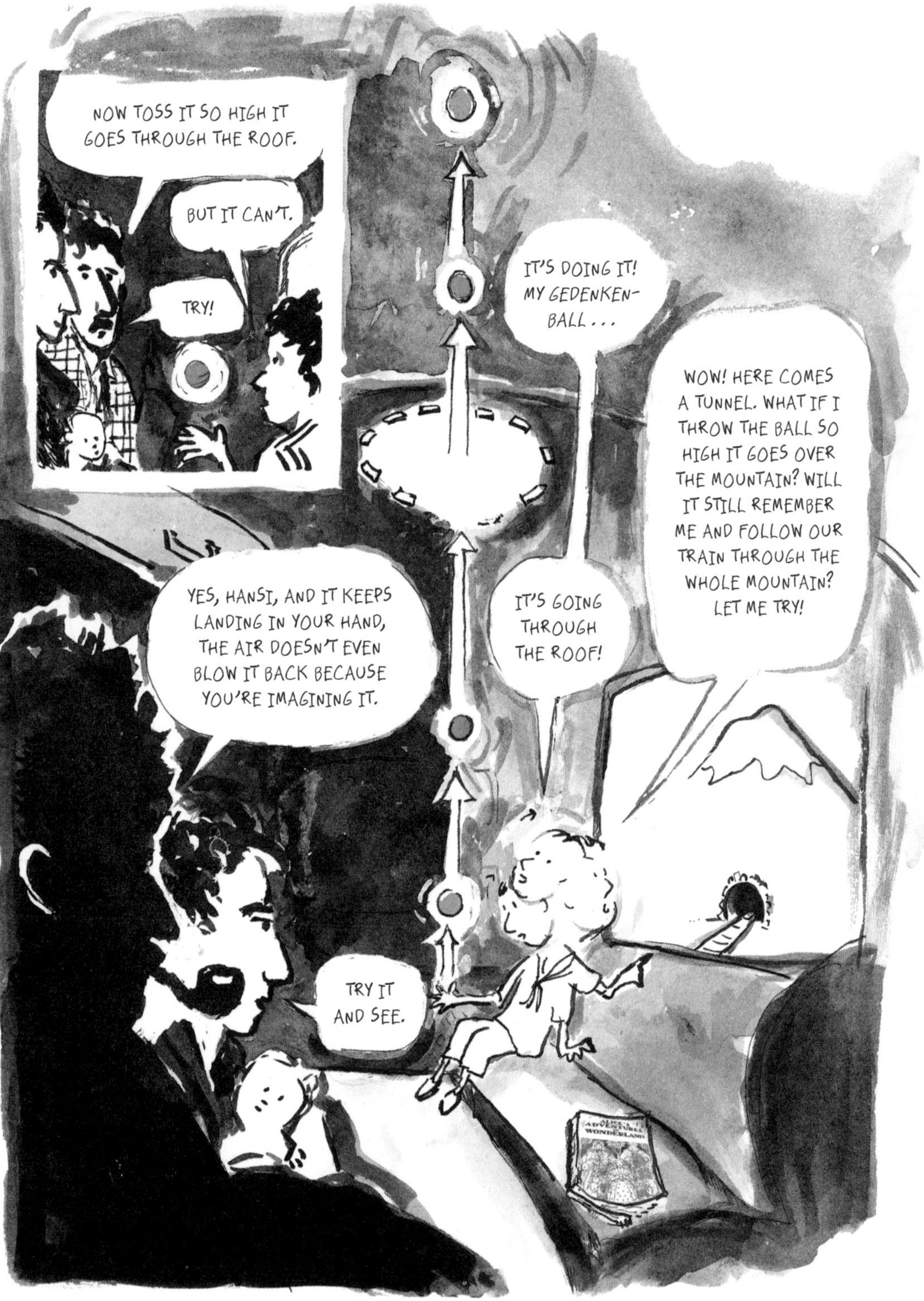
NOW TOSS IT SO HIGH IT GOES THROUGH THE ROOF.
BUT IT CAN'T.
TRY!
IT'S DOING IT! MY GEDENKEN-BALL . . .
WOW! HERE COMES A TUNNEL. WHAT IF I THROW THE BALL SO HIGH IT GOES OVER THE MOUNTAIN? WILL IT STILL REMEMBER ME AND FOLLOW OUR TRAIN THROUGH THE WHOLE MOUNTAIN? LET ME TRY!
YES, HANSI, AND IT KEEPS LANDING IN YOUR HAND, THE AIR DOESN'T EVEN BLOW IT BACK BECAUSE YOU'RE IMAGINING IT.
IT'S GOING THROUGH THE ROOF!
TRY IT AND SEE.
ALICE'S ADVENTURES IN WONDERLAND

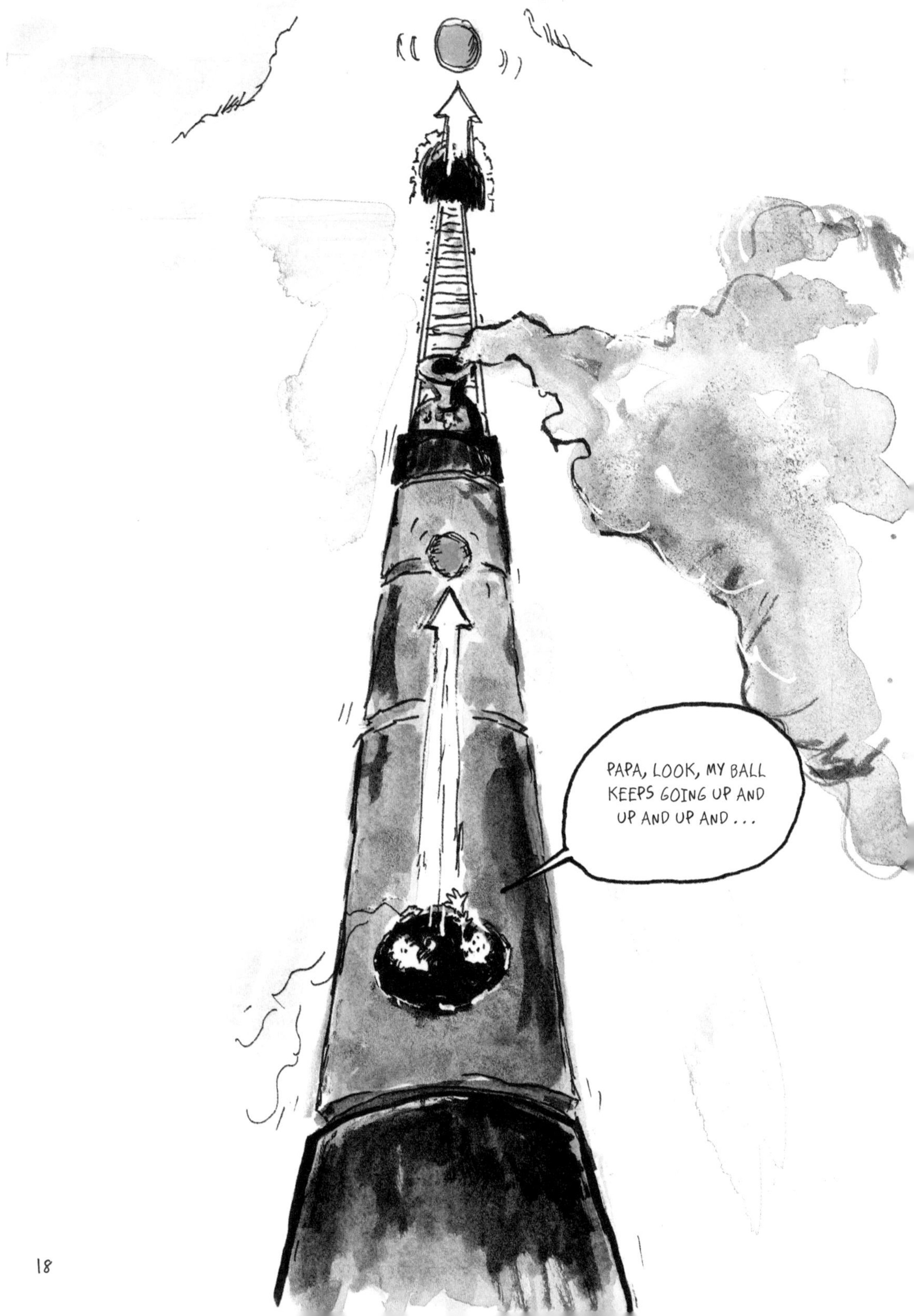
PAPA, LOOK, MY BALL KEEPS GOING UP AND UP AND UP AND . . .

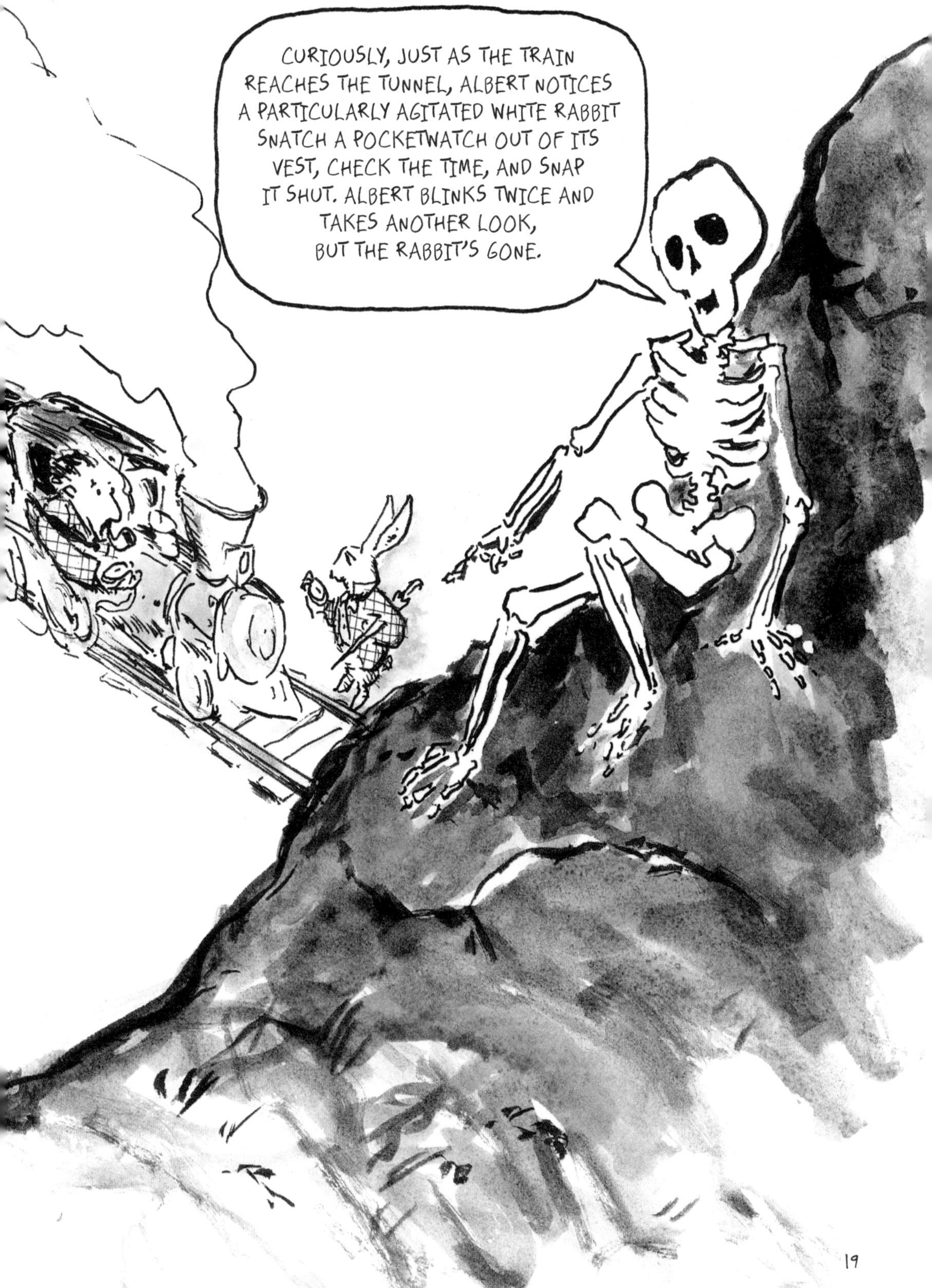
CURIOUSLY, JUST AS THE TRAIN REACHES THE TUNNEL, ALBERT NOTICES A PARTICULARLY AGITATED WHITE RABBIT SNATCH A POCKETWATCH OUT OF ITS VEST, CHECK THE TIME, AND SNAP IT SHUT. ALBERT BLINKS TWICE AND TAKES ANOTHER LOOK, BUT THE RABBIT'S GONE.

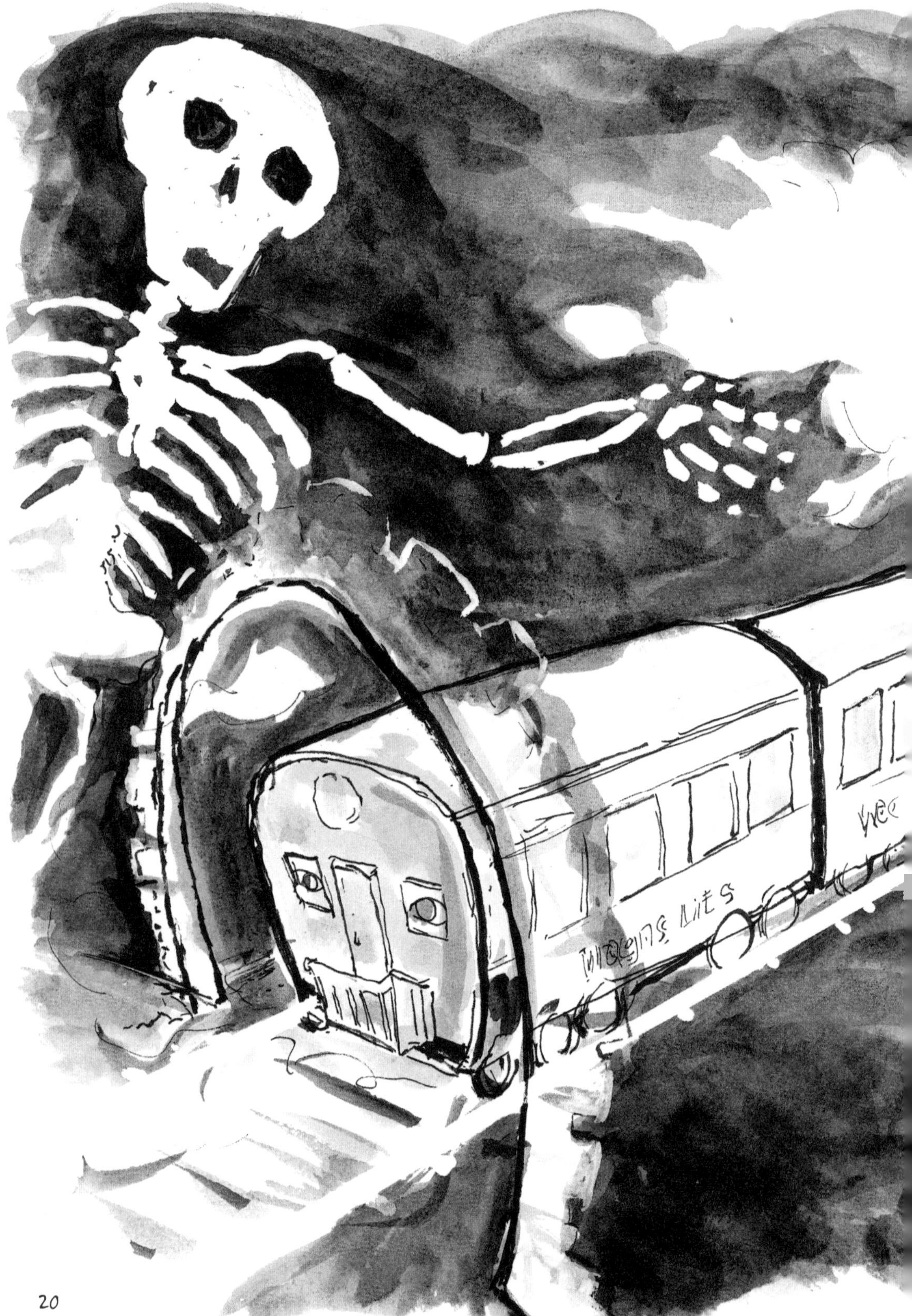

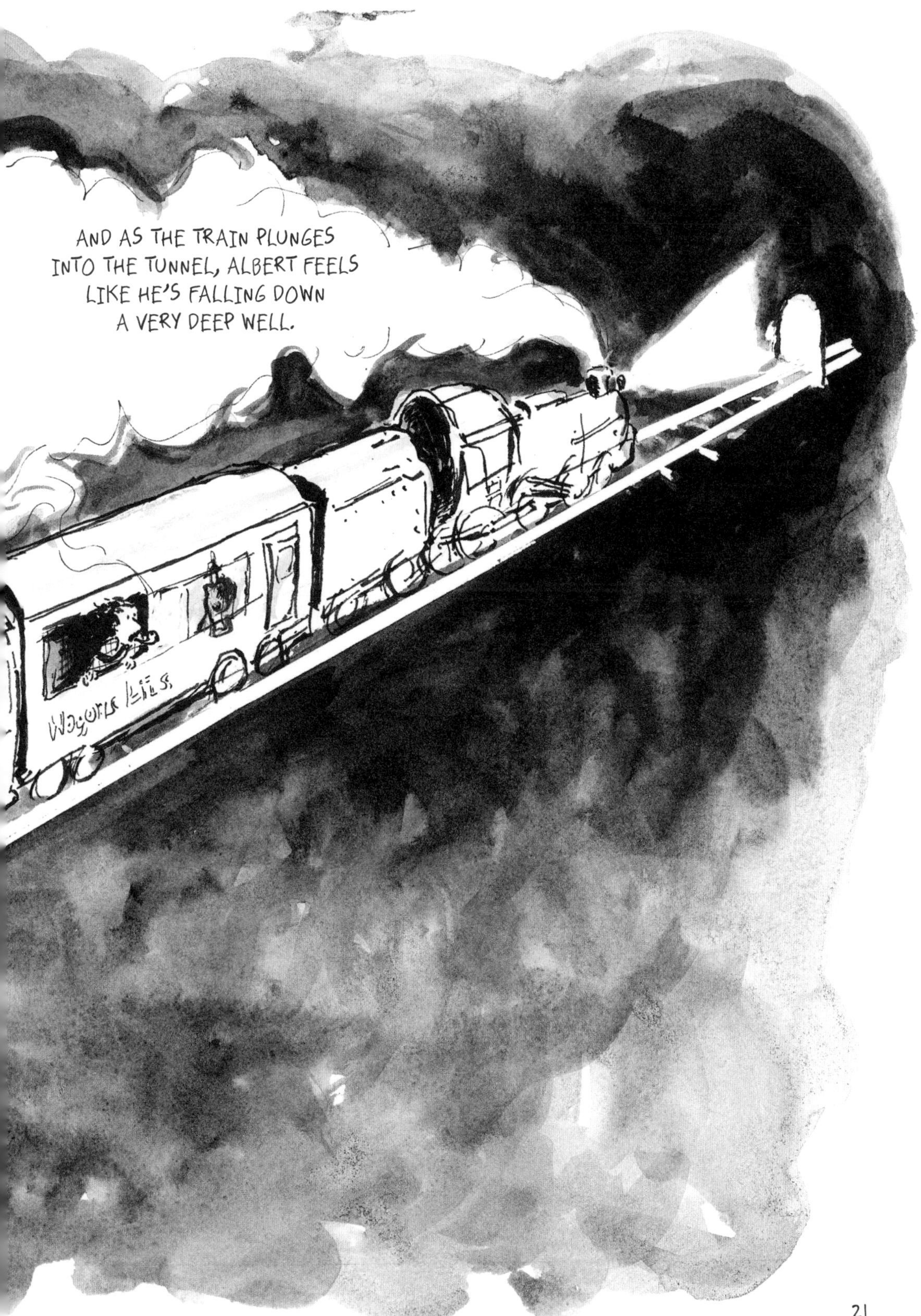
AND AS THE TRAIN PLUNGES INTO THE TUNNEL, ALBERT FEELS LIKE HE'S FALLING DOWN A VERY DEEP WELL.

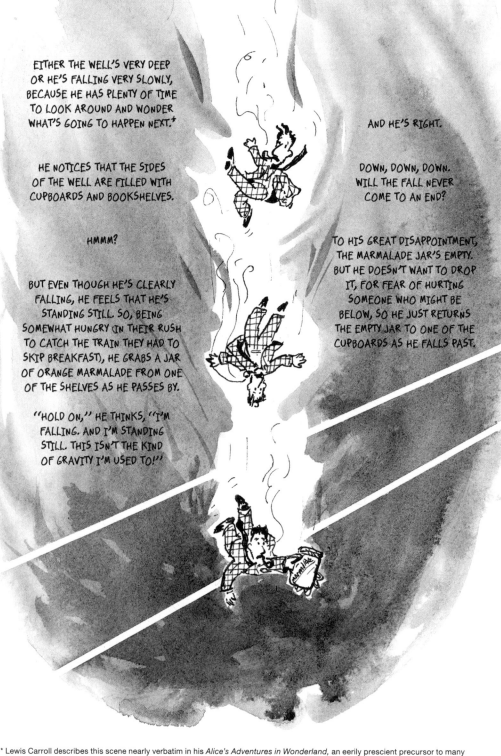

EITHER THE WELL'S VERY DEEP OR HE'S FALLING VERY SLOWLY, BECAUSE HE HAS PLENTY OF TIME TO LOOK AROUND AND WONDER WHAT'S GOING TO HAPPEN NEXT.*

HE NOTICES THAT THE SIDES OF THE WELL ARE FILLED WITH CUPBOARDS AND BOOKSHELVES.

HMMM?

BUT EVEN THOUGH HE'S CLEARLY FALLING, HE FEELS THAT HE'S STANDING STILL. SO, BEING SOMEWHAT HUNGRY (IN THEIR RUSH TO CATCH THE TRAIN THEY HAD TO SKIP BREAKFAST), HE GRABS A JAR OF ORANGE MARMALADE FROM ONE OF THE SHELVES AS HE PASSES BY.

"HOLD ON," HE THINKS, "I'M FALLING. AND I'M STANDING STILL. THIS ISN'T THE KIND OF GRAVITY I'M USED TO!"

AND HE'S RIGHT.

DOWN, DOWN, DOWN. WILL THE FALL NEVER COME TO AN END?

TO HIS GREAT DISAPPOINTMENT, THE MARMALADE JAR'S EMPTY. BUT HE DOESN'T WANT TO DROP IT, FOR FEAR OF HURTING SOMEONE WHO MIGHT BE BELOW, SO HE JUST RETURNS THE EMPTY JAR TO ONE OF THE CUPBOARDS AS HE FALLS PAST.

* Lewis Carroll describes this scene nearly verbatim in his *Alice's Adventures in Wonderland*, an eerily prescient precursor to many of Einstein's most salient thoughts during his sojourn in Prague—published in 1870, fourteen years before Einstein was even born.

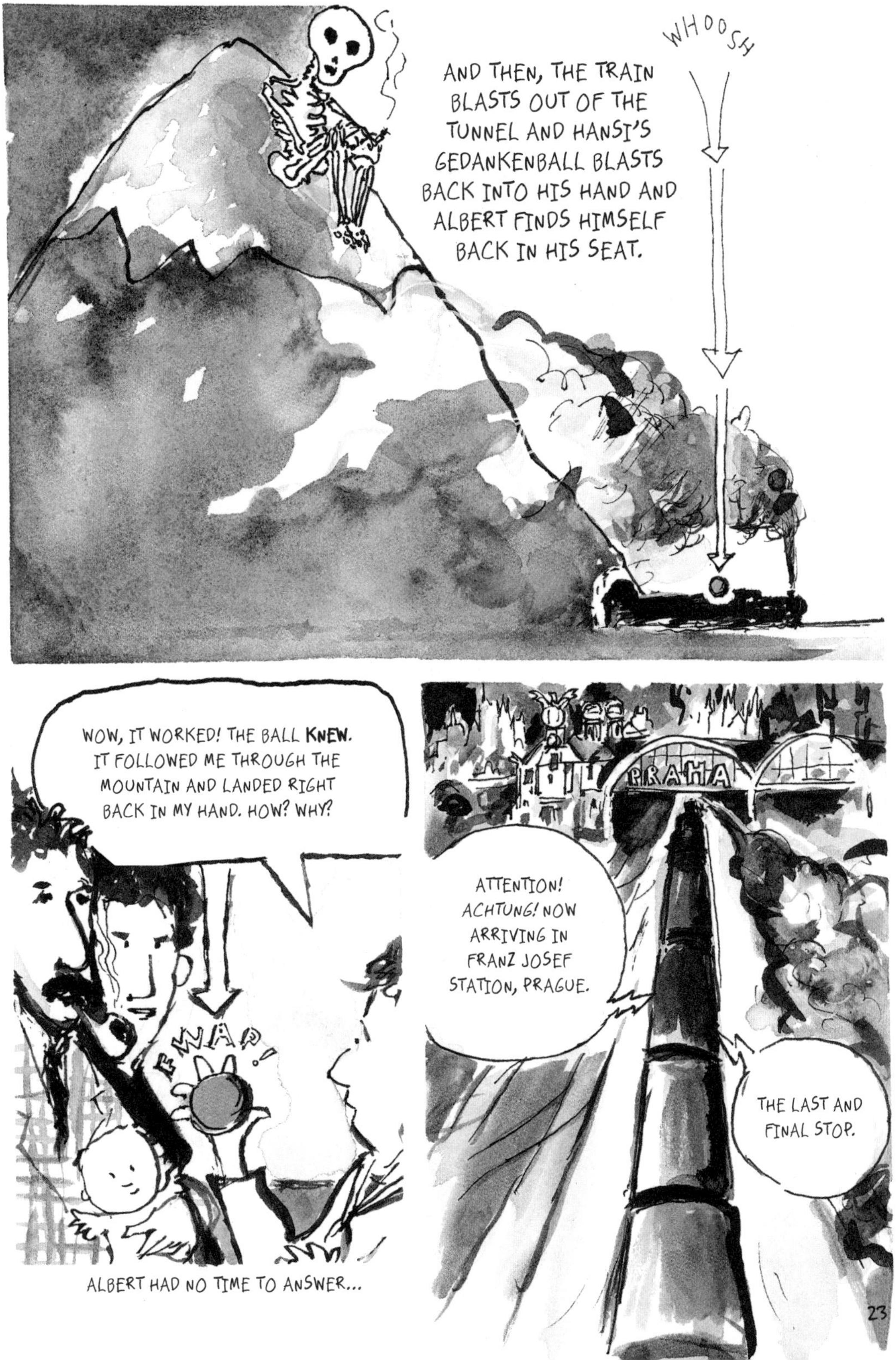

WHOOSH
AND THEN, THE TRAIN BLASTS OUT OF THE TUNNEL AND HANSI'S GEDANKENBALL BLASTS BACK INTO HIS HAND AND ALBERT FINDS HIMSELF BACK IN HIS SEAT.
WOW, IT WORKED! THE BALL **KNEW**. IT FOLLOWED ME THROUGH THE MOUNTAIN AND LANDED RIGHT BACK IN MY HAND. HOW? WHY?
FWAP!
ALBERT HAD NO TIME TO ANSWER...
PRAHA
ATTENTION! ACHTUNG! NOW ARRIVING IN FRANZ JOSEF STATION, PRAGUE.
THE LAST AND FINAL STOP.

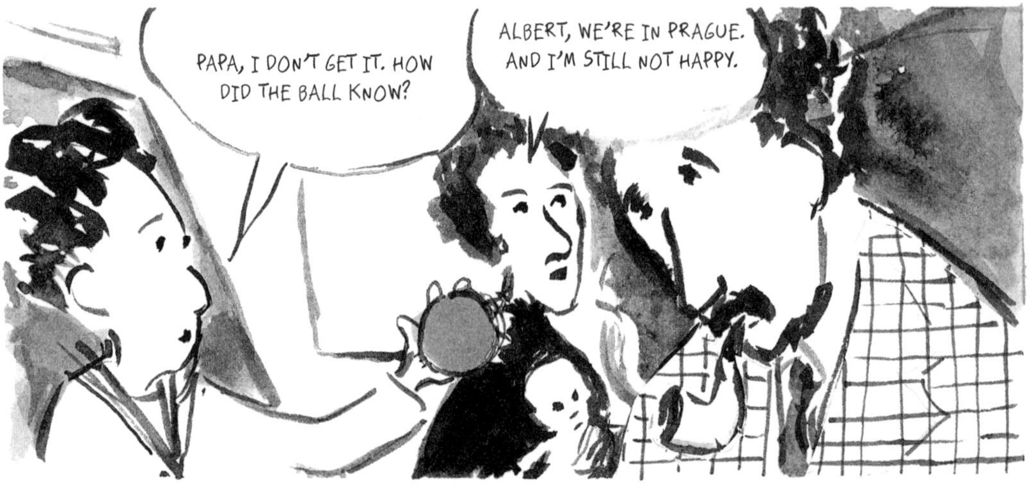
PAPA, I DON'T GET IT. HOW DID THE BALL KNOW?
ALBERT, WE'RE IN PRAGUE. AND I'M STILL NOT HAPPY.

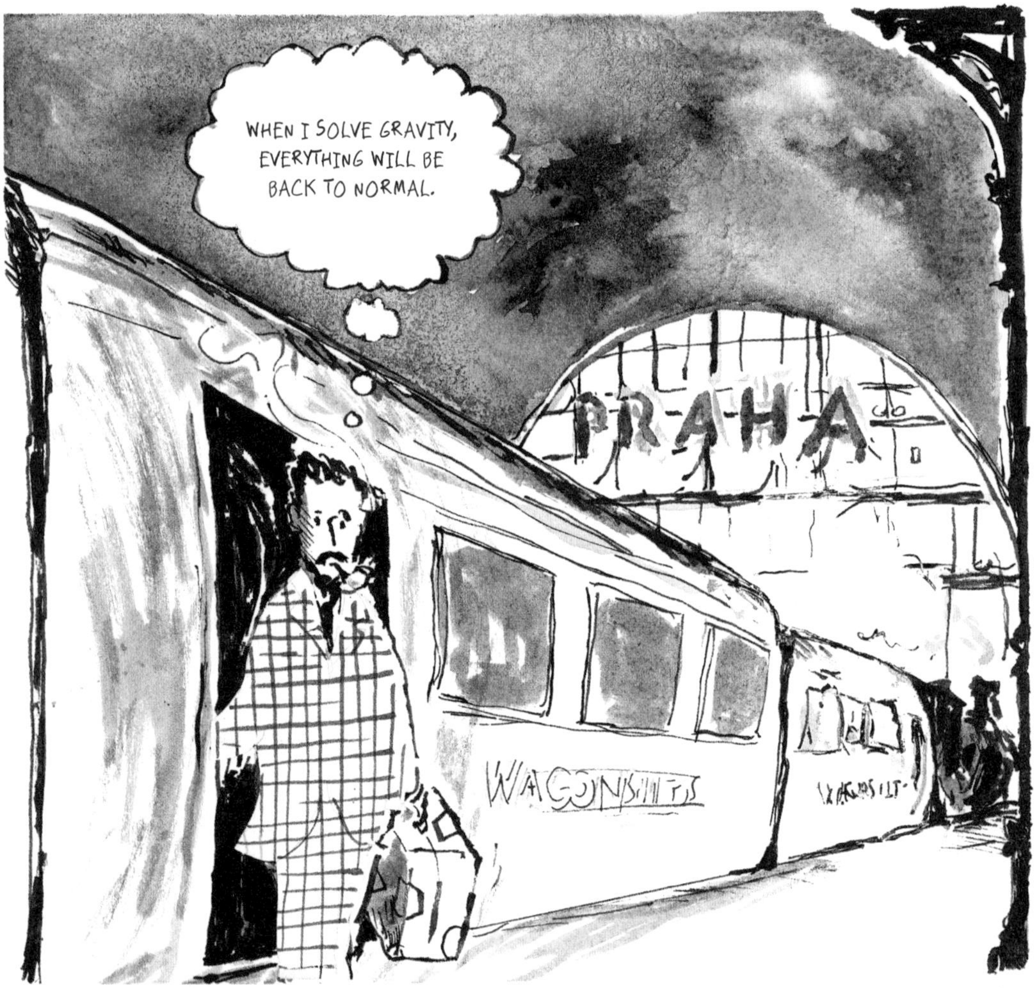
WHEN I SOLVE GRAVITY, EVERYTHING WILL BE BACK TO NORMAL.
PRAHA

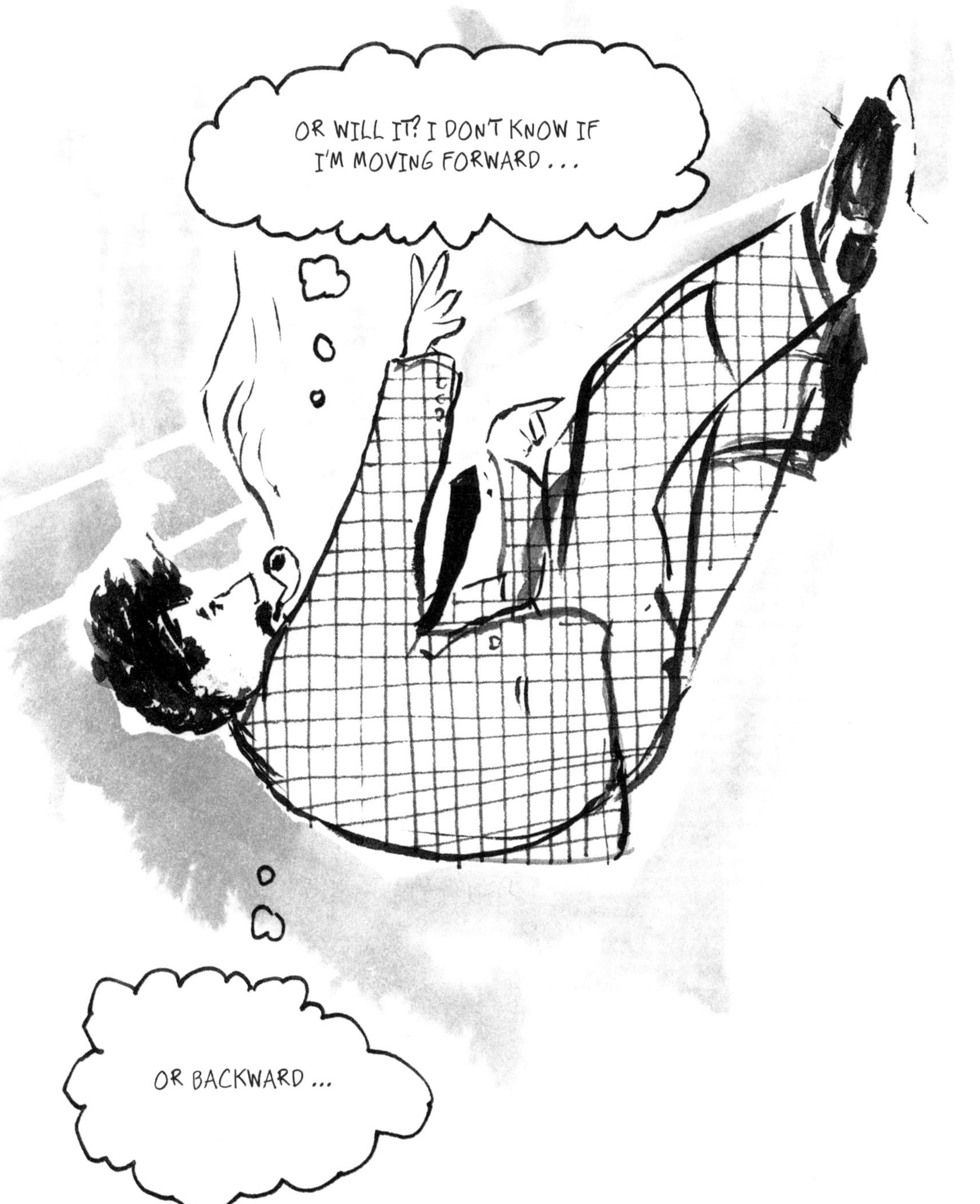
OR WILL IT? I DON'T KNOW IF I'M MOVING FORWARD . . .
OR BACKWARD ...

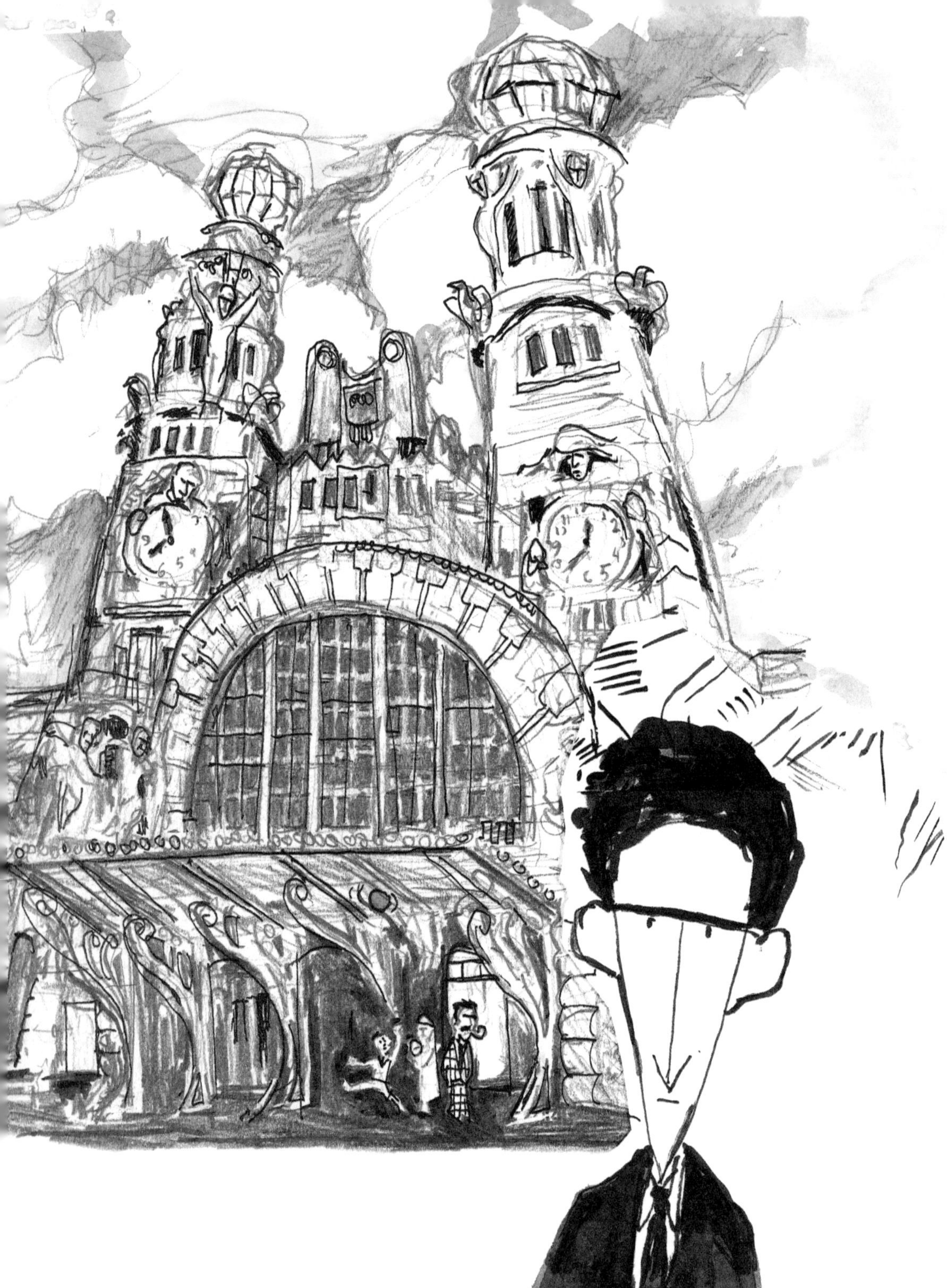

CHAPTER II

OBLAN A LOAY
CIVILIAN SWIMMING POOL

THE POOL OF TEARS

OF ALL MANKIND'S
AILMENTS, INSOMNIA IS
THE MOST HORRIBLE.

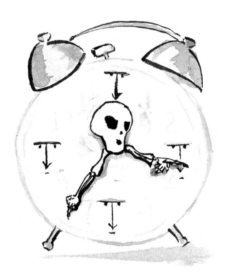

EYES WIDE OPEN, EACH
TICK OF THE CLOCK
A CANNON BLAST,
TICK, TICK, TICK, TICK
UNTIL DAWN FINALLY,
SOURFULLY BREAKS.

AND IF INSOMNIA
WERE AN OLYMPIC
EVENT, FRANZ KAFKA
WOULD BE ITS
MICHAEL PHELPS.

AS HE DOES MOST MORNINGS,
A BLEARY-EYED KAFKA HEADS
TO PRAGUE'S CIVILIAN
SWIMMING POOL* TO SWIM LAPS
WITH HIS BEST FRIEND AND
LITERARY GADFLY, MAX BROD.[5]

* Where as a child he used to swim daily with his father, Hermann, one of his scarce happy memories of him.

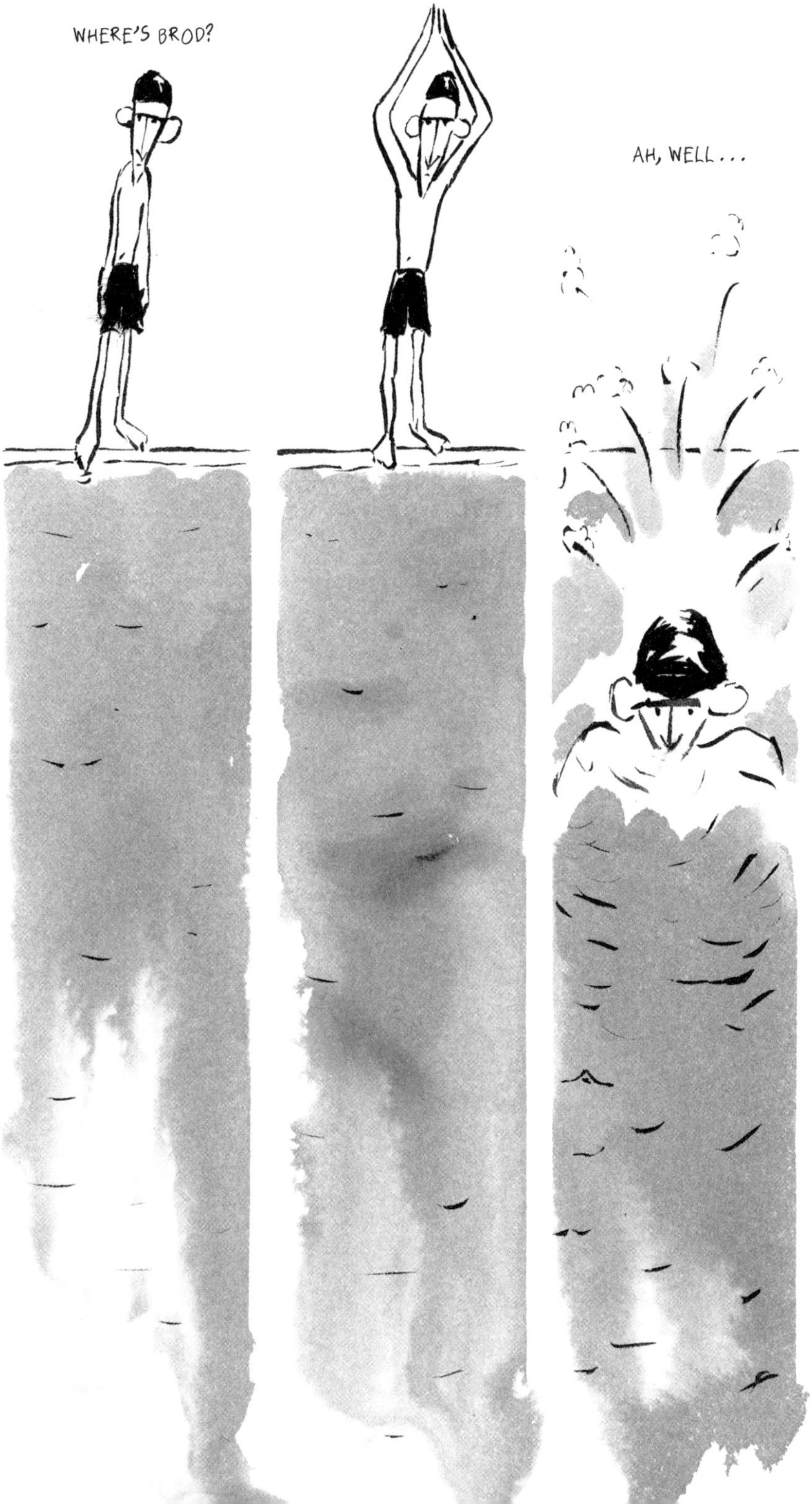
WHERE'S BROD?
AH, WELL...

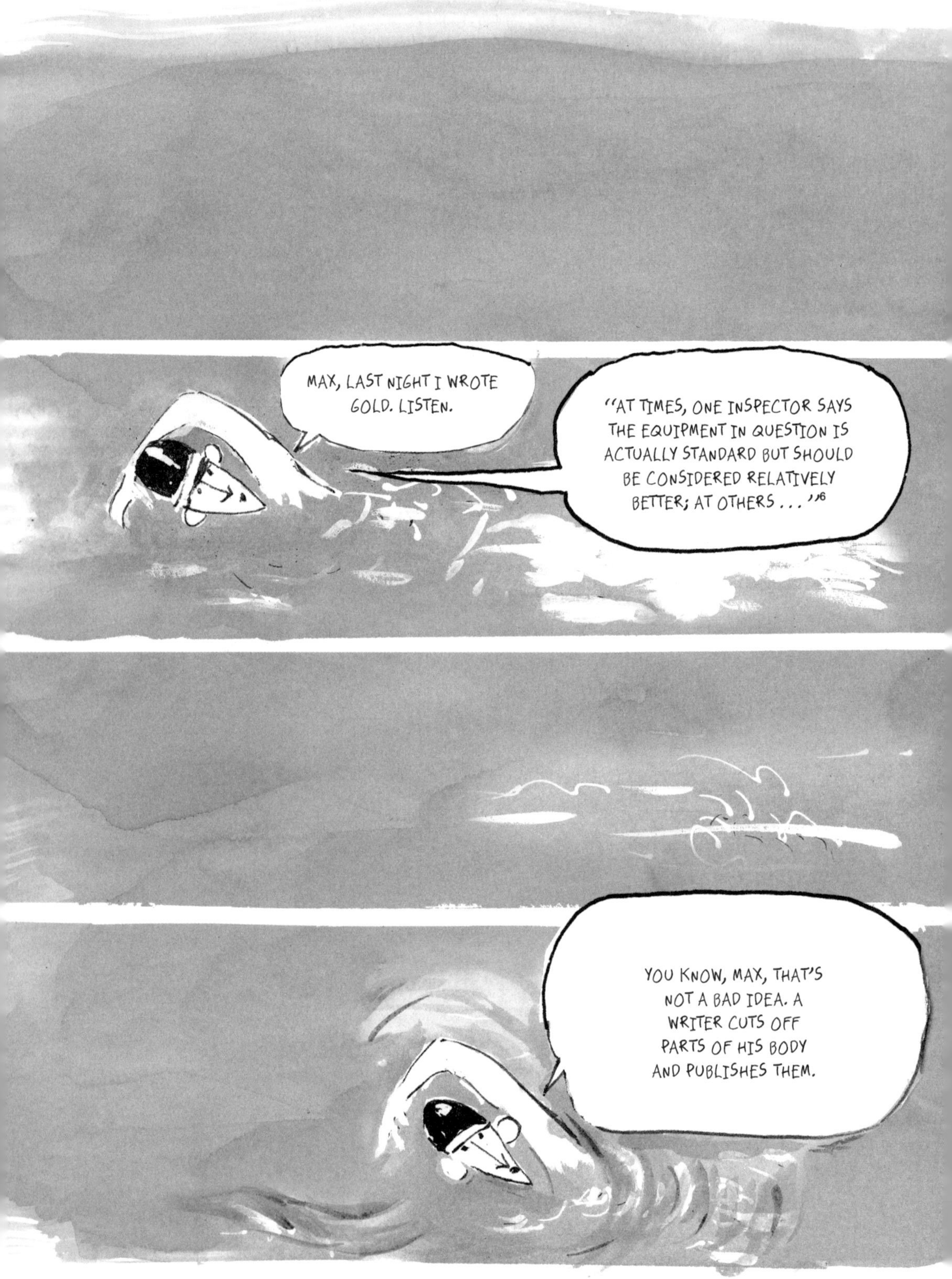
MAX, LAST NIGHT I WROTE GOLD. LISTEN.
"AT TIMES, ONE INSPECTOR SAYS THE EQUIPMENT IN QUESTION IS ACTUALLY STANDARD BUT SHOULD BE CONSIDERED RELATIVELY BETTER; AT OTHERS . . ."16
YOU KNOW, MAX, THAT'S NOT A BAD IDEA. A WRITER CUTS OFF PARTS OF HIS BODY AND PUBLISHES THEM.

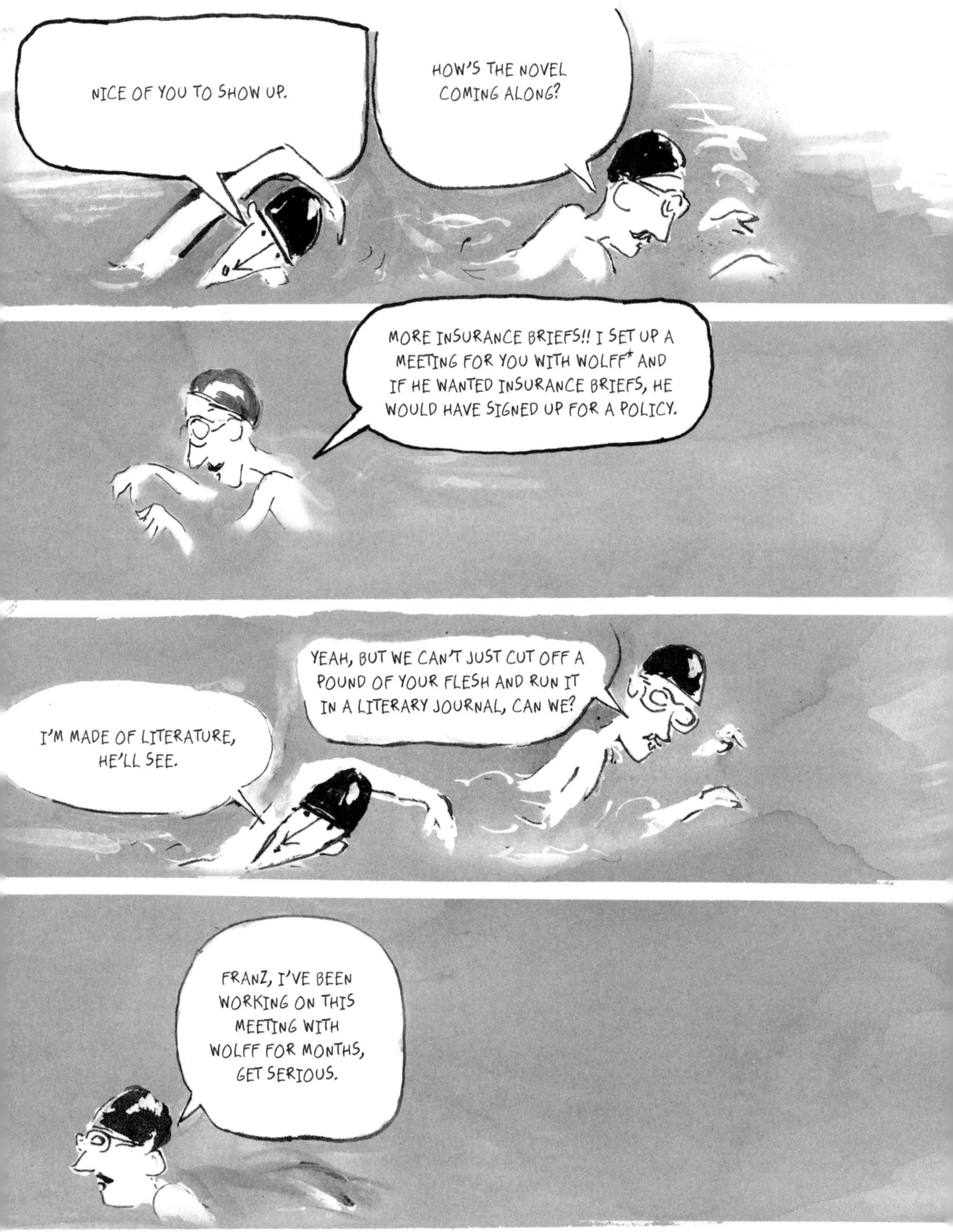
NICE OF YOU TO SHOW UP.
HOW'S THE NOVEL COMING ALONG?
MORE INSURANCE BRIEFS!! I SET UP A MEETING FOR YOU WITH WOLFF* AND IF HE WANTED INSURANCE BRIEFS, HE WOULD HAVE SIGNED UP FOR A POLICY.
I'M MADE OF LITERATURE, HE'LL SEE.
YEAH, BUT WE CAN'T JUST CUT OFF A POUND OF YOUR FLESH AND RUN IT IN A LITERARY JOURNAL, CAN WE?
FRANZ, I'VE BEEN WORKING ON THIS MEETING WITH WOLFF FOR MONTHS, GET SERIOUS.

* Kurt Wolff (1887–1963). German publisher, editor, and writer

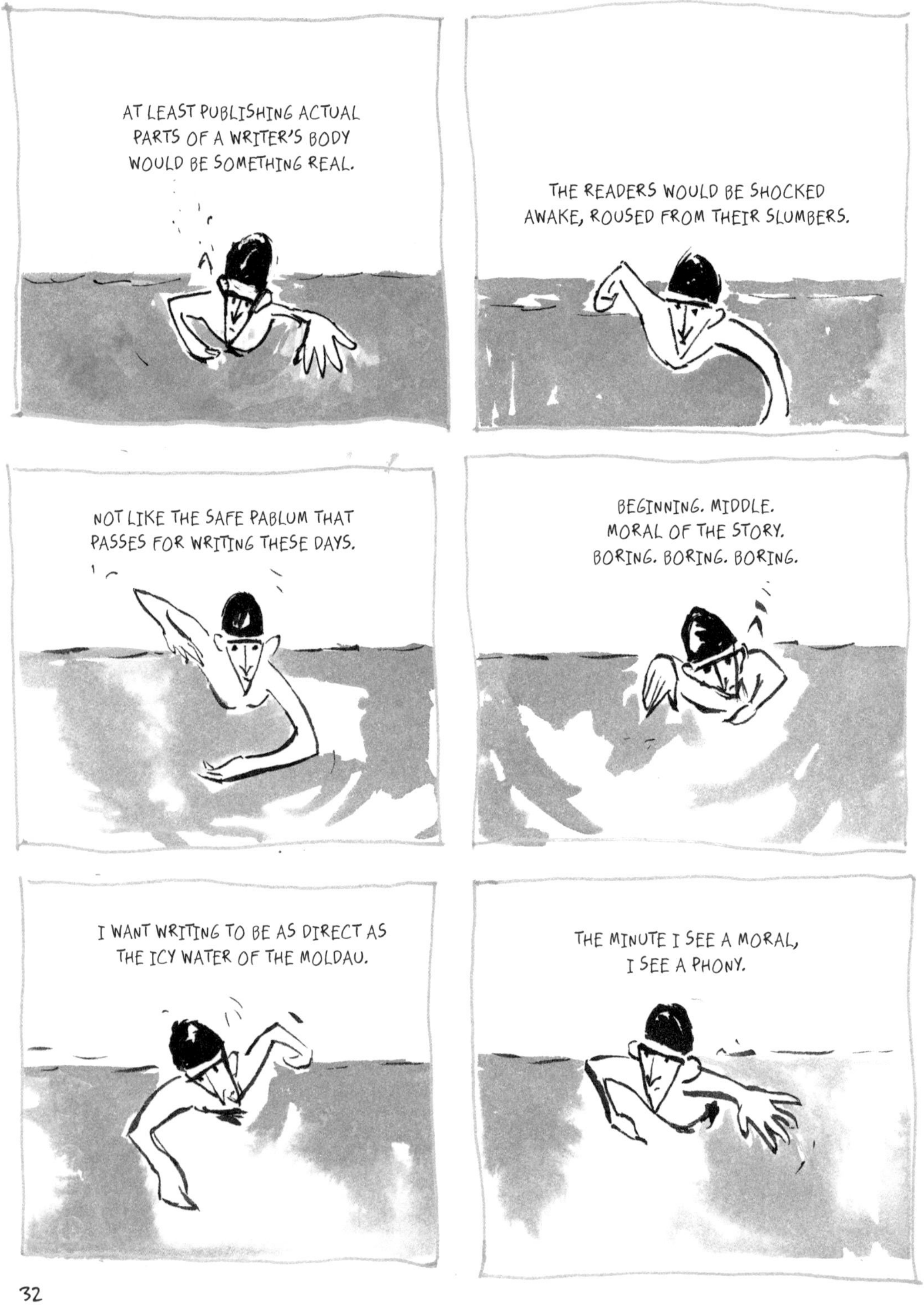
AT LEAST PUBLISHING ACTUAL PARTS OF A WRITER'S BODY WOULD BE SOMETHING REAL.
THE READERS WOULD BE SHOCKED AWAKE, ROUSED FROM THEIR SLUMBERS.
NOT LIKE THE SAFE PABLUM THAT PASSES FOR WRITING THESE DAYS.
BEGINNING. MIDDLE. MORAL OF THE STORY. BORING. BORING. BORING.
I WANT WRITING TO BE AS DIRECT AS THE ICY WATER OF THE MOLDAU.
THE MINUTE I SEE A MORAL, I SEE A PHONY.

READING SHOULD
BE UPSETTING,
DISRUPTIVE,
DISJOINTING.

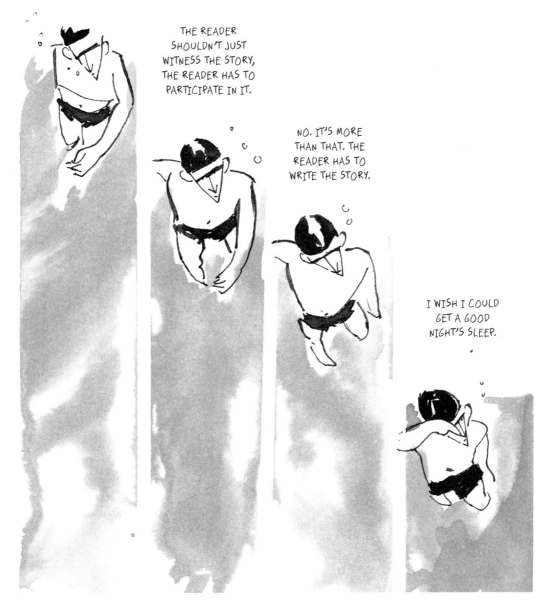

THE READER
SHOULDN'T JUST
WITNESS THE STORY,
THE READER HAS TO
PARTICIPATE IN IT.

NO. IT'S MORE
THAN THAT. THE
READER HAS TO
WRITE THE STORY.

I WISH I COULD
GET A GOOD
NIGHT'S SLEEP.

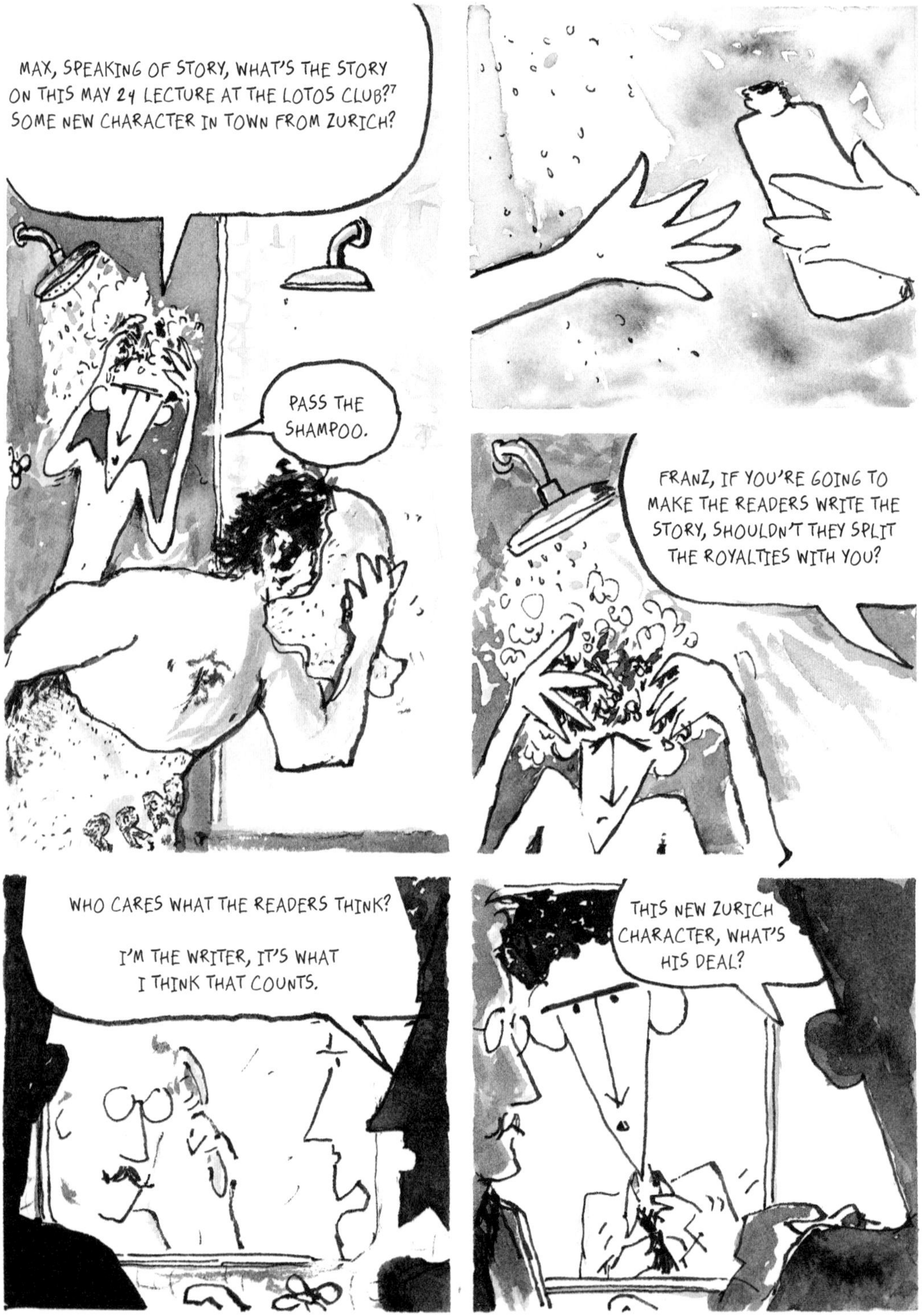
MAX, SPEAKING OF STORY, WHAT'S THE STORY ON THIS MAY 24 LECTURE AT THE LOTOS CLUB?[7] SOME NEW CHARACTER IN TOWN FROM ZURICH?
PASS THE SHAMPOO.
FRANZ, IF YOU'RE GOING TO MAKE THE READERS WRITE THE STORY, SHOULDN'T THEY SPLIT THE ROYALTIES WITH YOU?
WHO CARES WHAT THE READERS THINK?
I'M THE WRITER, IT'S WHAT I THINK THAT COUNTS.
THIS NEW ZURICH CHARACTER, WHAT'S HIS DEAL?

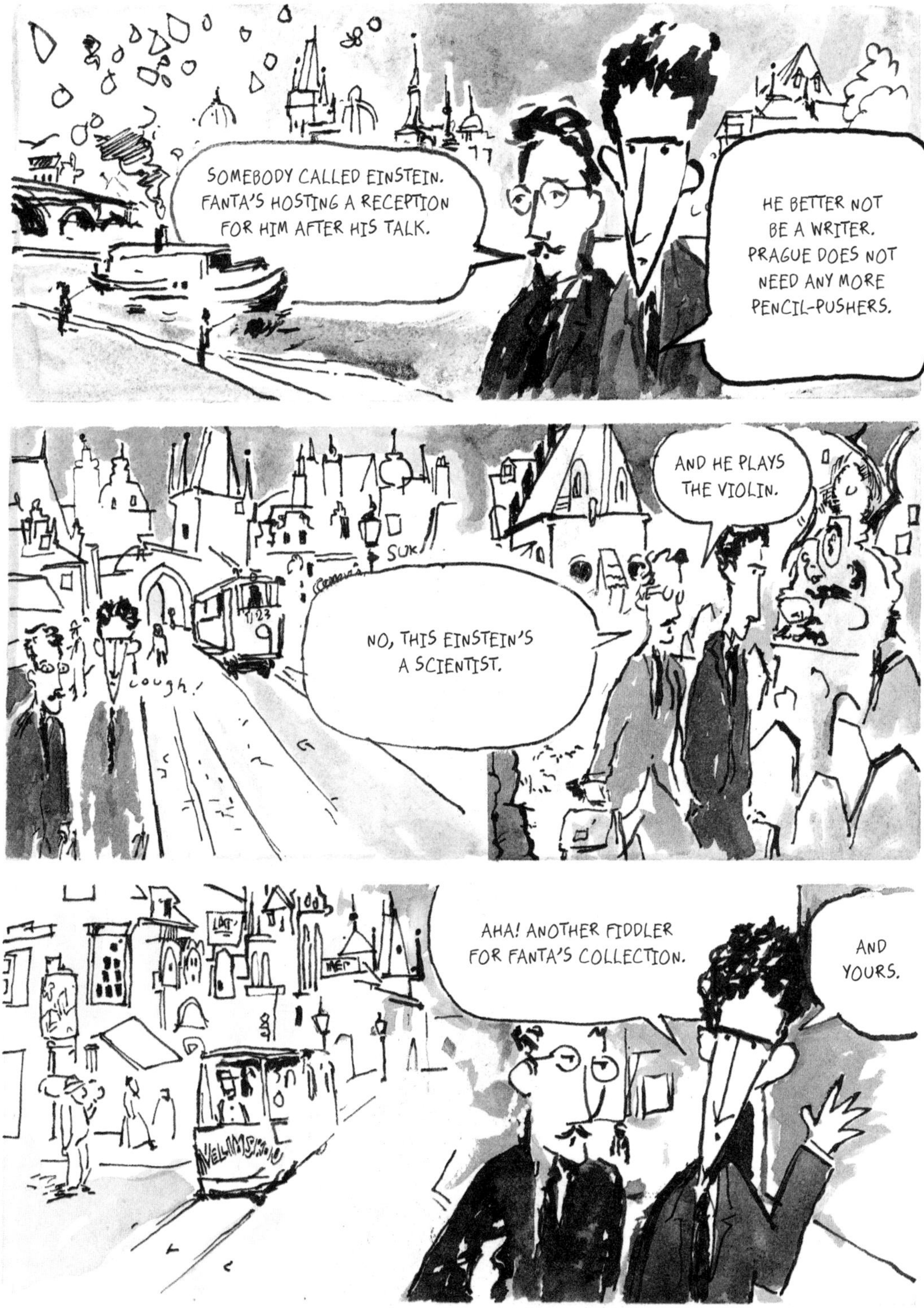
SOMEBODY CALLED EINSTEIN. FANTA'S HOSTING A RECEPTION FOR HIM AFTER HIS TALK.
HE BETTER NOT BE A WRITER. PRAGUE DOES NOT NEED ANY MORE PENCIL-PUSHERS.
cough!
NO, THIS EINSTEIN'S A SCIENTIST.
AND HE PLAYS THE VIOLIN.
AHA! ANOTHER FIDDLER FOR FANTA'S COLLECTION.
AND YOURS.

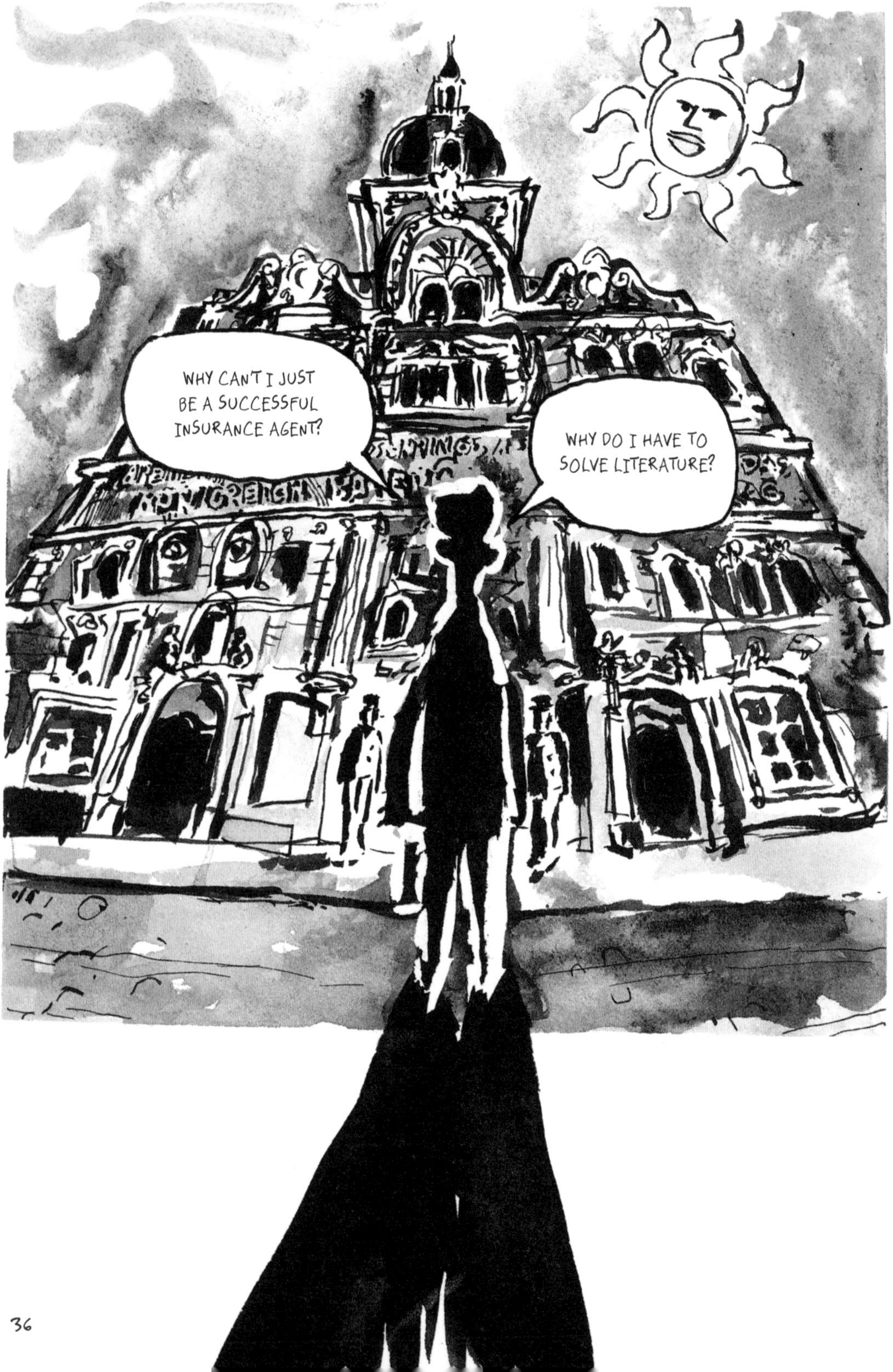
WHY CAN'T I JUST BE A SUCCESSFUL INSURANCE AGENT?
WHY DO I HAVE TO SOLVE LITERATURE?

CHAPTER III

MEET MAX ABRAHAM

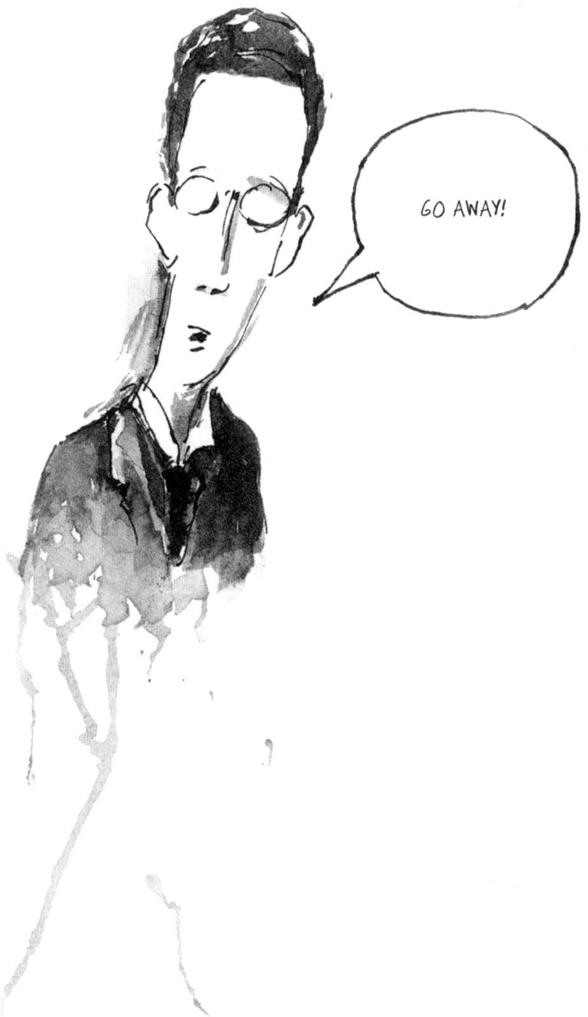
GO AWAY!

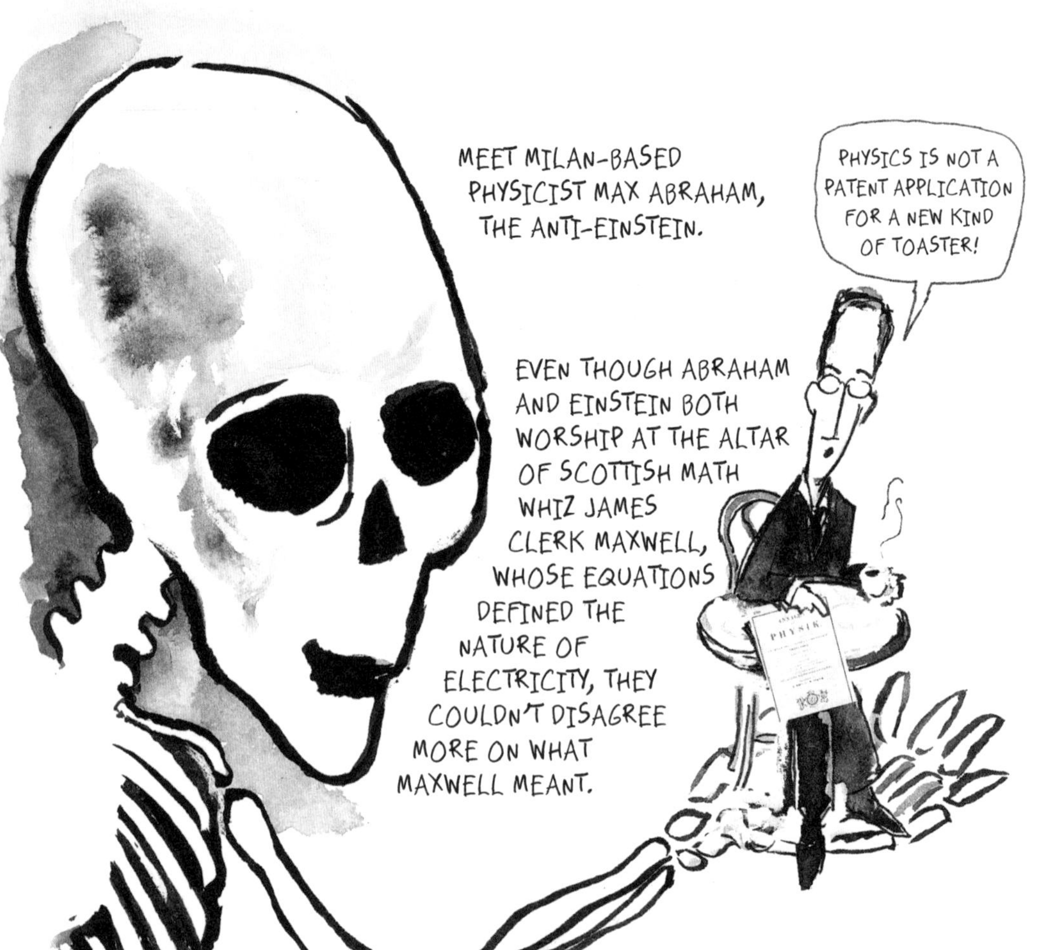
MEET MILAN-BASED PHYSICIST MAX ABRAHAM, THE ANTI-EINSTEIN.
PHYSICS IS NOT A PATENT APPLICATION FOR A NEW KIND OF TOASTER!
EVEN THOUGH ABRAHAM AND EINSTEIN BOTH WORSHIP AT THE ALTAR OF SCOTTISH MATH WHIZ JAMES CLERK MAXWELL, WHOSE EQUATIONS DEFINED THE NATURE OF ELECTRICITY, THEY COULDN'T DISAGREE MORE ON WHAT MAXWELL MEANT.

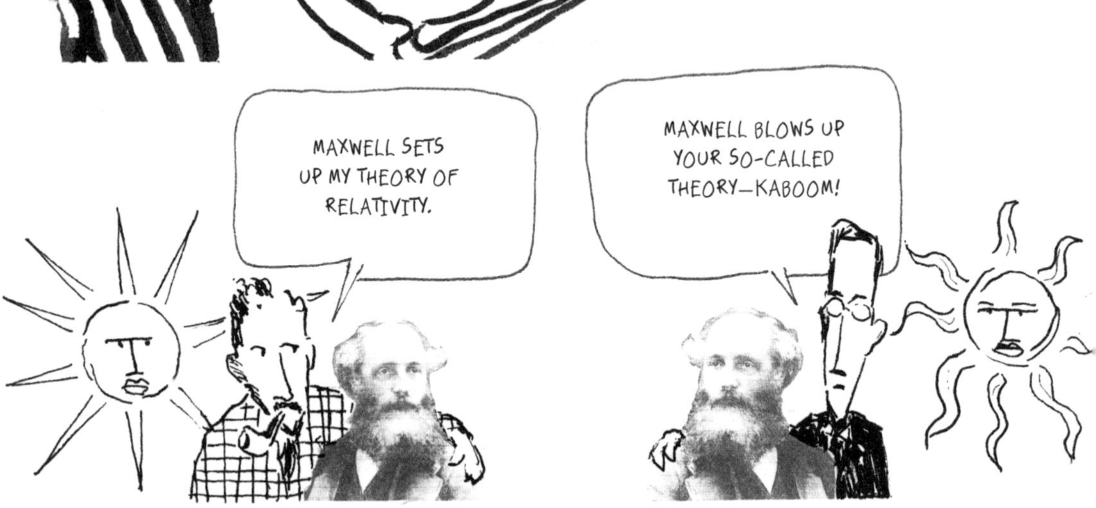
MAXWELL SETS UP MY THEORY OF RELATIVITY.
MAXWELL BLOWS UP YOUR SO-CALLED THEORY—KABOOM!

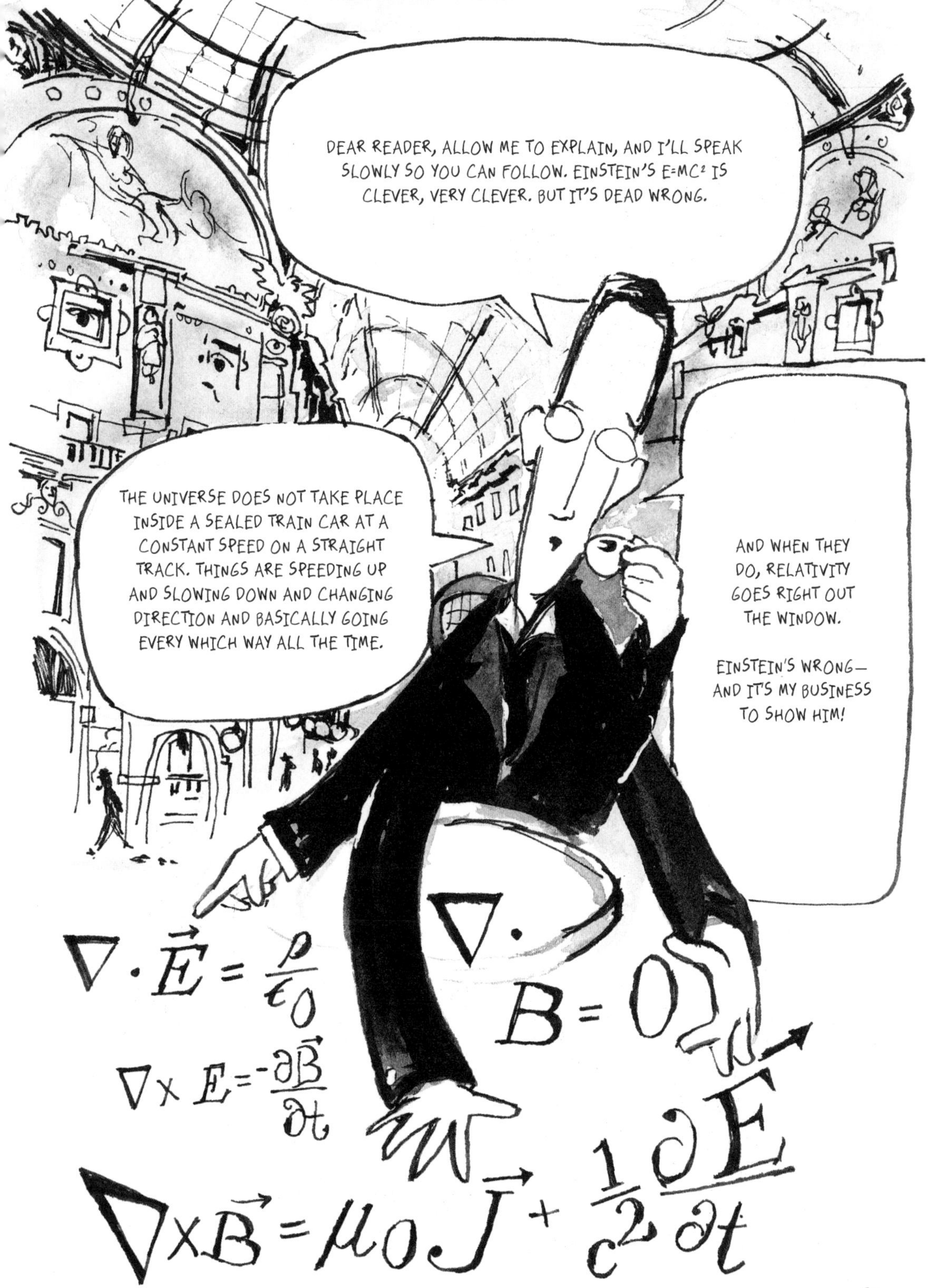
DEAR READER, ALLOW ME TO EXPLAIN, AND I'LL SPEAK SLOWLY SO YOU CAN FOLLOW. EINSTEIN'S E=MC² IS CLEVER, VERY CLEVER. BUT IT'S DEAD WRONG.
THE UNIVERSE DOES NOT TAKE PLACE INSIDE A SEALED TRAIN CAR AT A CONSTANT SPEED ON A STRAIGHT TRACK. THINGS ARE SPEEDING UP AND SLOWING DOWN AND CHANGING DIRECTION AND BASICALLY GOING EVERY WHICH WAY ALL THE TIME.
AND WHEN THEY DO, RELATIVITY GOES RIGHT OUT THE WINDOW.
EINSTEIN'S WRONG—AND IT'S MY BUSINESS TO SHOW HIM!
∇·E = ρ/ε0
∇·B = 0
∇×E = -∂B/∂t
∇×B = μ0 J + 1/c² ∂E/∂t

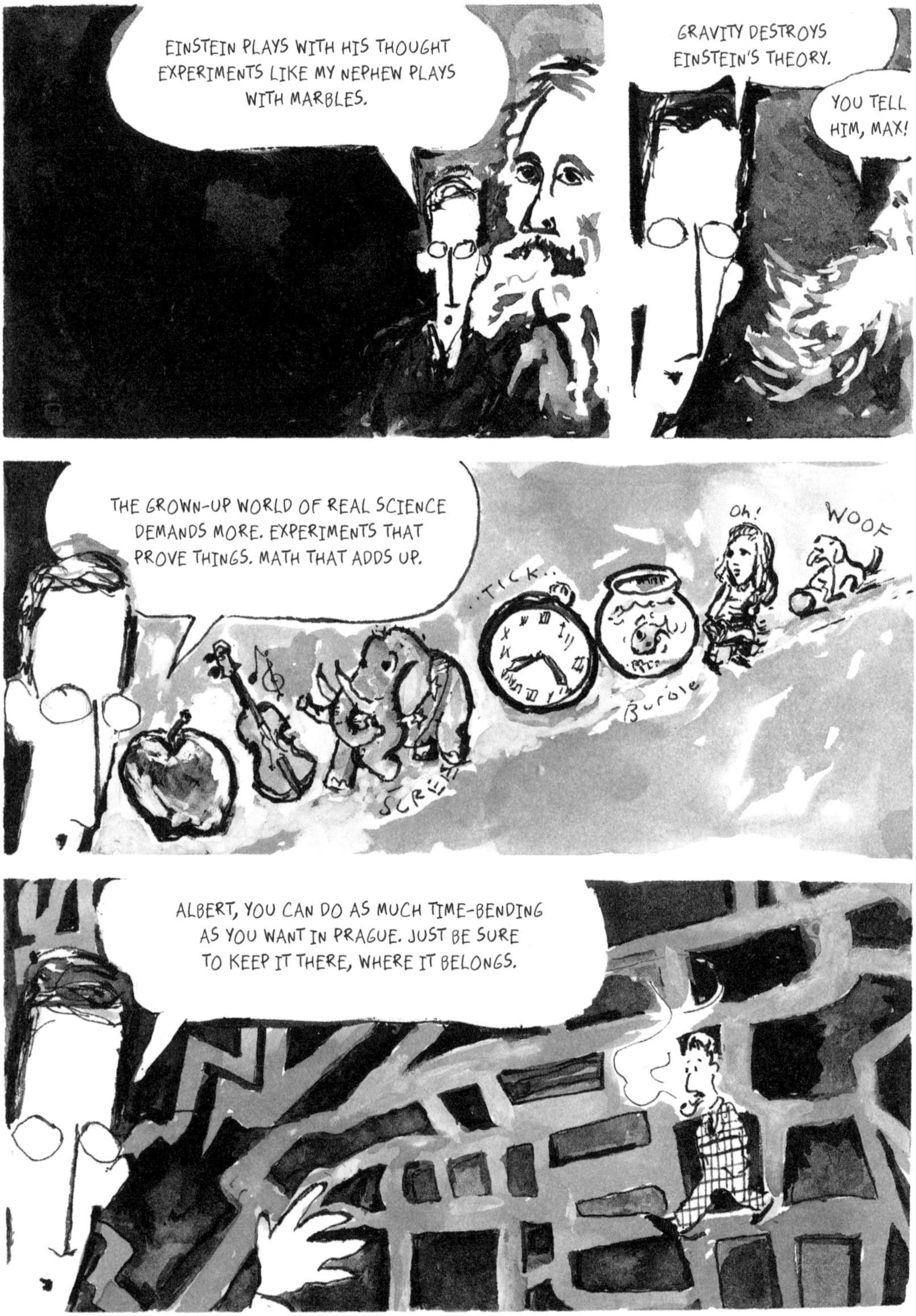
EINSTEIN PLAYS WITH HIS THOUGHT EXPERIMENTS LIKE MY NEPHEW PLAYS WITH MARBLES.
GRAVITY DESTROYS EINSTEIN'S THEORY.
YOU TELL HIM, MAX!
THE GROWN-UP WORLD OF REAL SCIENCE DEMANDS MORE. EXPERIMENTS THAT PROVE THINGS. MATH THAT ADDS UP.
TICK..
Oh!
WOOF
Burble
ALBERT, YOU CAN DO AS MUCH TIME-BENDING AS YOU WANT IN PRAGUE. JUST BE SURE TO KEEP IT THERE, WHERE IT BELONGS.

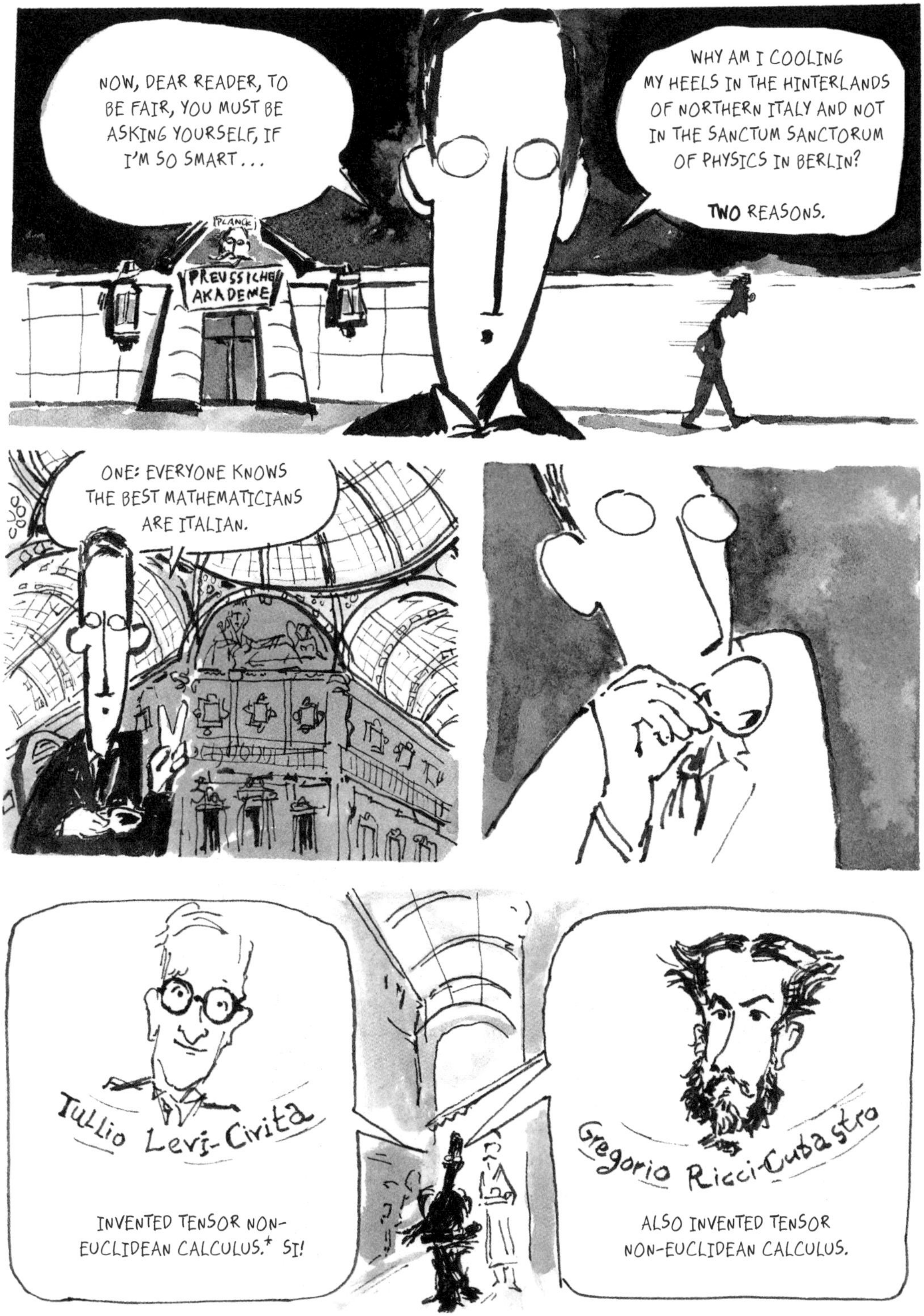
NOW, DEAR READER, TO BE FAIR, YOU MUST BE ASKING YOURSELF, IF I'M SO SMART . . .
WHY AM I COOLING MY HEELS IN THE HINTERLANDS OF NORTHERN ITALY AND NOT IN THE SANCTUM SANCTORUM OF PHYSICS IN BERLIN?
TWO REASONS.
PLANCK
PREUSSICHE AKADEME
ONE: EVERYONE KNOWS THE BEST MATHEMATICIANS ARE ITALIAN.
Tullio Levi-Civita
INVENTED TENSOR NON-EUCLIDEAN CALCULUS.* SI!
Gregorio Ricci-Curbastro
ALSO INVENTED TENSOR NON-EUCLIDEAN CALCULUS.

* Whatever that is?

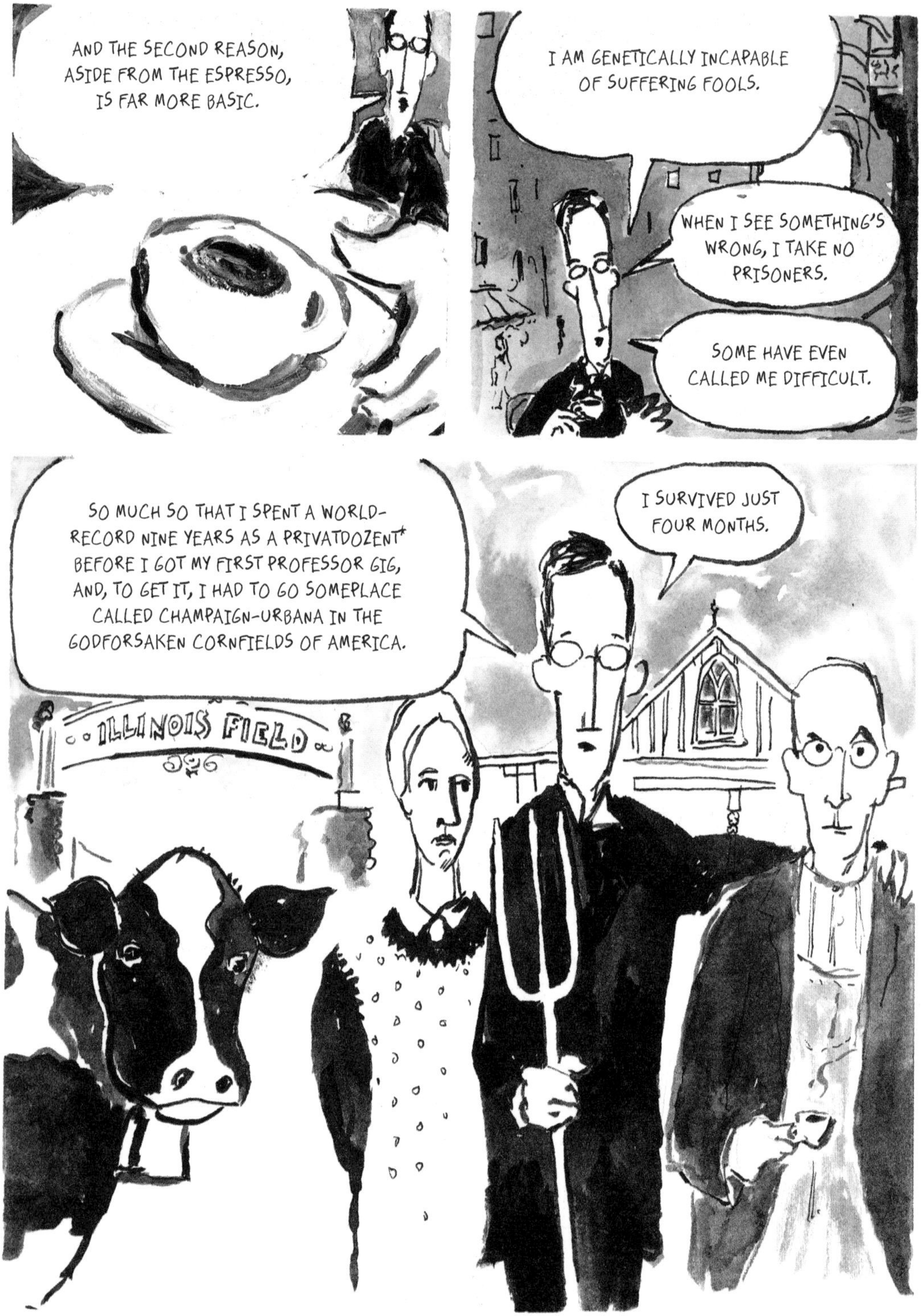
AND THE SECOND REASON, ASIDE FROM THE ESPRESSO, IS FAR MORE BASIC.
I AM GENETICALLY INCAPABLE OF SUFFERING FOOLS.
WHEN I SEE SOMETHING'S WRONG, I TAKE NO PRISONERS.
SOME HAVE EVEN CALLED ME DIFFICULT.
SO MUCH SO THAT I SPENT A WORLD-RECORD NINE YEARS AS A PRIVATDOZENT* BEFORE I GOT MY FIRST PROFESSOR GIG, AND, TO GET IT, I HAD TO GO SOMEPLACE CALLED CHAMPAIGN-URBANA IN THE GODFORSAKEN CORNFIELDS OF AMERICA.
I SURVIVED JUST FOUR MONTHS.
ILLINOIS FIELD

*A second-tier university lecturer, kind of a stand-in for a real professor, without the respect or the salary

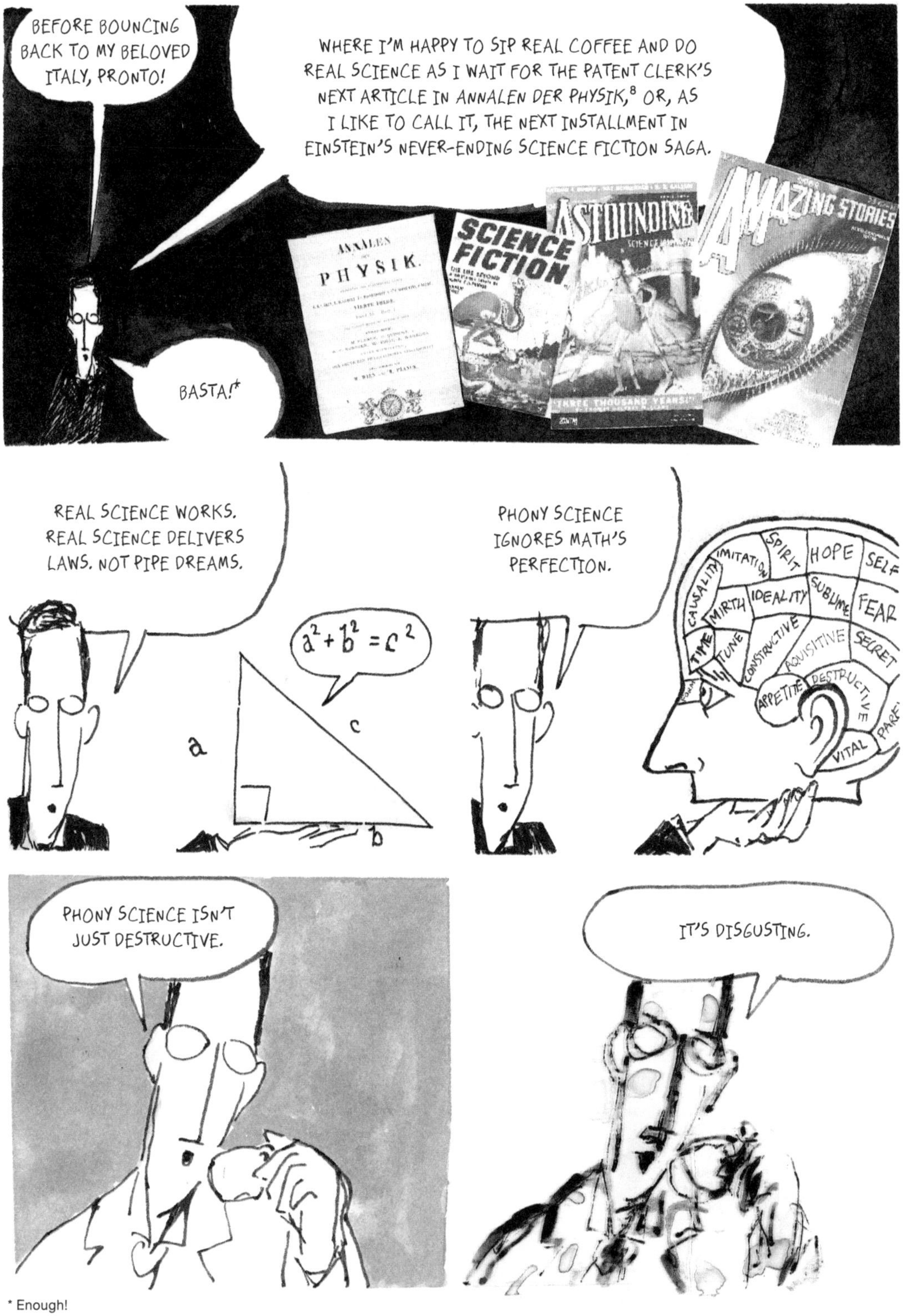
BEFORE BOUNCING BACK TO MY BELOVED ITALY, PRONTO!
WHERE I'M HAPPY TO SIP REAL COFFEE AND DO REAL SCIENCE AS I WAIT FOR THE PATENT CLERK'S NEXT ARTICLE IN *ANNALEN DER PHYSIK*,[8] OR, AS I LIKE TO CALL IT, THE NEXT INSTALLMENT IN EINSTEIN'S NEVER-ENDING SCIENCE FICTION SAGA.
ANNALEN PHYSIK
SCIENCE FICTION
ASTOUNDING
AMAZING STORIES
"THREE THOUSAND YEARS!"
BASTA!*
REAL SCIENCE WORKS. REAL SCIENCE DELIVERS LAWS. NOT PIPE DREAMS.
$a^2+b^2=c^2$
a
b
c
PHONY SCIENCE IGNORES MATH'S PERFECTION.
IMITATION
SPIRIT
HOPE
CAUSALITY
MIRTH
IDEALITY
SUBLIME
FEAR
TIME
TUNE
CONSTRUCTIVE
SECRET
APPETITE
DESTRUCTIVE
VITAL
PHONY SCIENCE ISN'T JUST DESTRUCTIVE.
IT'S DISGUSTING.

* Enough!

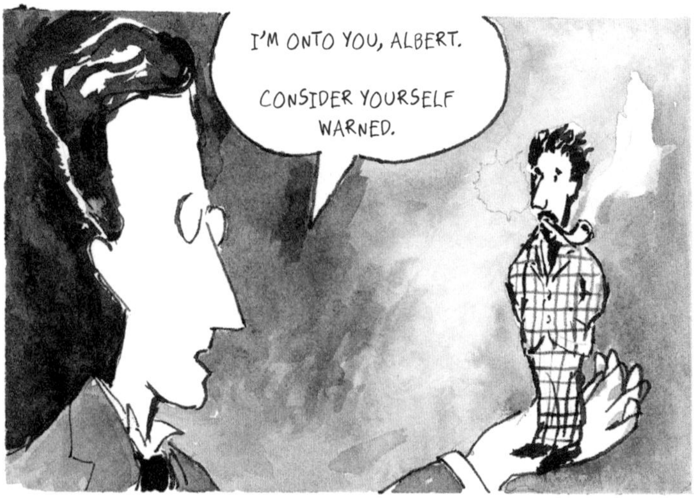
I'M ONTO YOU, ALBERT.
CONSIDER YOURSELF WARNED.

CHAPTER IV
"The People Are So Happy Here!"

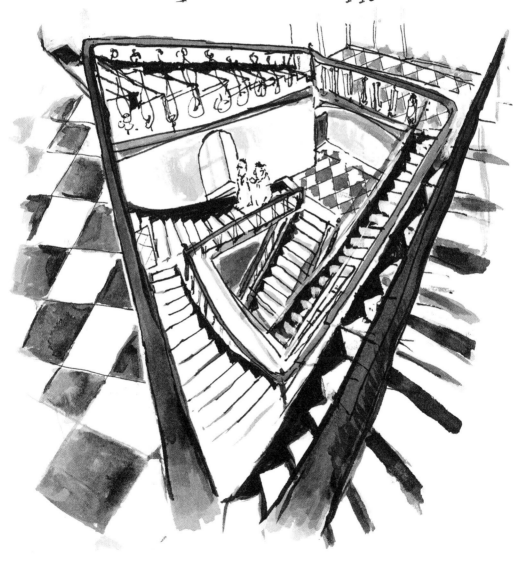

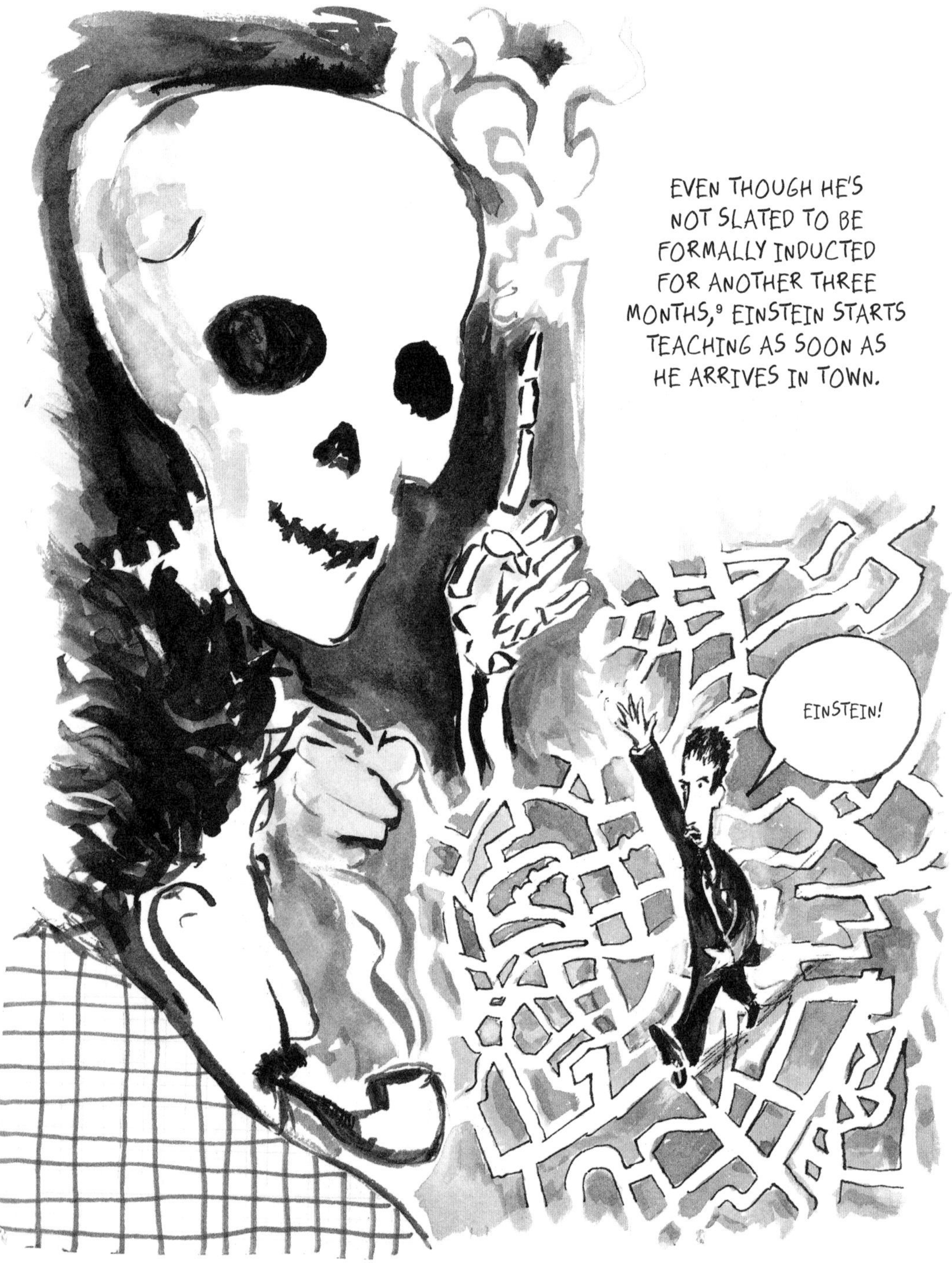
EVEN THOUGH HE'S NOT SLATED TO BE FORMALLY INDUCTED FOR ANOTHER THREE MONTHS,[9] EINSTEIN STARTS TEACHING AS SOON AS HE ARRIVES IN TOWN.
EINSTEIN!

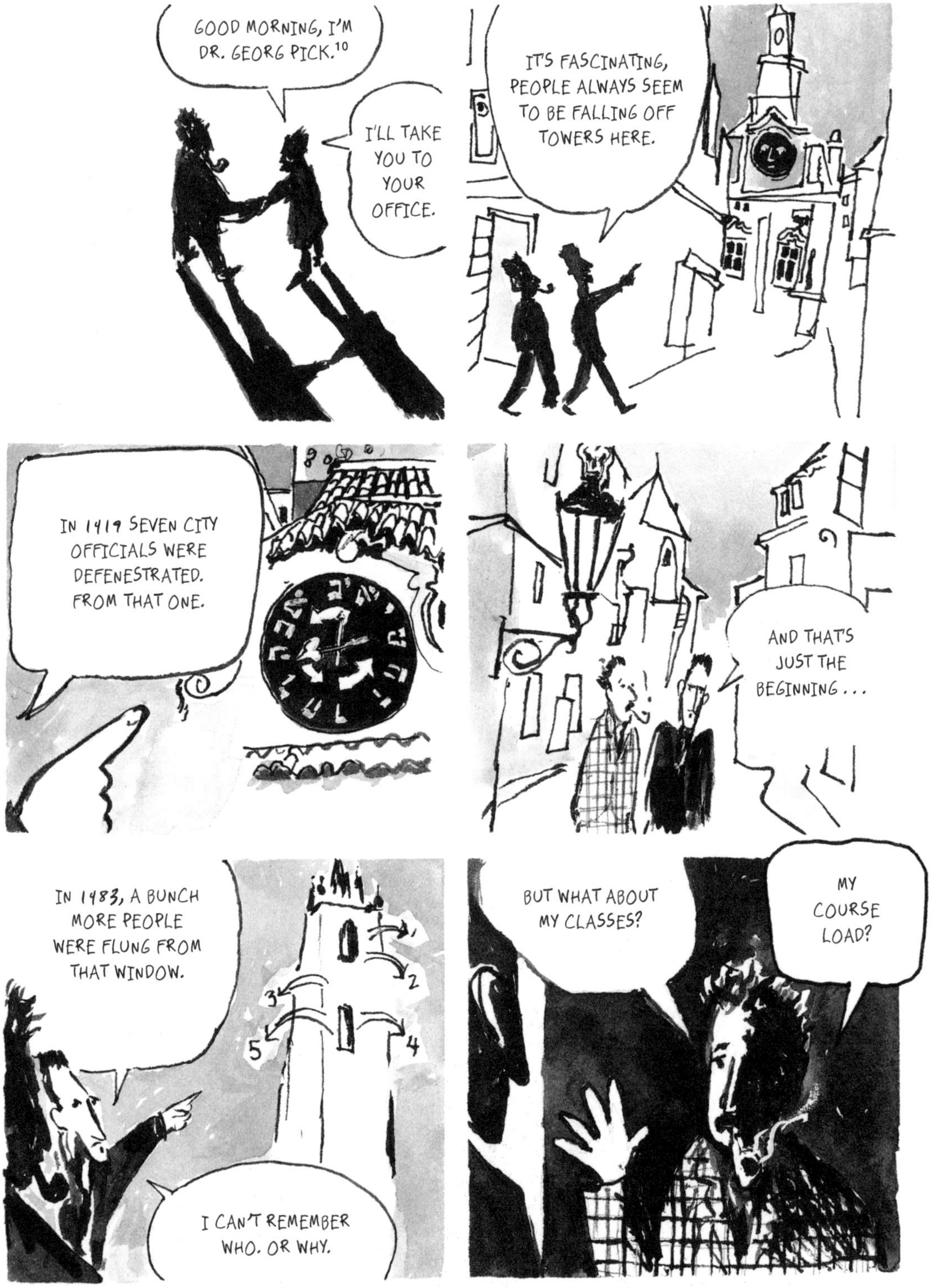
GOOD MORNING, I'M DR. GEORG PICK.10
I'LL TAKE YOU TO YOUR OFFICE.
IT'S FASCINATING, PEOPLE ALWAYS SEEM TO BE FALLING OFF TOWERS HERE.
IN 1419 SEVEN CITY OFFICIALS WERE DEFENESTRATED. FROM THAT ONE.
AND THAT'S JUST THE BEGINNING . . .
IN 1483, A BUNCH MORE PEOPLE WERE FLUNG FROM THAT WINDOW.
1
2
3
4
5
I CAN'T REMEMBER WHO. OR WHY.
BUT WHAT ABOUT MY CLASSES?
MY COURSE LOAD?

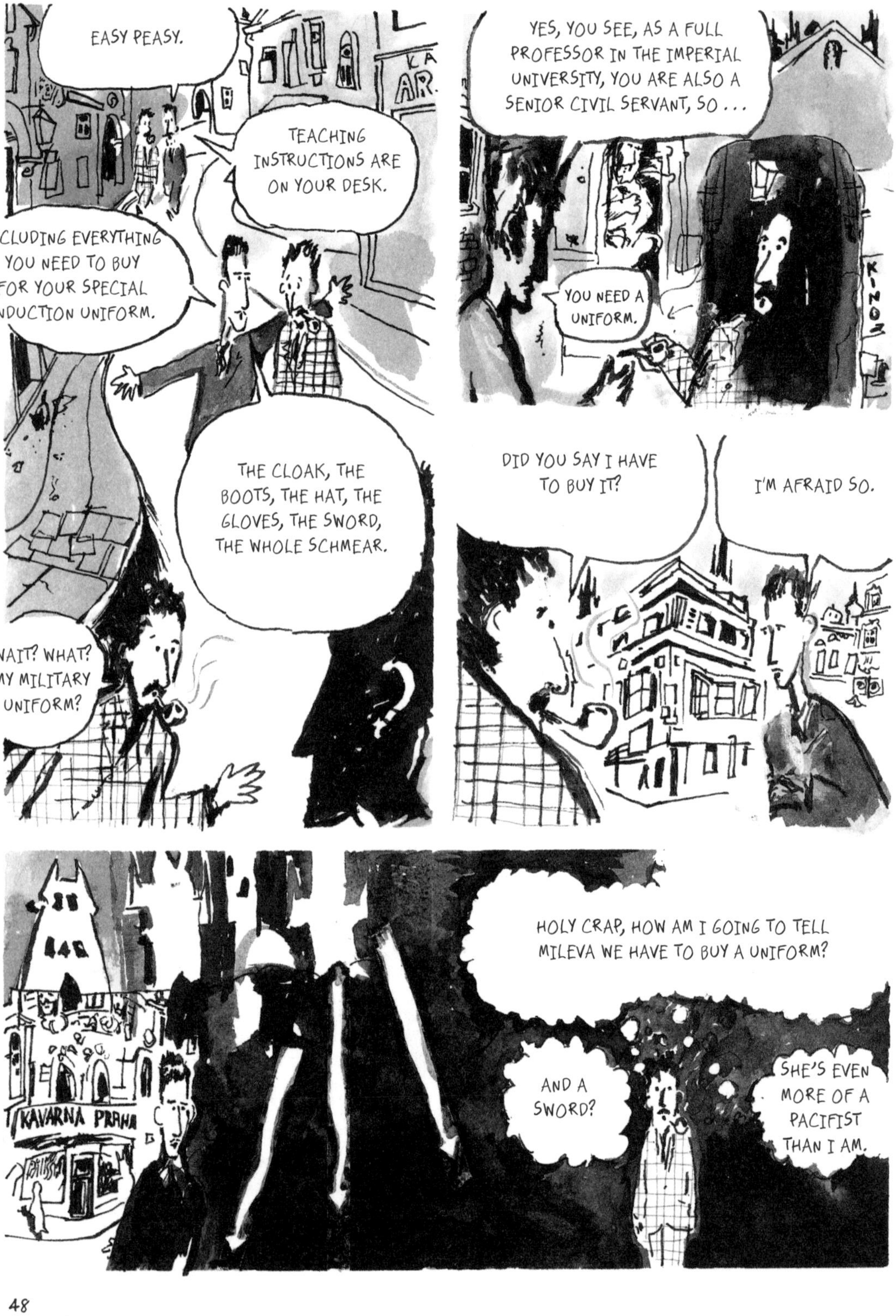
EASY PEASY.
TEACHING INSTRUCTIONS ARE ON YOUR DESK.
INCLUDING EVERYTHING YOU NEED TO BUY FOR YOUR SPECIAL INDUCTION UNIFORM.
THE CLOAK, THE BOOTS, THE HAT, THE GLOVES, THE SWORD, THE WHOLE SCHMEAR.
WAIT? WHAT? MY MILITARY UNIFORM?
YES, YOU SEE, AS A FULL PROFESSOR IN THE IMPERIAL UNIVERSITY, YOU ARE ALSO A SENIOR CIVIL SERVANT, SO . . .
YOU NEED A UNIFORM.
KINO
DID YOU SAY I HAVE TO BUY IT?
I'M AFRAID SO.
HOLY CRAP, HOW AM I GOING TO TELL MILEVA WE HAVE TO BUY A UNIFORM?
AND A SWORD?
SHE'S EVEN MORE OF A PACIFIST THAN I AM.
KAVARNA PRAHA

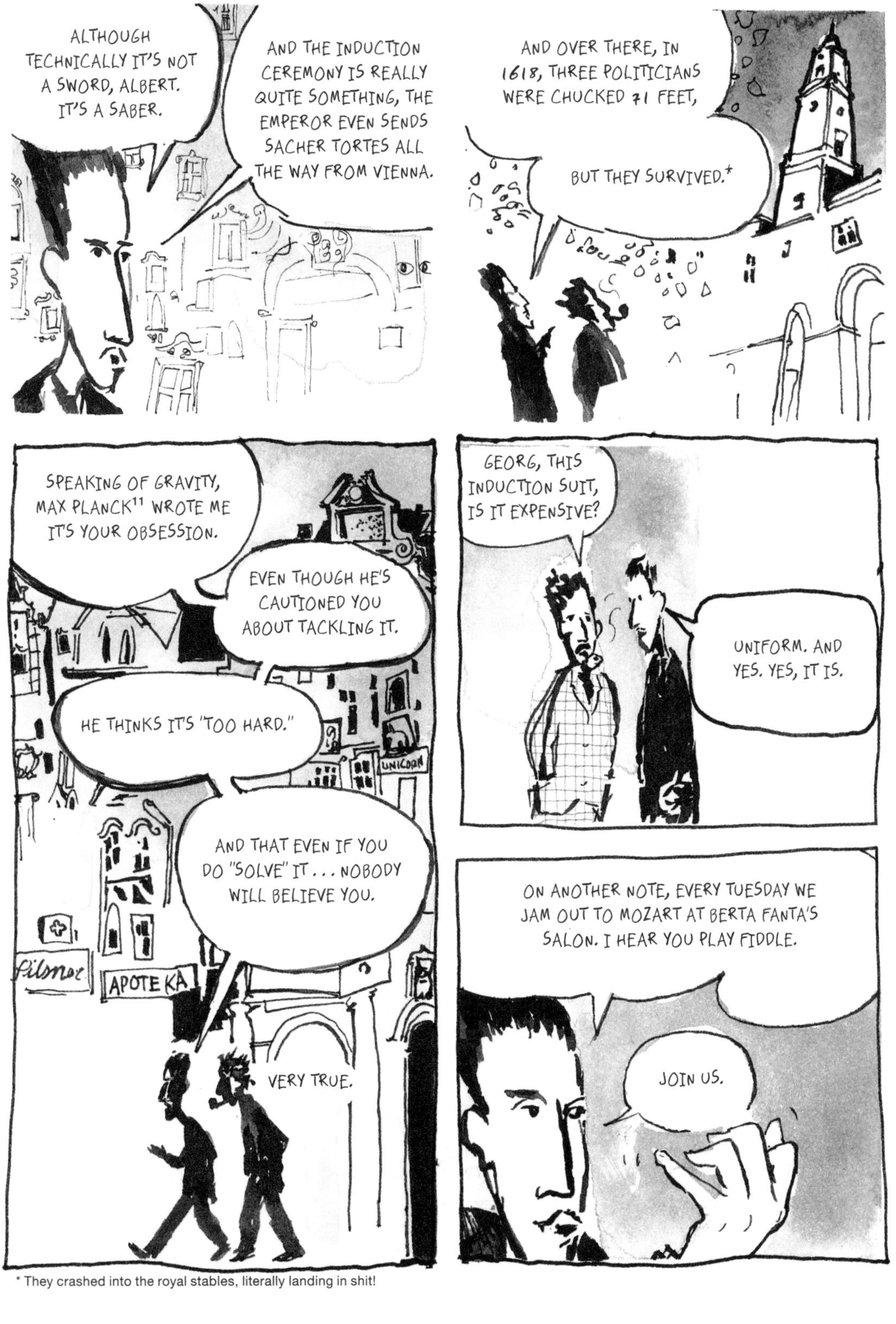
ALTHOUGH TECHNICALLY IT'S NOT A SWORD, ALBERT. IT'S A SABER.
AND THE INDUCTION CEREMONY IS REALLY QUITE SOMETHING, THE EMPEROR EVEN SENDS SACHER TORTES ALL THE WAY FROM VIENNA.
AND OVER THERE, IN 1618, THREE POLITICIANS WERE CHUCKED 71 FEET,
BUT THEY SURVIVED.*
SPEAKING OF GRAVITY, MAX PLANCK[11] WROTE ME IT'S YOUR OBSESSION.
EVEN THOUGH HE'S CAUTIONED YOU ABOUT TACKLING IT.
HE THINKS IT'S "TOO HARD."
AND THAT EVEN IF YOU DO "SOLVE" IT . . . NOBODY WILL BELIEVE YOU.
UNICORN
Pilsner
APOTEKA
VERY TRUE.
GEORG, THIS INDUCTION SUIT, IS IT EXPENSIVE?
UNIFORM. AND YES. YES, IT IS.
ON ANOTHER NOTE, EVERY TUESDAY WE JAM OUT TO MOZART AT BERTA FANTA'S SALON. I HEAR YOU PLAY FIDDLE.
JOIN US.

* They crashed into the royal stables, literally landing in shit!

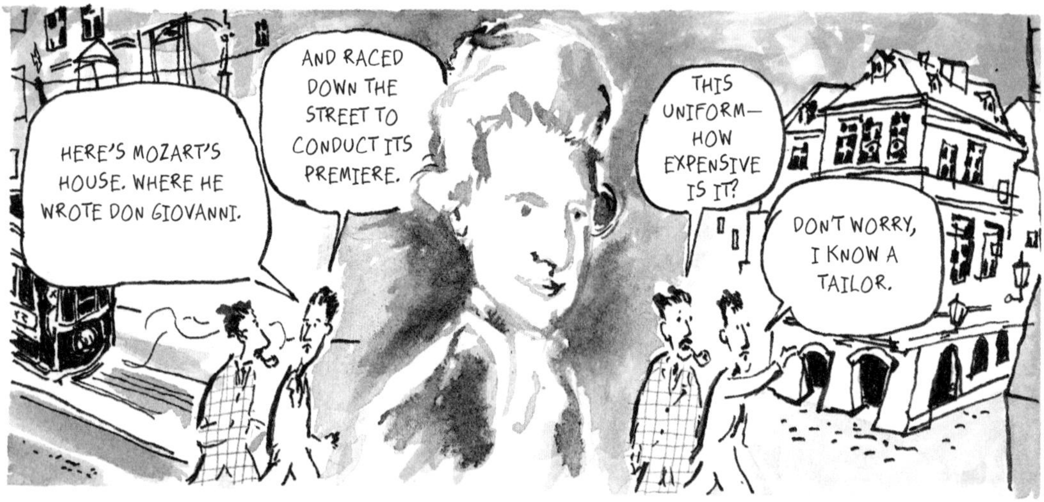
HERE'S MOZART'S HOUSE. WHERE HE WROTE DON GIOVANNI.
AND RACED DOWN THE STREET TO CONDUCT ITS PREMIERE.
THIS UNIFORM— HOW EXPENSIVE IS IT?
DON'T WORRY, I KNOW A TAILOR.

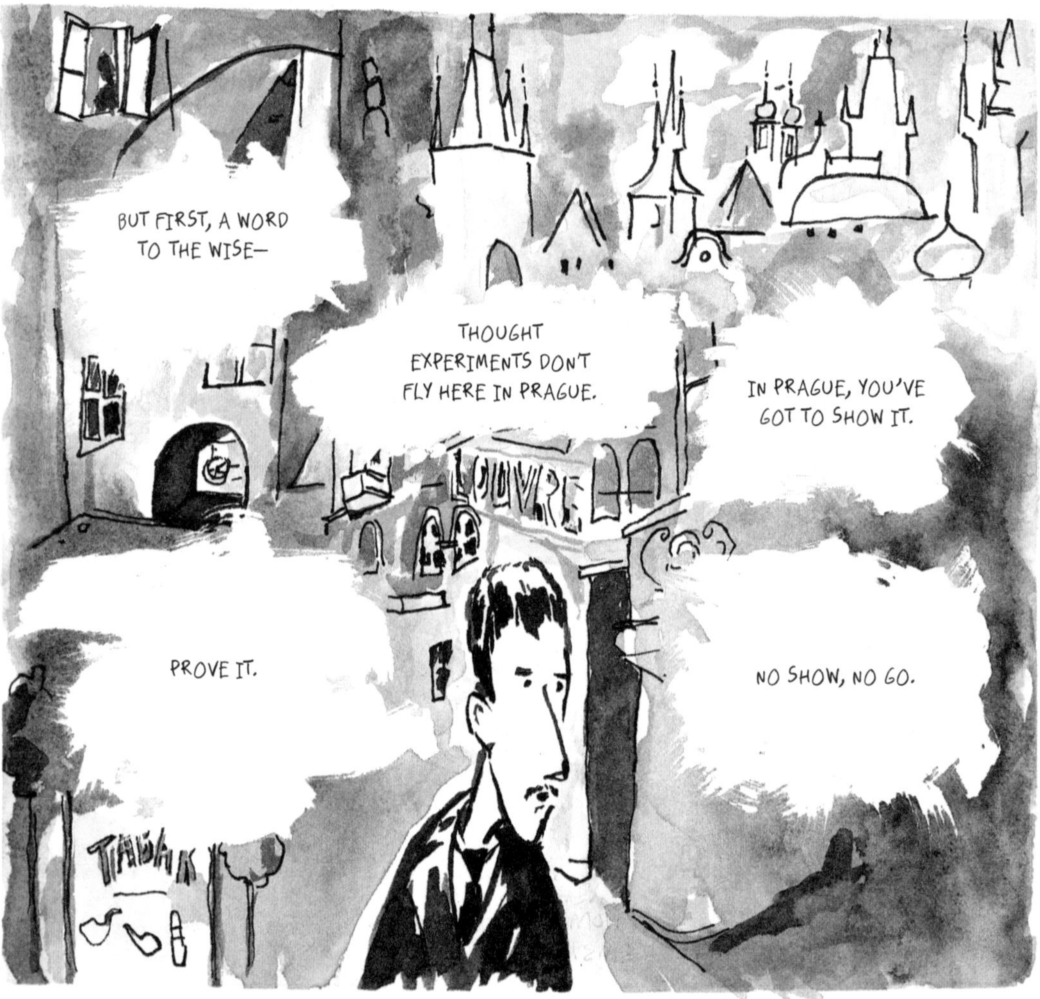
BUT FIRST, A WORD TO THE WISE—
THOUGHT EXPERIMENTS DON'T FLY HERE IN PRAGUE.
IN PRAGUE, YOU'VE GOT TO SHOW IT.
LOUVRE
PROVE IT.
NO SHOW, NO GO.

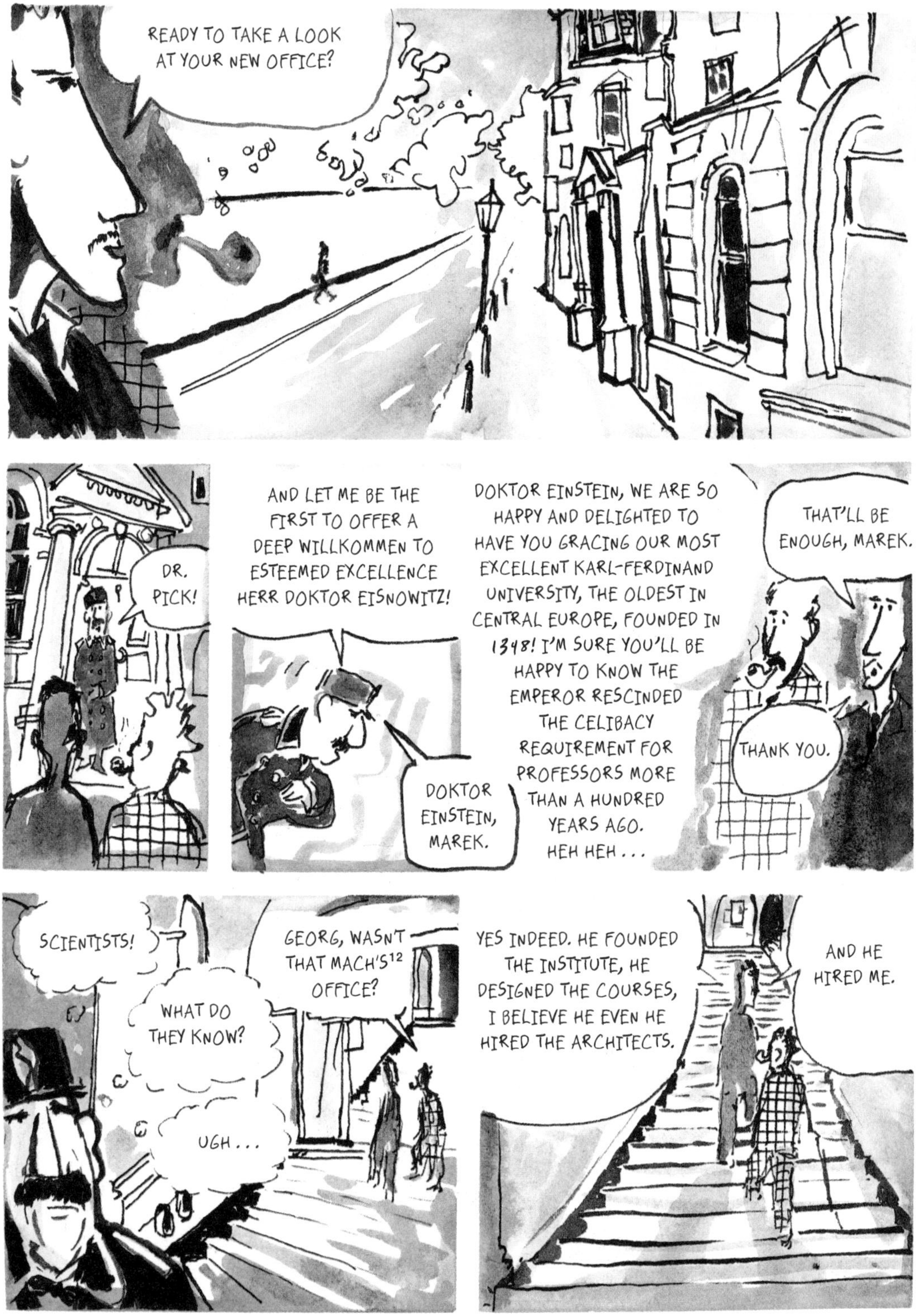
READY TO TAKE A LOOK AT YOUR NEW OFFICE?
DR. PICK!
AND LET ME BE THE FIRST TO OFFER A DEEP WILLKOMMEN TO ESTEEMED EXCELLENCE HERR DOKTOR EISNOWITZ!
DOKTOR EINSTEIN, MAREK.
DOKTOR EINSTEIN, WE ARE SO HAPPY AND DELIGHTED TO HAVE YOU GRACING OUR MOST EXCELLENT KARL-FERDINAND UNIVERSITY, THE OLDEST IN CENTRAL EUROPE, FOUNDED IN 1348! I'M SURE YOU'LL BE HAPPY TO KNOW THE EMPEROR RESCINDED THE CELIBACY REQUIREMENT FOR PROFESSORS MORE THAN A HUNDRED YEARS AGO. HEH HEH . . .
THAT'LL BE ENOUGH, MAREK.
THANK YOU.
SCIENTISTS!
WHAT DO THEY KNOW?
UGH . . .
GEORG, WASN'T THAT MACH'S[12] OFFICE?
YES INDEED. HE FOUNDED THE INSTITUTE, HE DESIGNED THE COURSES, I BELIEVE HE EVEN HE HIRED THE ARCHITECTS.
AND HE HIRED ME.

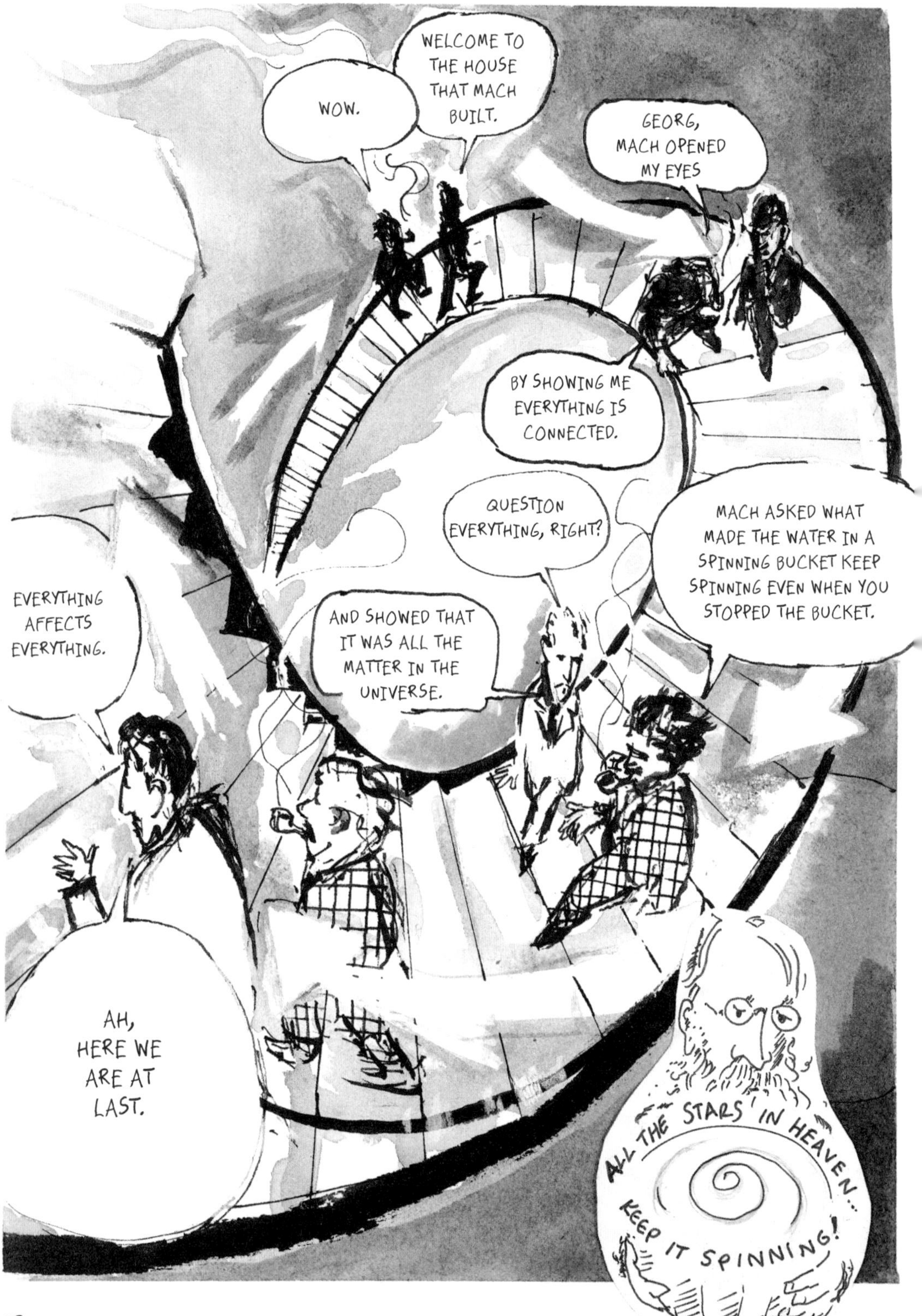
WOW.
WELCOME TO THE HOUSE THAT MACH BUILT.
GEORG, MACH OPENED MY EYES
BY SHOWING ME EVERYTHING IS CONNECTED.
QUESTION EVERYTHING, RIGHT?
MACH ASKED WHAT MADE THE WATER IN A SPINNING BUCKET KEEP SPINNING EVEN WHEN YOU STOPPED THE BUCKET.
EVERYTHING AFFECTS EVERYTHING.
AND SHOWED THAT IT WAS ALL THE MATTER IN THE UNIVERSE.
AH, HERE WE ARE AT LAST.
ALL THE STARS IN HEAVEN...
KEEP IT SPINNING!

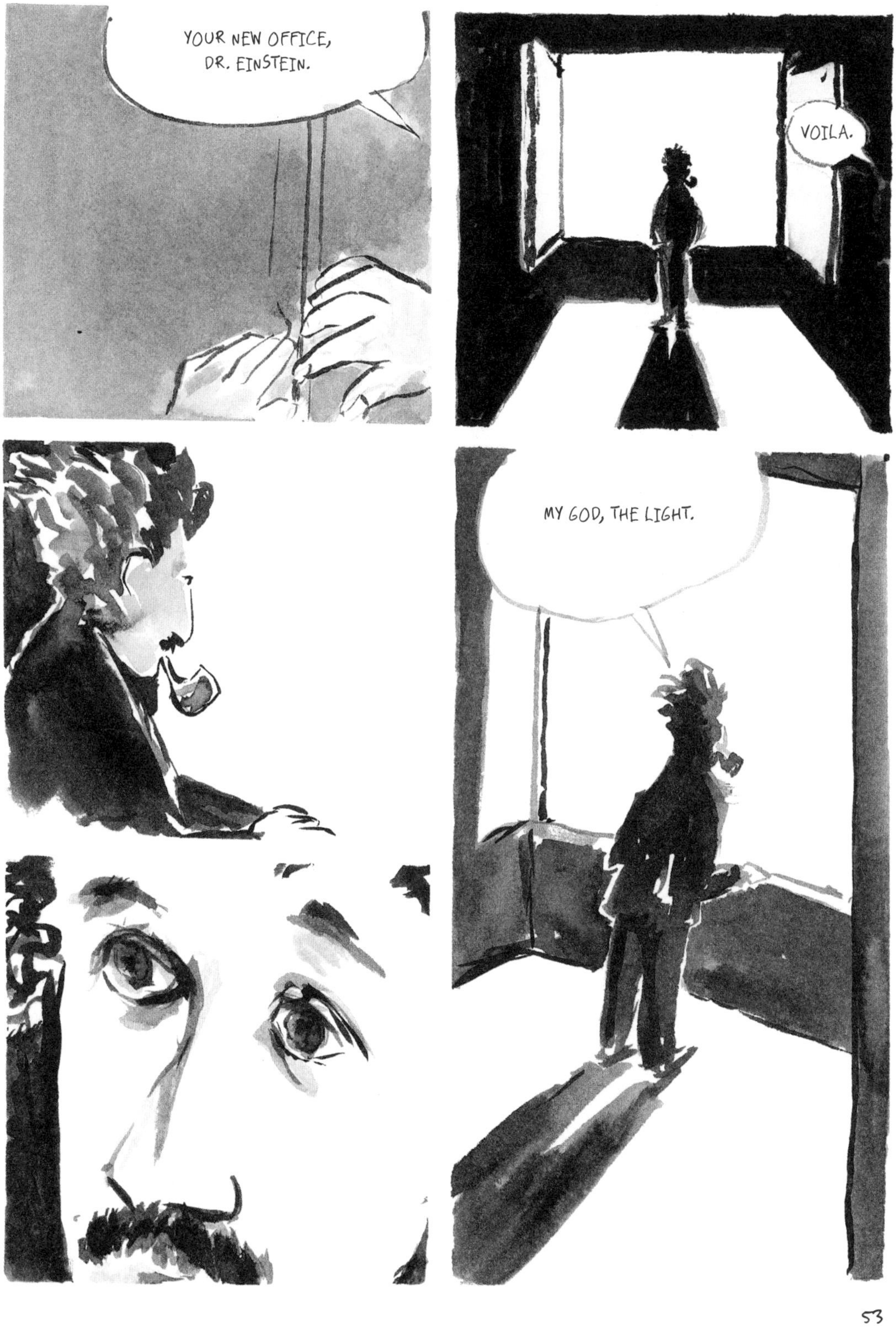
YOUR NEW OFFICE,
DR. EINSTEIN.
VOILA.
MY GOD, THE LIGHT.

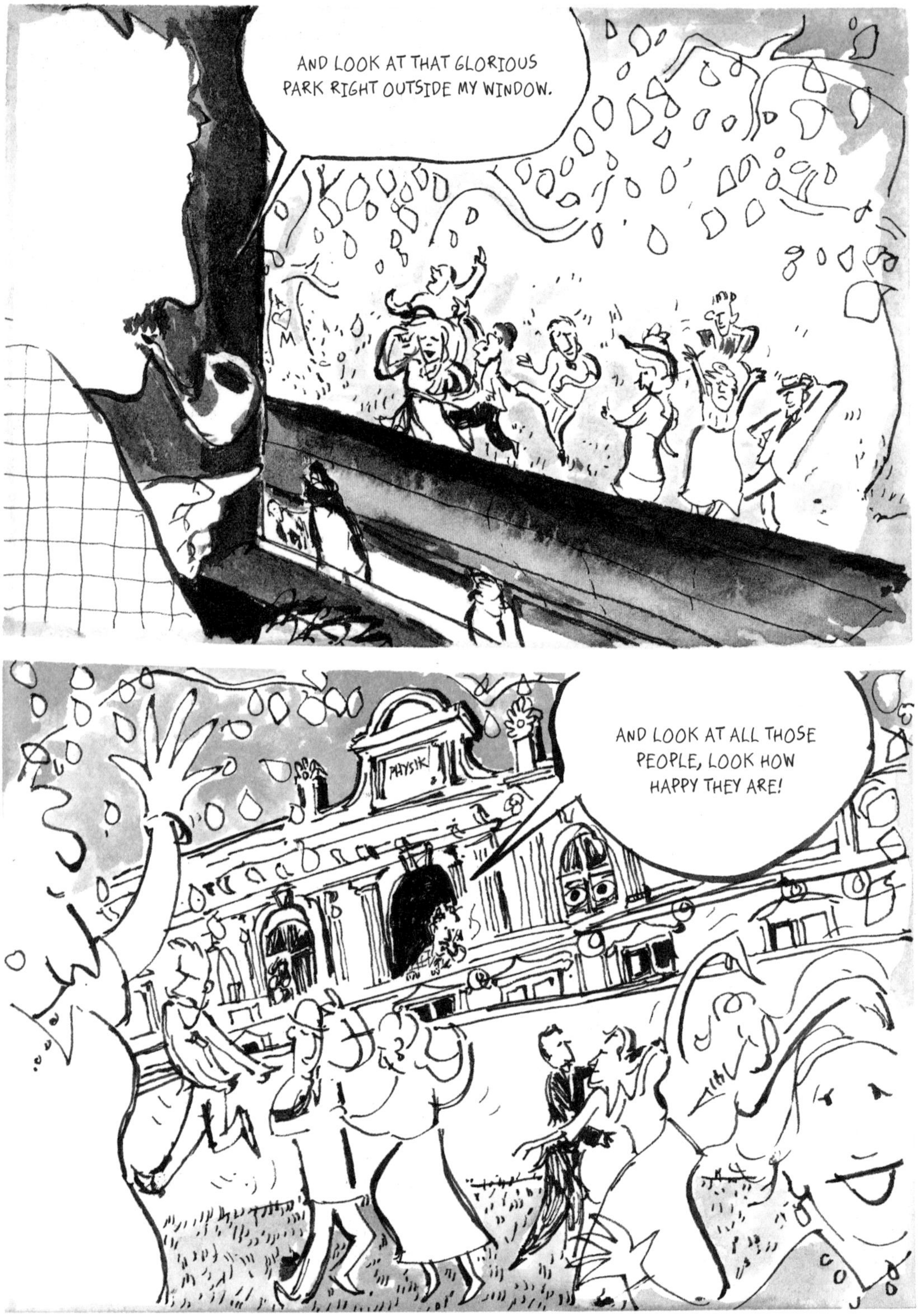
AND LOOK AT THAT GLORIOUS
PARK RIGHT OUTSIDE MY WINDOW.
PHYSIK!
AND LOOK AT ALL THOSE
PEOPLE, LOOK HOW
HAPPY THEY ARE!

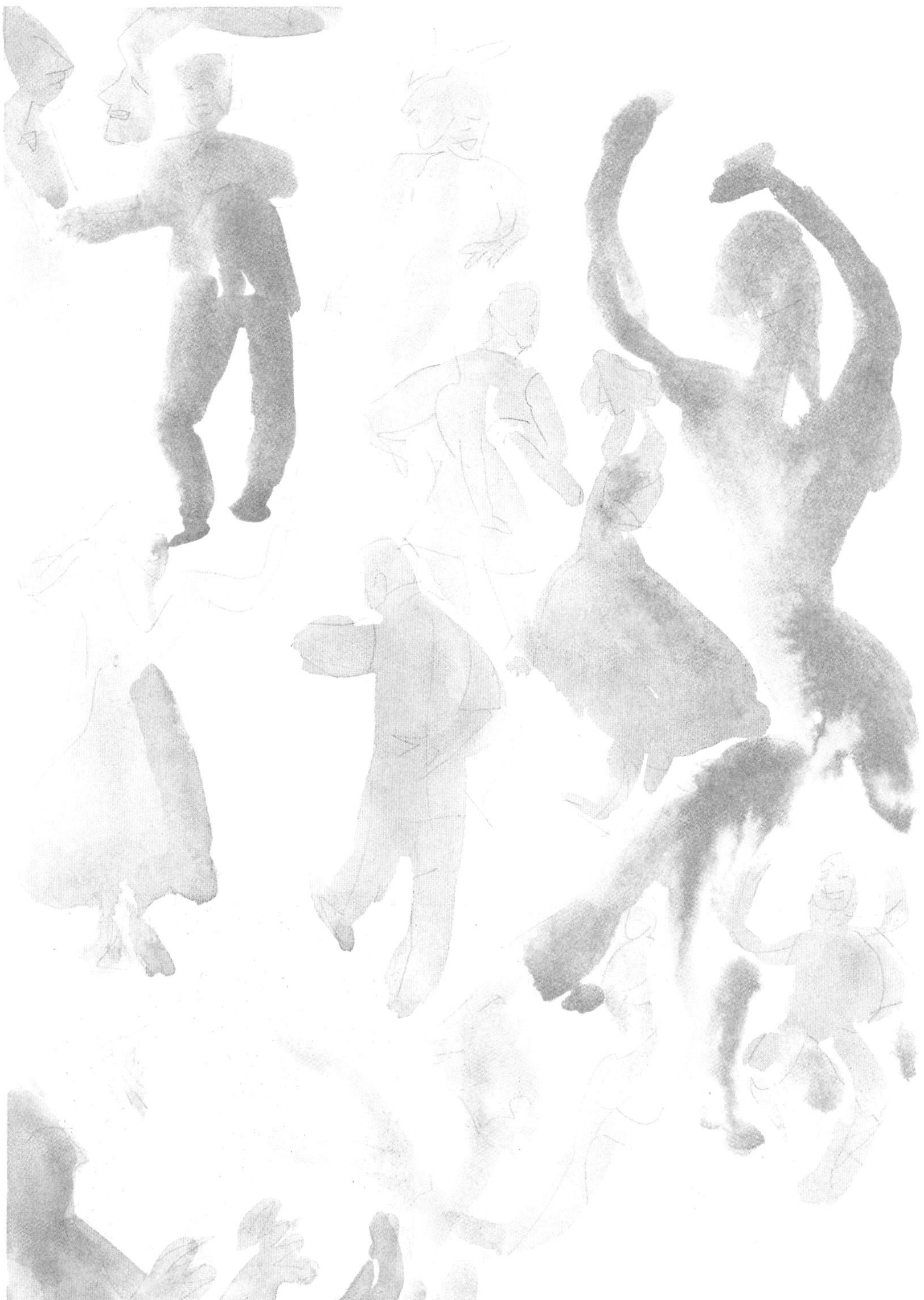

55

JUST LOOK AT THEM DANCE!

MILEVA WILL LOVE IT HERE.

I JUST KNOW IT.

SHE'LL SEE, SHE'LL SEE WE MADE THE RIGHT MOVE.

UM...

ERR...

UH, ALBERT.

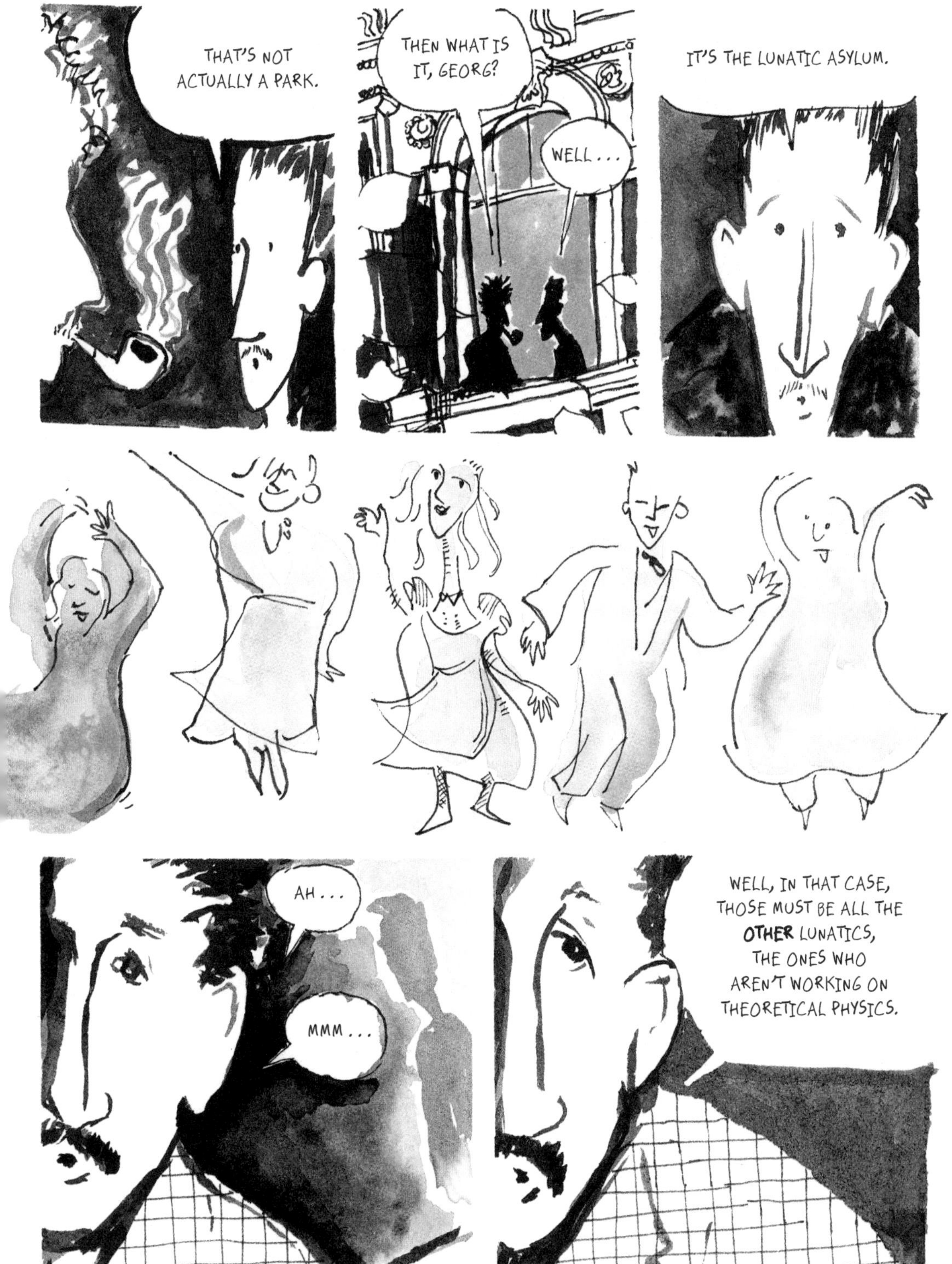

THAT'S NOT ACTUALLY A PARK.
THEN WHAT IS IT, GEORG?
WELL . . .
IT'S THE LUNATIC ASYLUM.
AH . . .
MMM . . .
WELL, IN THAT CASE, THOSE MUST BE ALL THE **OTHER** LUNATICS, THE ONES WHO AREN'T WORKING ON THEORETICAL PHYSICS.

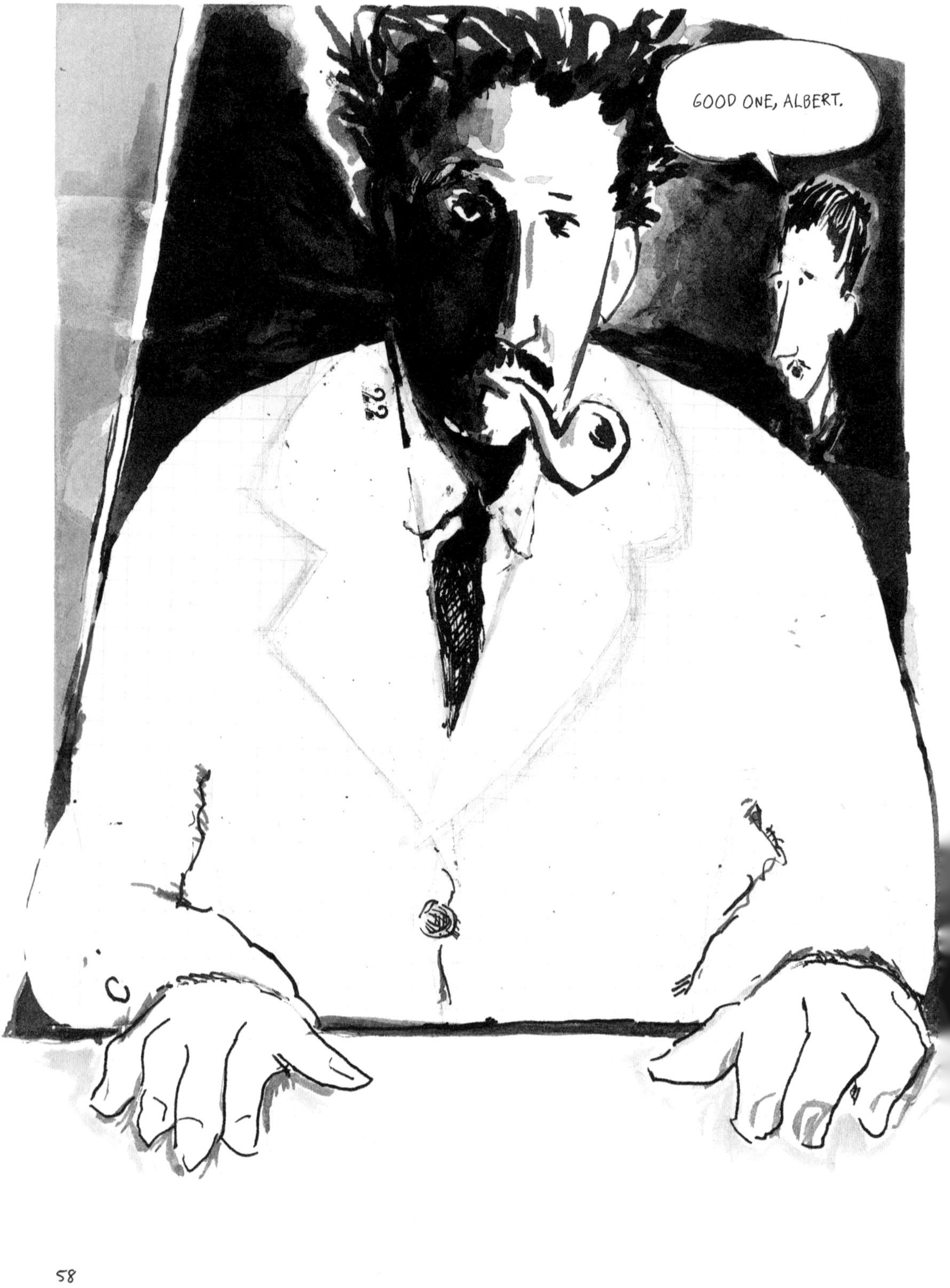
GOOD ONE, ALBERT.

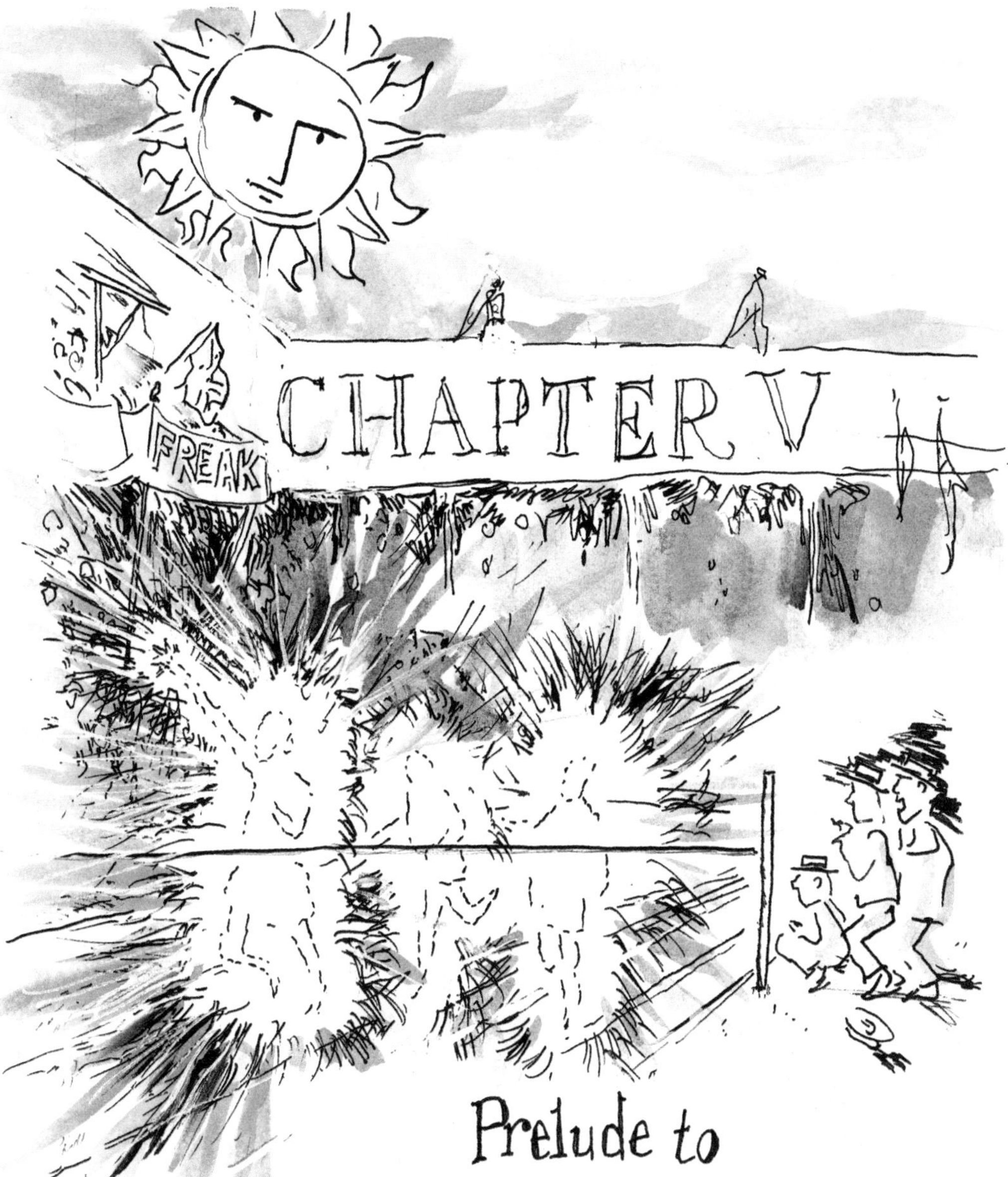
CHAPTER V
FREAK

Prelude to Berta Fanta's Mad Tea Party

May 24, 1911

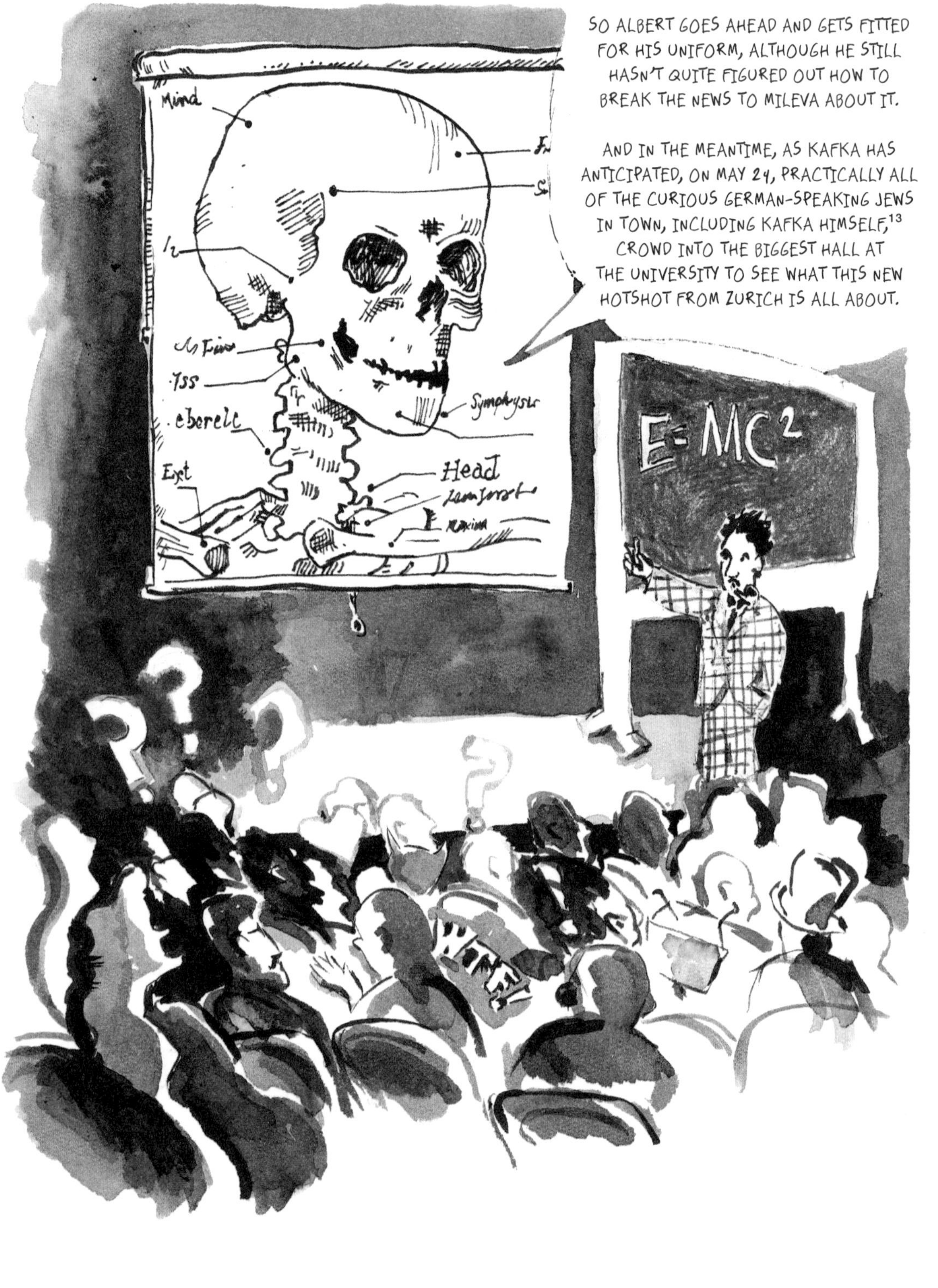
SO ALBERT GOES AHEAD AND GETS FITTED FOR HIS UNIFORM, ALTHOUGH HE STILL HASN'T QUITE FIGURED OUT HOW TO BREAK THE NEWS TO MILEVA ABOUT IT.
AND IN THE MEANTIME, AS KAFKA HAS ANTICIPATED, ON MAY 24, PRACTICALLY ALL OF THE CURIOUS GERMAN-SPEAKING JEWS IN TOWN, INCLUDING KAFKA HIMSELF,[13] CROWD INTO THE BIGGEST HALL AT THE UNIVERSITY TO SEE WHAT THIS NEW HOTSHOT FROM ZURICH IS ALL ABOUT.
Mind
Head
E=MC2

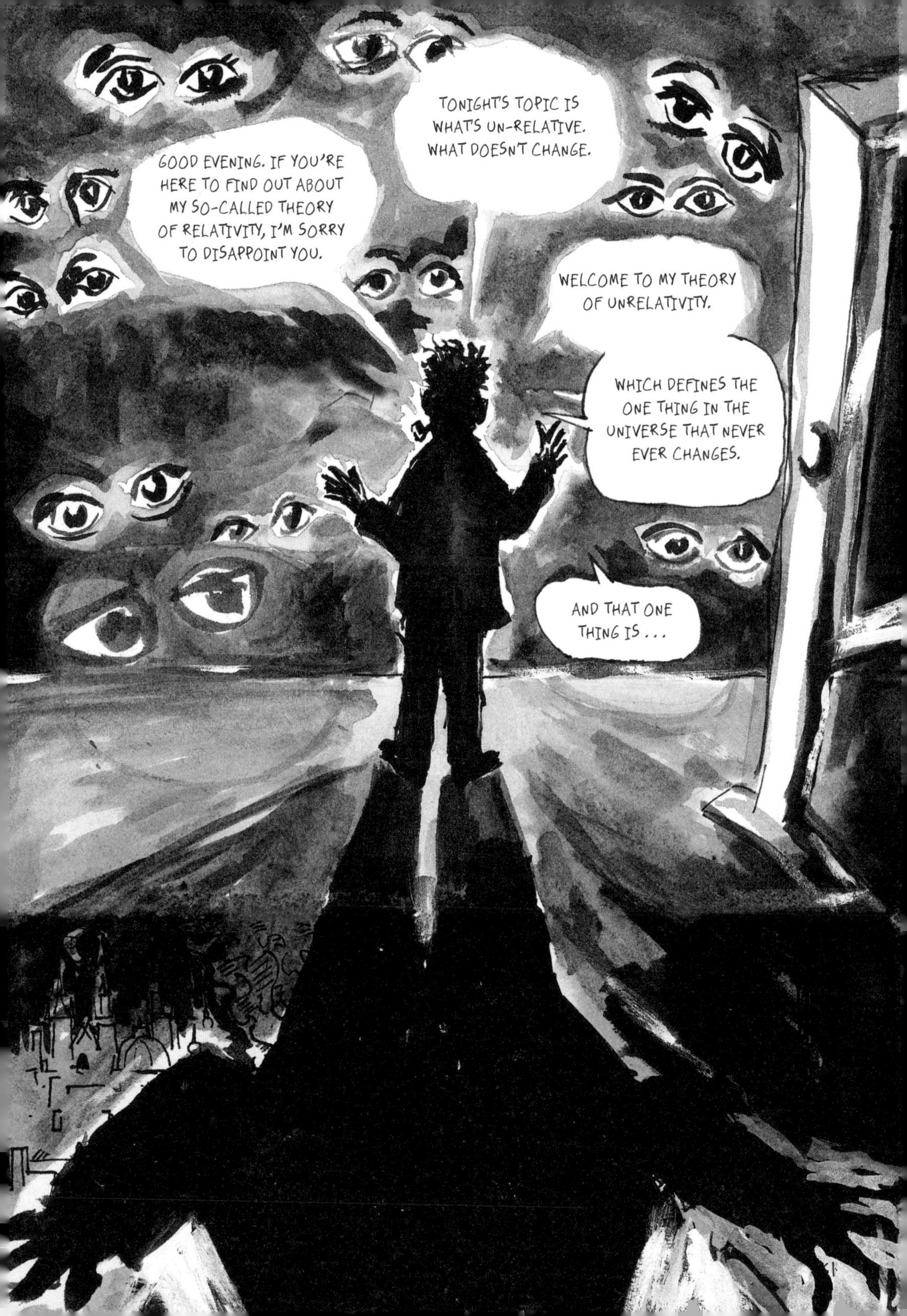
GOOD EVENING. IF YOU'RE HERE TO FIND OUT ABOUT MY SO-CALLED THEORY OF RELATIVITY, I'M SORRY TO DISAPPOINT YOU.
TONIGHT'S TOPIC IS WHAT'S UN-RELATIVE. WHAT DOESN'T CHANGE.
WELCOME TO MY THEORY OF UNRELATIVITY.
WHICH DEFINES THE ONE THING IN THE UNIVERSE THAT NEVER EVER CHANGES.
AND THAT ONE THING IS . . .

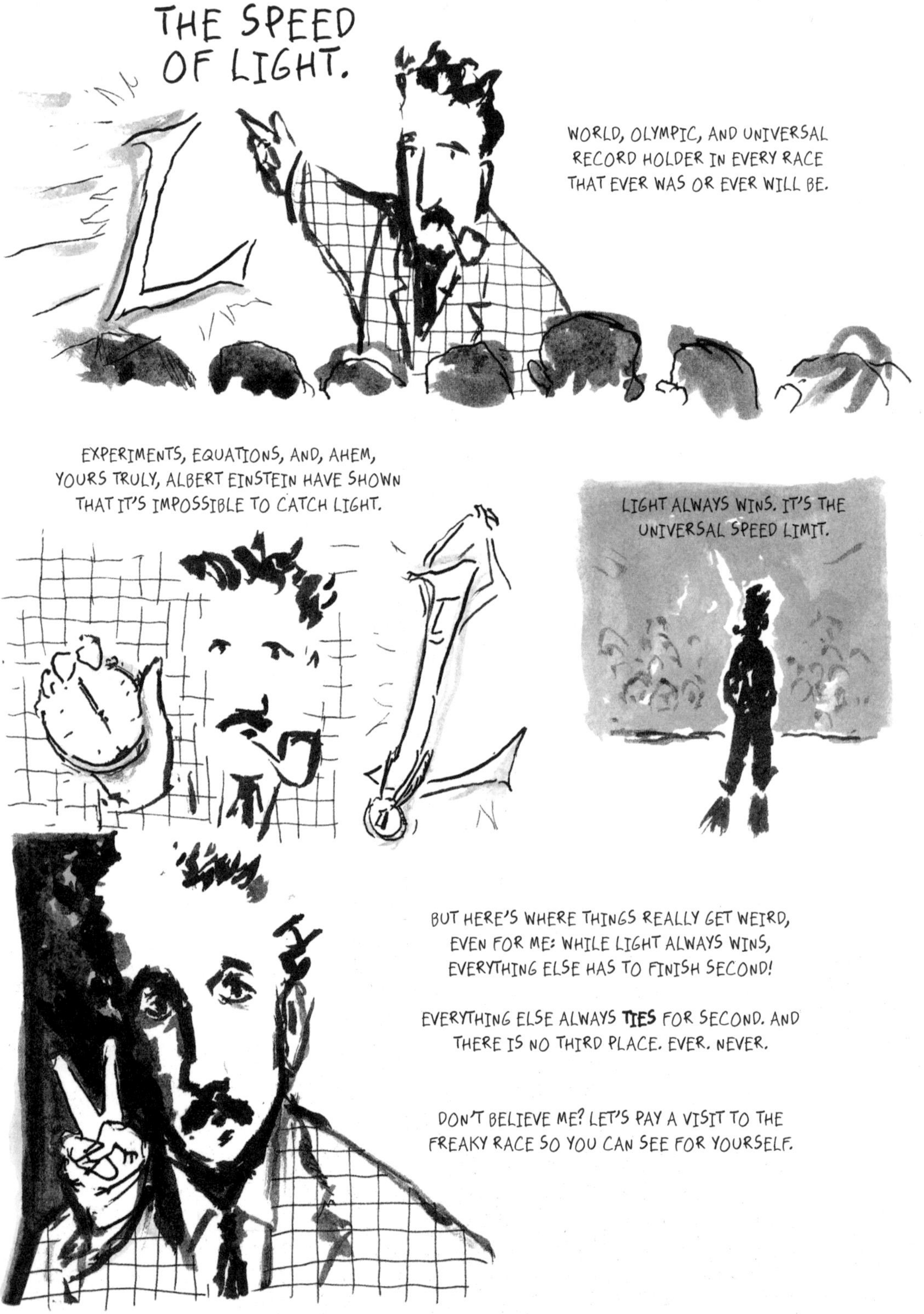
THE SPEED OF LIGHT.
WORLD, OLYMPIC, AND UNIVERSAL RECORD HOLDER IN EVERY RACE THAT EVER WAS OR EVER WILL BE.
EXPERIMENTS, EQUATIONS, AND, AHEM, YOURS TRULY, ALBERT EINSTEIN HAVE SHOWN THAT IT'S IMPOSSIBLE TO CATCH LIGHT.
LIGHT ALWAYS WINS. IT'S THE UNIVERSAL SPEED LIMIT.
BUT HERE'S WHERE THINGS REALLY GET WEIRD, EVEN FOR ME: WHILE LIGHT ALWAYS WINS, EVERYTHING ELSE HAS TO FINISH SECOND!
EVERYTHING ELSE ALWAYS **TIES** FOR SECOND. AND THERE IS NO THIRD PLACE. EVER. NEVER.
DON'T BELIEVE ME? LET'S PAY A VISIT TO THE FREAKY RACE SO YOU CAN SEE FOR YOURSELF.

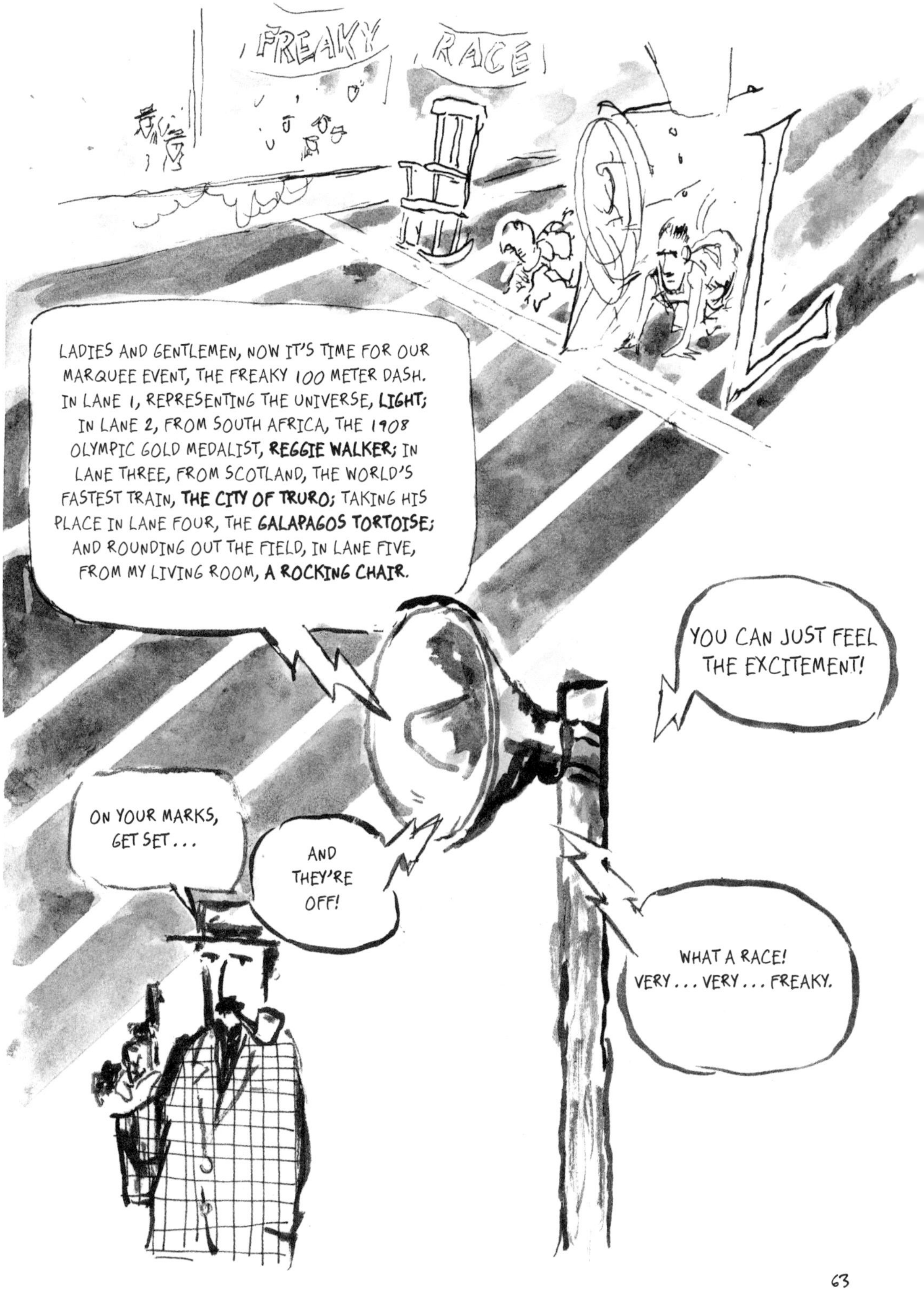
FREAKY RACE
LADIES AND GENTLEMEN, NOW IT'S TIME FOR OUR MARQUEE EVENT, THE FREAKY 100 METER DASH. IN LANE 1, REPRESENTING THE UNIVERSE, **LIGHT**; IN LANE 2, FROM SOUTH AFRICA, THE 1908 OLYMPIC GOLD MEDALIST, **REGGIE WALKER**; IN LANE THREE, FROM SCOTLAND, THE WORLD'S FASTEST TRAIN, **THE CITY OF TRURO**; TAKING HIS PLACE IN LANE FOUR, THE **GALAPAGOS TORTOISE**; AND ROUNDING OUT THE FIELD, IN LANE FIVE, FROM MY LIVING ROOM, **A ROCKING CHAIR**.
YOU CAN JUST FEEL THE EXCITEMENT!
ON YOUR MARKS, GET SET . . .
AND THEY'RE OFF!
WHAT A RACE! VERY . . . VERY . . . FREAKY.

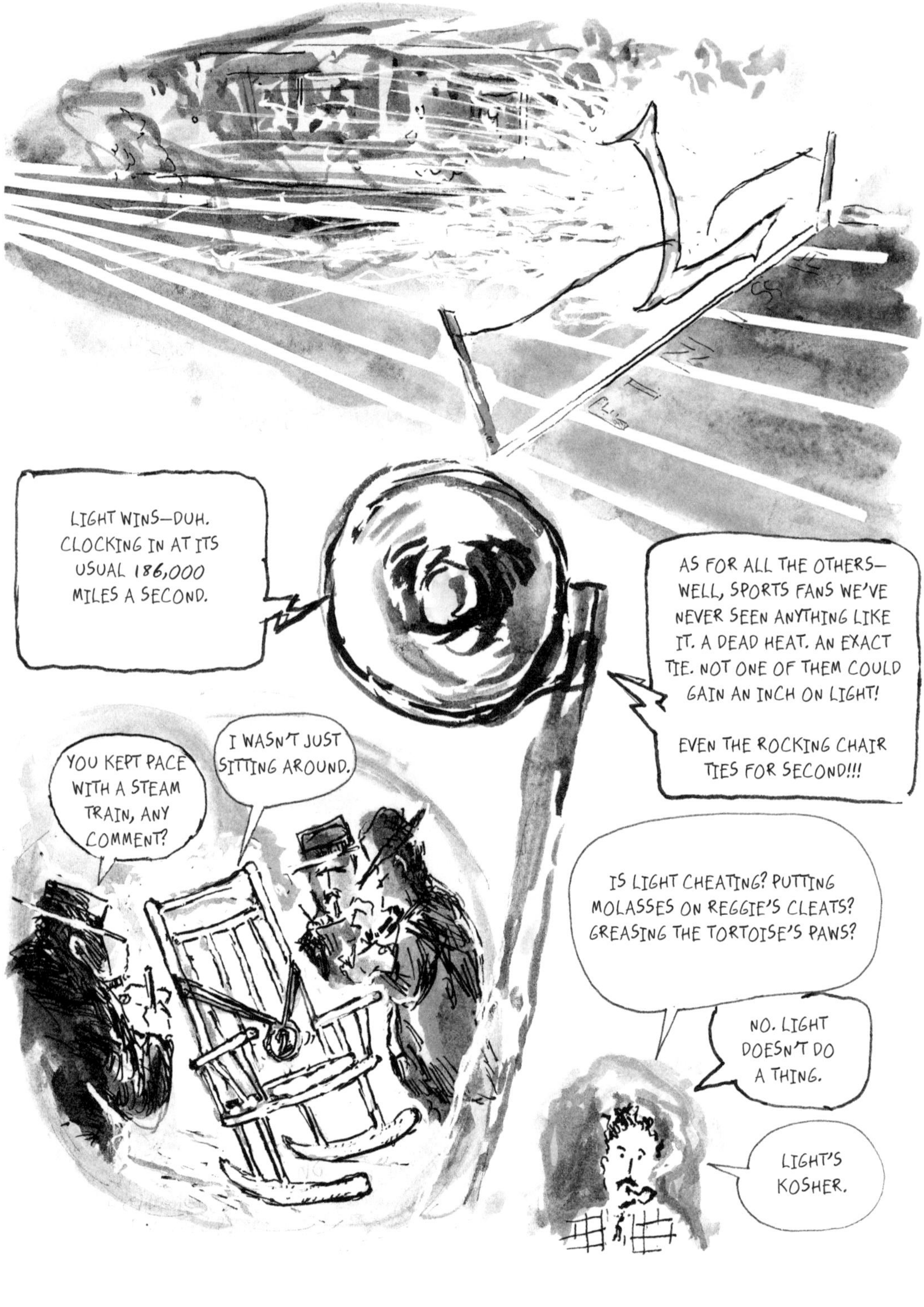
LIGHT WINS—DUH. CLOCKING IN AT ITS USUAL 186,000 MILES A SECOND.
AS FOR ALL THE OTHERS— WELL, SPORTS FANS WE'VE NEVER SEEN ANYTHING LIKE IT. A DEAD HEAT. AN EXACT TIE. NOT ONE OF THEM COULD GAIN AN INCH ON LIGHT!
EVEN THE ROCKING CHAIR TIES FOR SECOND!!!
YOU KEPT PACE WITH A STEAM TRAIN, ANY COMMENT?
I WASN'T JUST SITTING AROUND.
IS LIGHT CHEATING? PUTTING MOLASSES ON REGGIE'S CLEATS? GREASING THE TORTOISE'S PAWS?
NO. LIGHT DOESN'T DO A THING.
LIGHT'S KOSHER.

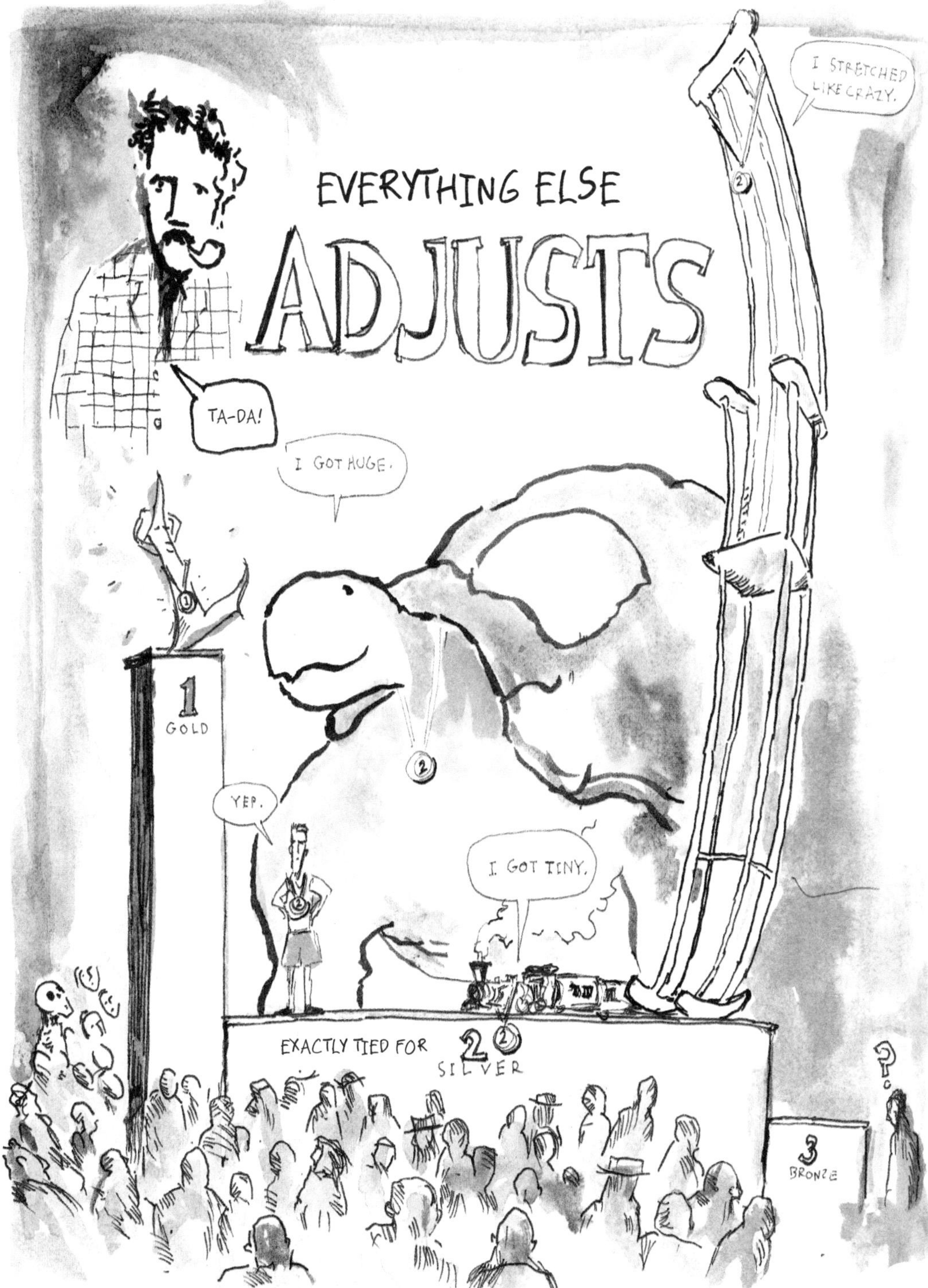
EVERYTHING ELSE
ADJUSTS
TA-DA!
I GOT HUGE.
I STRETCHED LIKE CRAZY.
1
GOLD
YEP.
I. GOT TINY.
EXACTLY TIED FOR 2
SILVER
3
BRONZE

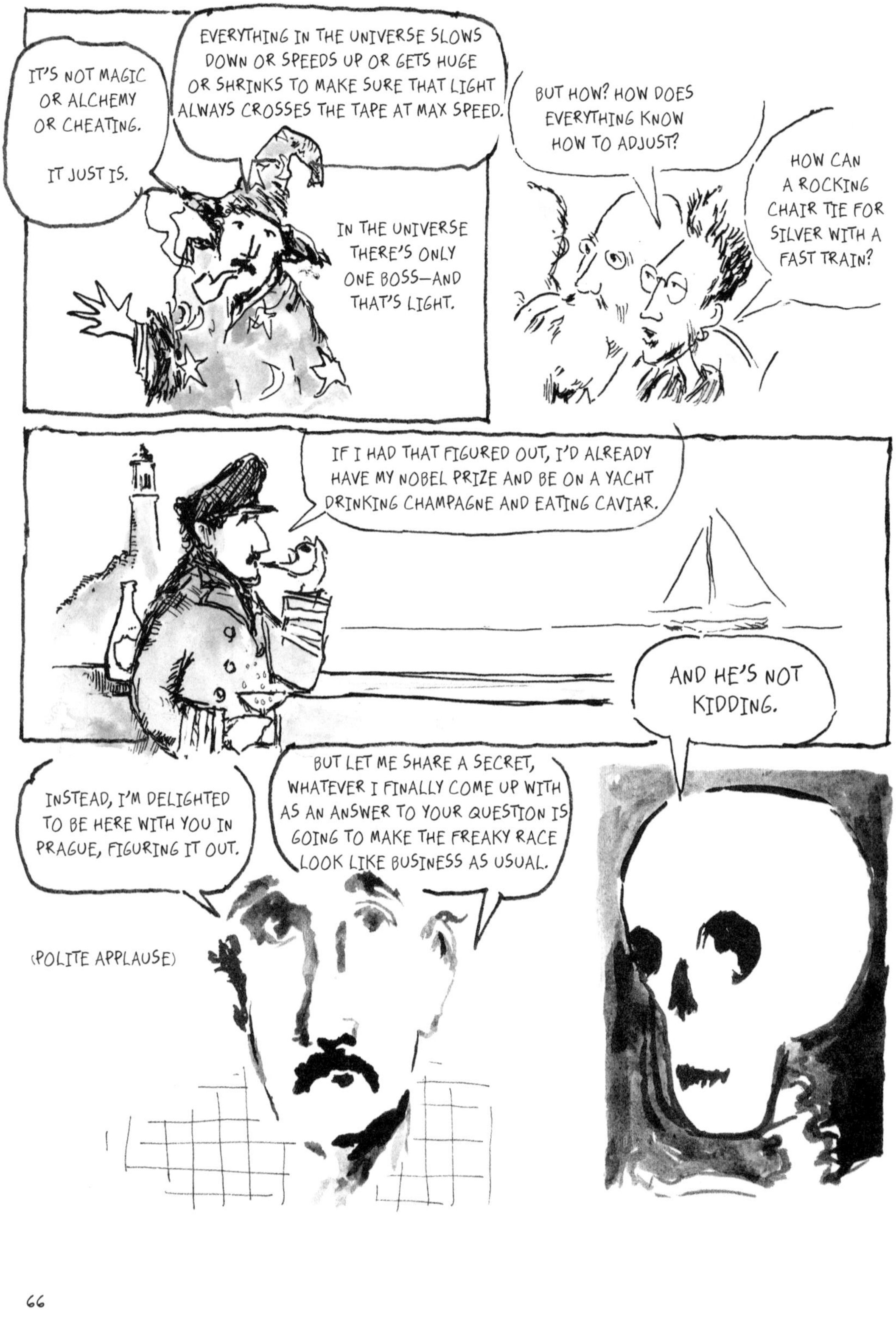
IT'S NOT MAGIC OR ALCHEMY OR CHEATING.
IT JUST IS.
EVERYTHING IN THE UNIVERSE SLOWS DOWN OR SPEEDS UP OR GETS HUGE OR SHRINKS TO MAKE SURE THAT LIGHT ALWAYS CROSSES THE TAPE AT MAX SPEED.
IN THE UNIVERSE THERE'S ONLY ONE BOSS—AND THAT'S LIGHT.
BUT HOW? HOW DOES EVERYTHING KNOW HOW TO ADJUST?
HOW CAN A ROCKING CHAIR TIE FOR SILVER WITH A FAST TRAIN?
IF I HAD THAT FIGURED OUT, I'D ALREADY HAVE MY NOBEL PRIZE AND BE ON A YACHT DRINKING CHAMPAGNE AND EATING CAVIAR.
AND HE'S NOT KIDDING.
INSTEAD, I'M DELIGHTED TO BE HERE WITH YOU IN PRAGUE, FIGURING IT OUT.
BUT LET ME SHARE A SECRET, WHATEVER I FINALLY COME UP WITH AS AN ANSWER TO YOUR QUESTION IS GOING TO MAKE THE FREAKY RACE LOOK LIKE BUSINESS AS USUAL.
(POLITE APPLAUSE)

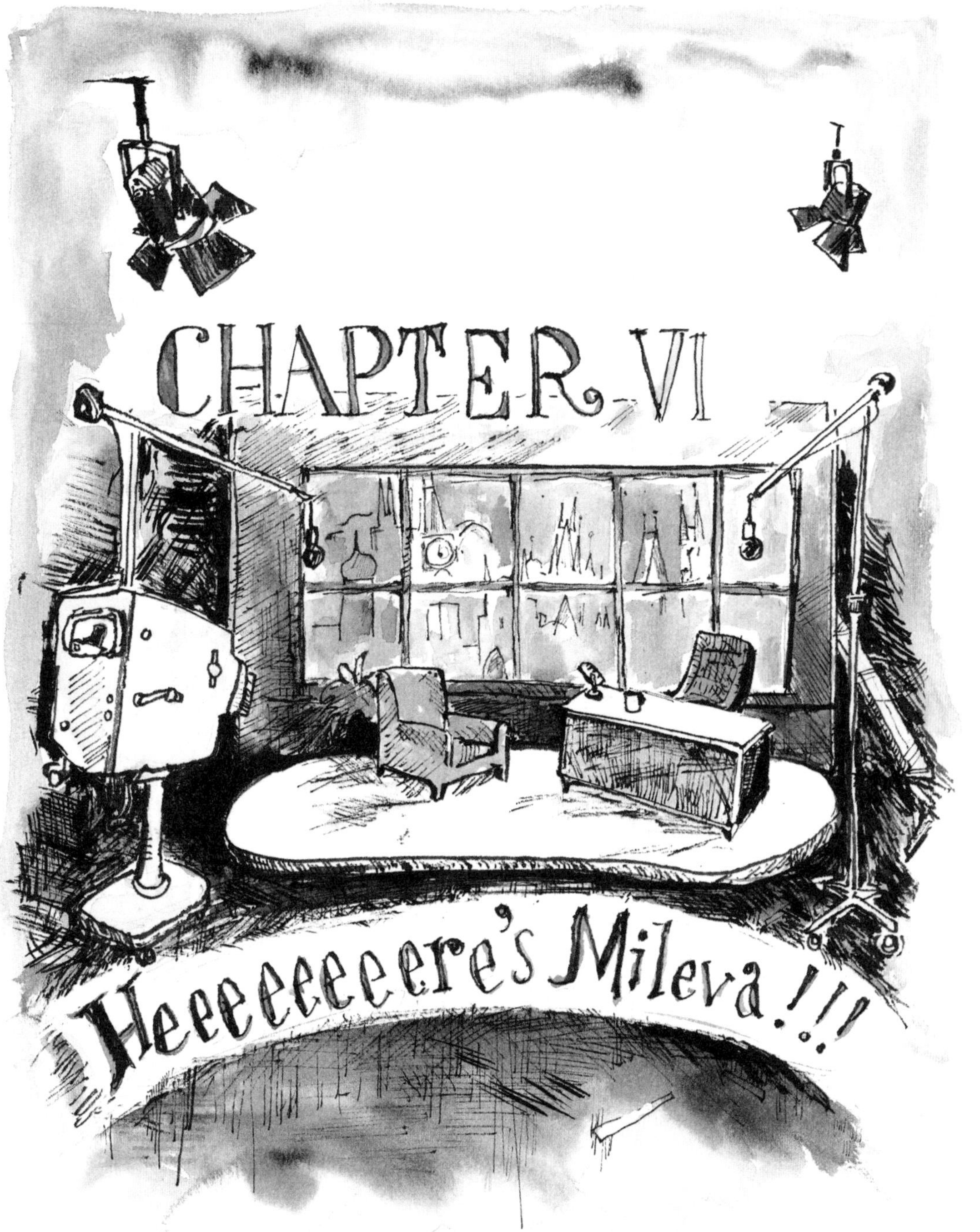
CHAPTER VI
Heeeeeeeere's Mileva!!!

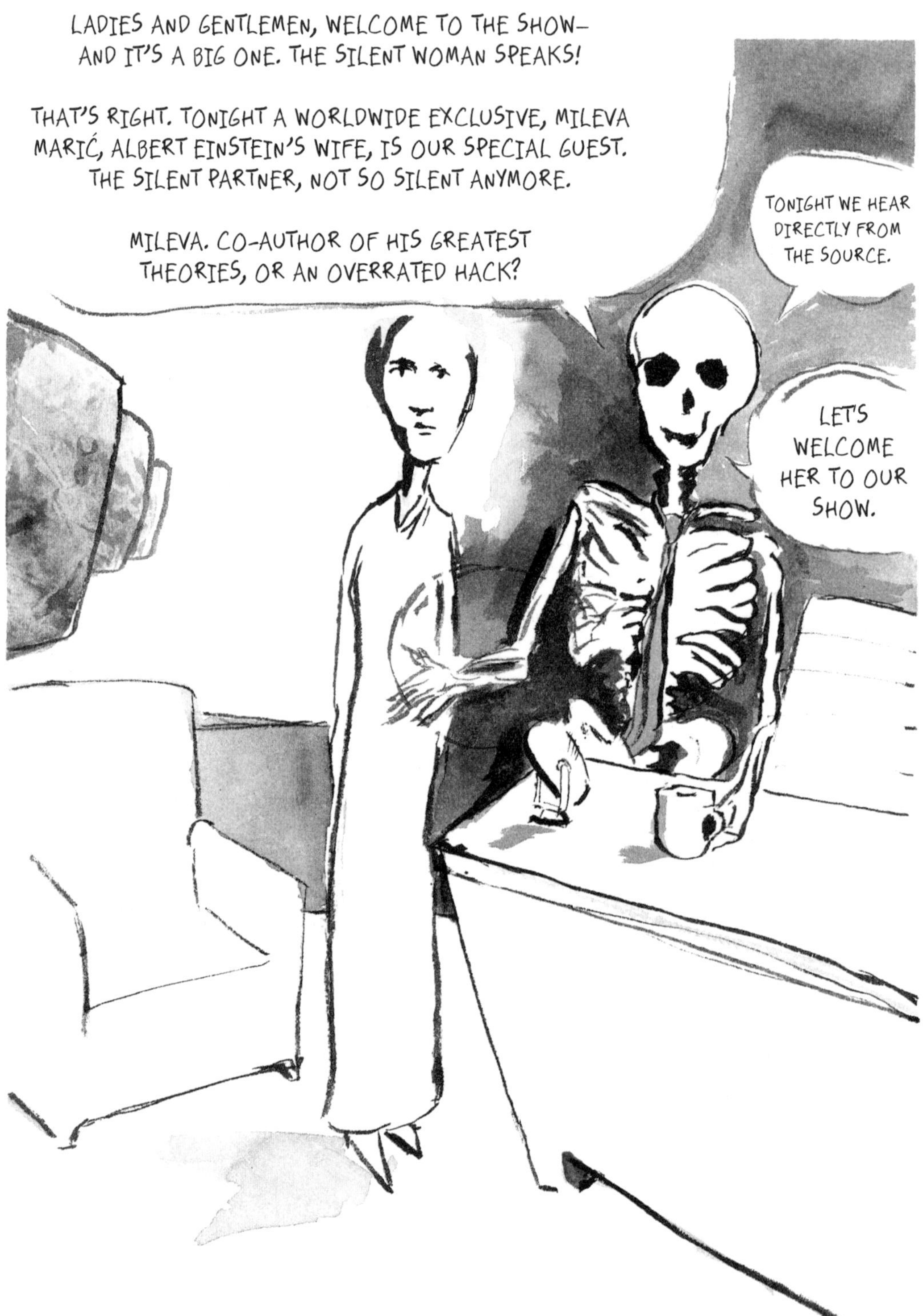
LADIES AND GENTLEMEN, WELCOME TO THE SHOW–
AND IT'S A BIG ONE. THE SILENT WOMAN SPEAKS!
THAT'S RIGHT. TONIGHT A WORLDWIDE EXCLUSIVE, MILEVA MARIĆ, ALBERT EINSTEIN'S WIFE, IS OUR SPECIAL GUEST. THE SILENT PARTNER, NOT SO SILENT ANYMORE.
MILEVA. CO-AUTHOR OF HIS GREATEST THEORIES, OR AN OVERRATED HACK?
TONIGHT WE HEAR DIRECTLY FROM THE SOURCE.
LET'S WELCOME HER TO OUR SHOW.

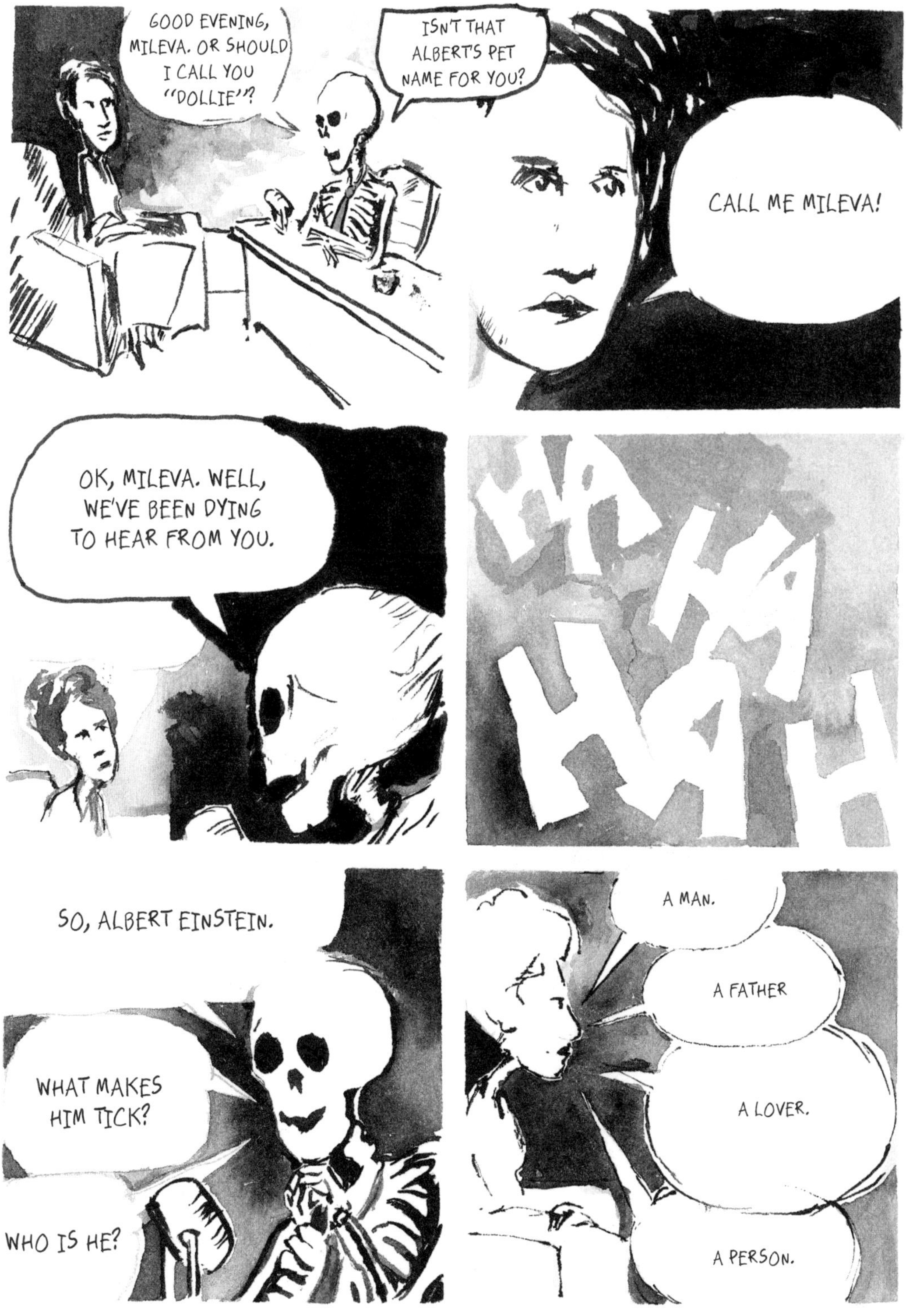
GOOD EVENING, MILEVA. OR SHOULD I CALL YOU "DOLLIE"?
ISN'T THAT ALBERT'S PET NAME FOR YOU?
CALL ME MILEVA!
OK, MILEVA. WELL, WE'VE BEEN DYING TO HEAR FROM YOU.
HA HA HA HA!
SO, ALBERT EINSTEIN.
WHAT MAKES HIM TICK?
WHO IS HE?
A MAN.
A FATHER
A LOVER.
A PERSON.

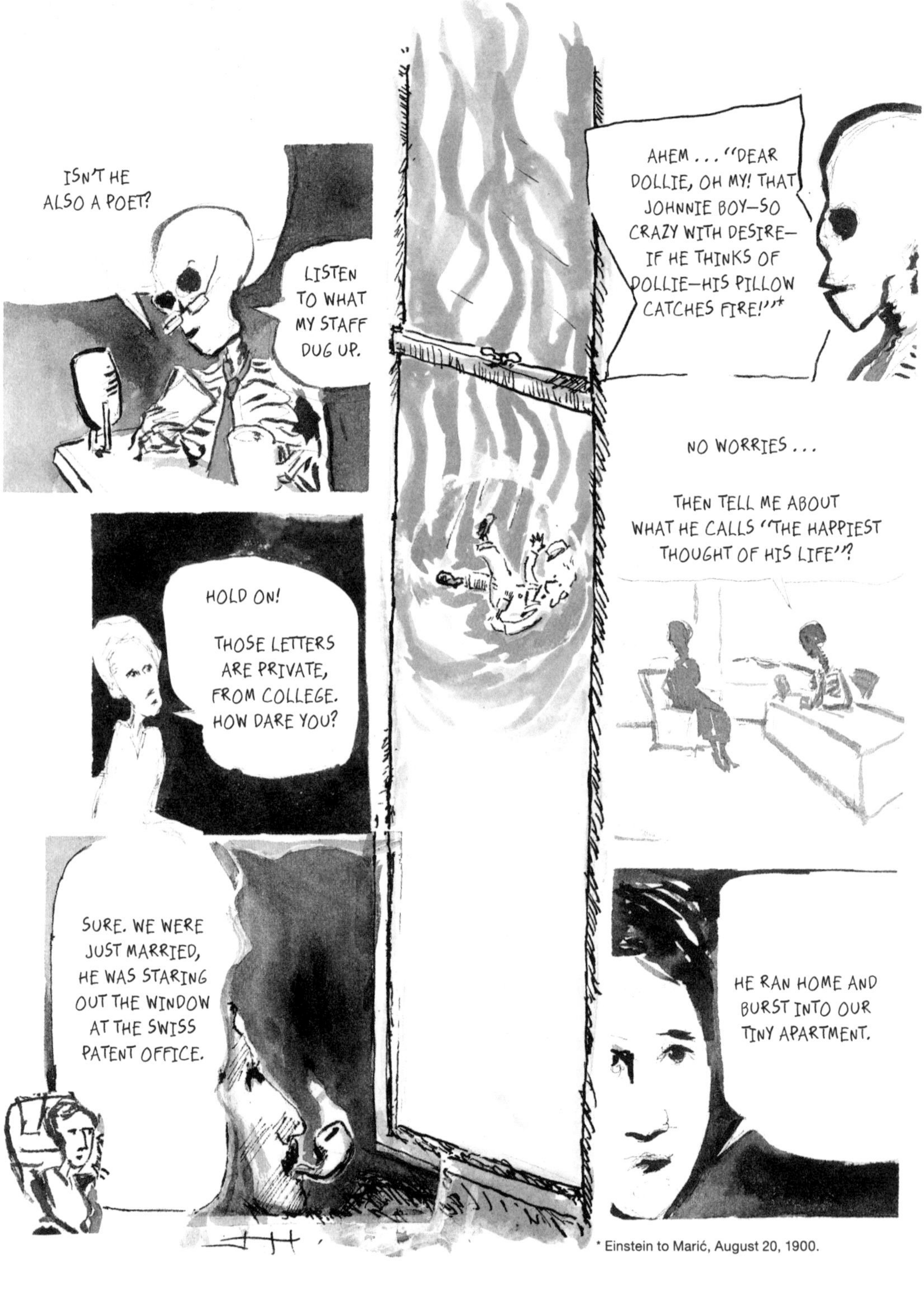
ISN'T HE ALSO A POET?
LISTEN TO WHAT MY STAFF DUG UP.
AHEM . . . "DEAR DOLLIE, OH MY! THAT JOHNNIE BOY—SO CRAZY WITH DESIRE—IF HE THINKS OF DOLLIE—HIS PILLOW CATCHES FIRE!"*
HOLD ON!
THOSE LETTERS ARE PRIVATE, FROM COLLEGE. HOW DARE YOU?
NO WORRIES . . .
THEN TELL ME ABOUT WHAT HE CALLS "THE HAPPIEST THOUGHT OF HIS LIFE"?
SURE. WE WERE JUST MARRIED, HE WAS STARING OUT THE WINDOW AT THE SWISS PATENT OFFICE.
HE RAN HOME AND BURST INTO OUR TINY APARTMENT.

* Einstein to Marić, August 20, 1900.

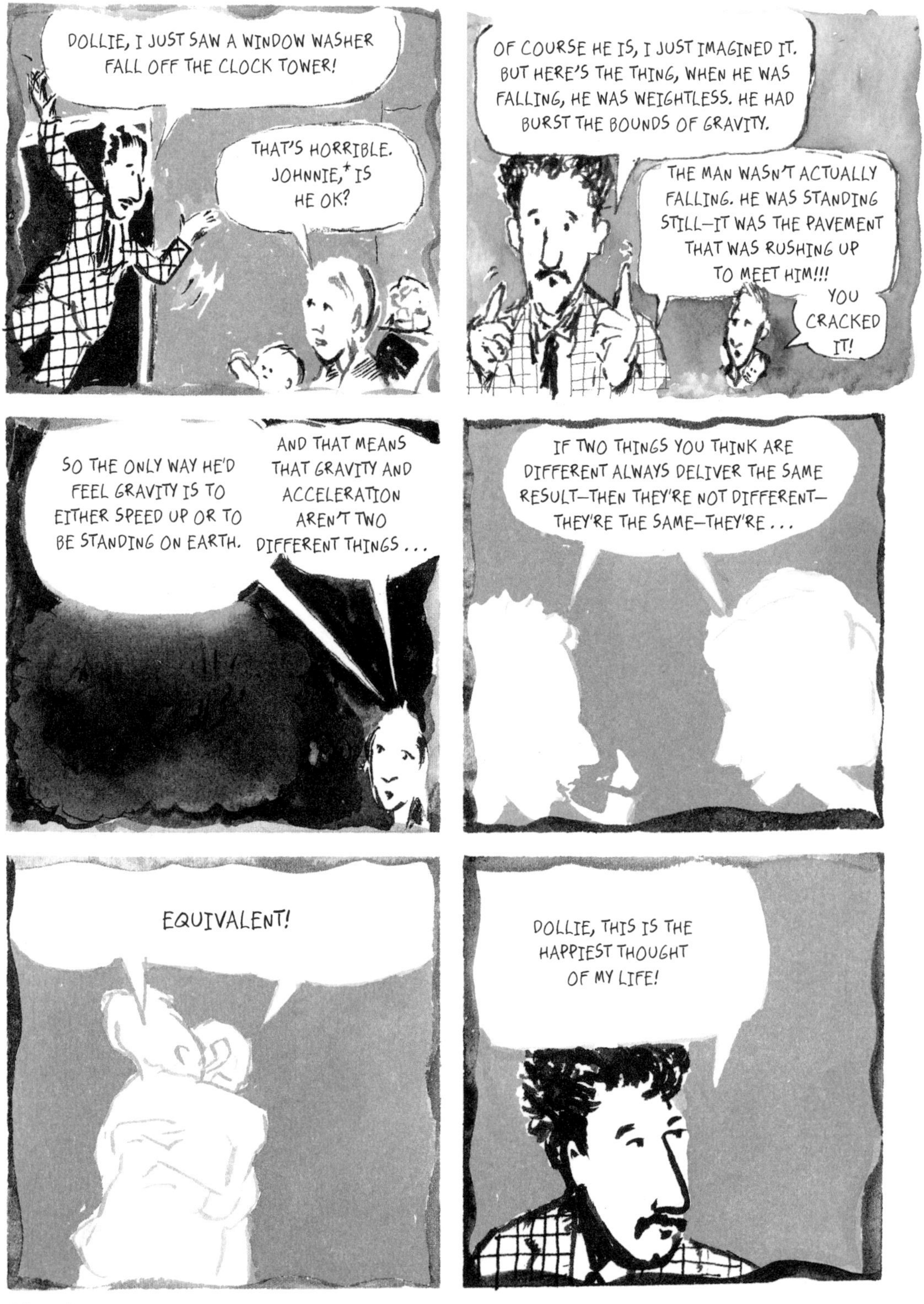
DOLLIE, I JUST SAW A WINDOW WASHER FALL OFF THE CLOCK TOWER!
THAT'S HORRIBLE. JOHNNIE,* IS HE OK?
OF COURSE HE IS, I JUST IMAGINED IT. BUT HERE'S THE THING, WHEN HE WAS FALLING, HE WAS WEIGHTLESS. HE HAD BURST THE BOUNDS OF GRAVITY.
THE MAN WASN'T ACTUALLY FALLING. HE WAS STANDING STILL—IT WAS THE PAVEMENT THAT WAS RUSHING UP TO MEET HIM!!!
YOU CRACKED IT!
SO THE ONLY WAY HE'D FEEL GRAVITY IS TO EITHER SPEED UP OR TO BE STANDING ON EARTH.
AND THAT MEANS THAT GRAVITY AND ACCELERATION AREN'T TWO DIFFERENT THINGS . . .
IF TWO THINGS YOU THINK ARE DIFFERENT ALWAYS DELIVER THE SAME RESULT—THEN THEY'RE NOT DIFFERENT—THEY'RE THE SAME—THEY'RE . . .
EQUIVALENT!
DOLLIE, THIS IS THE HAPPIEST THOUGHT OF MY LIFE!

*Mileva's (Dollie) pet name for Albert as revealed in long-hidden private letters.

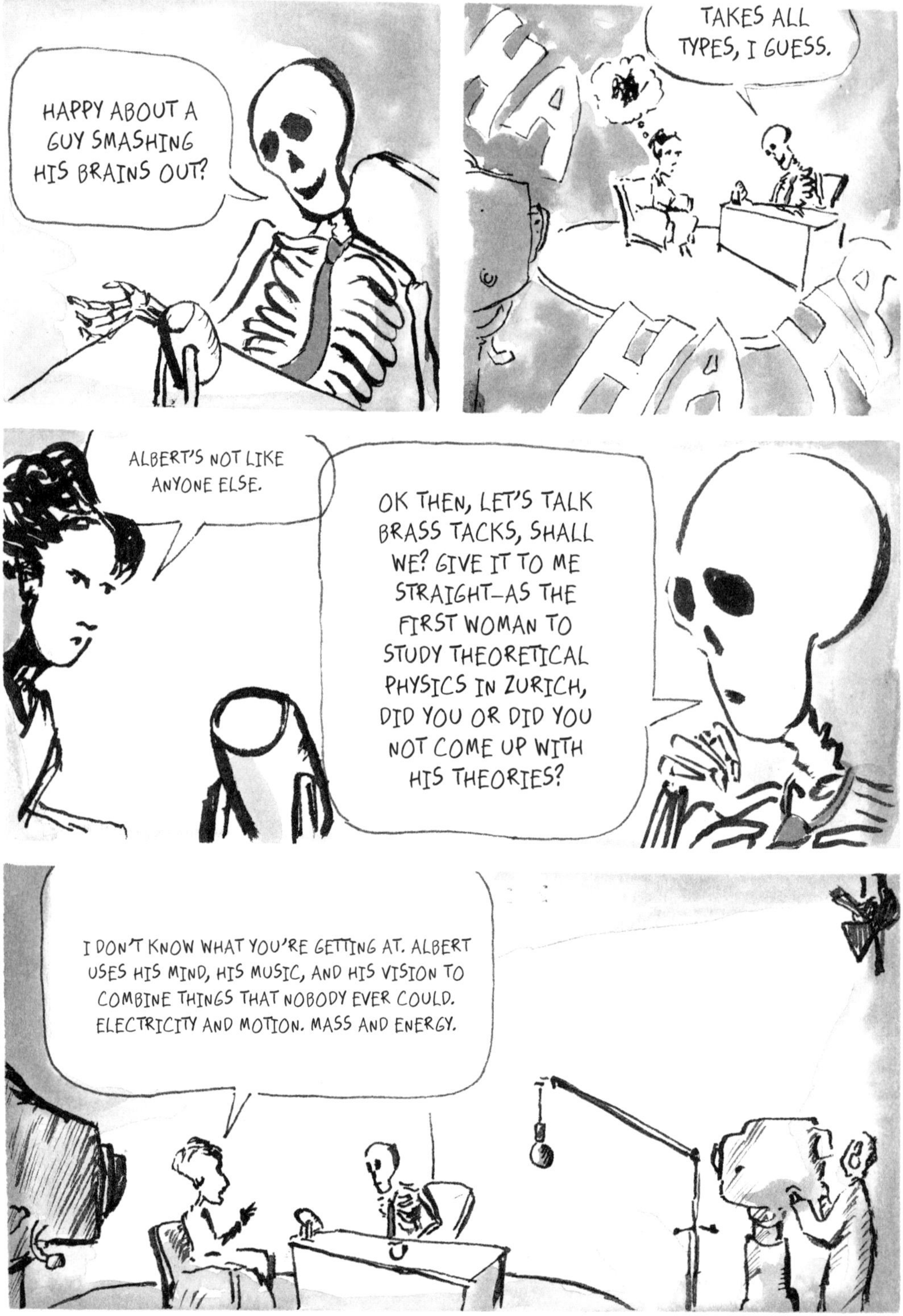
HAPPY ABOUT A GUY SMASHING HIS BRAINS OUT?
HA
TAKES ALL TYPES, I GUESS.
HA HA HA
ALBERT'S NOT LIKE ANYONE ELSE.
OK THEN, LET'S TALK BRASS TACKS, SHALL WE? GIVE IT TO ME STRAIGHT—AS THE FIRST WOMAN TO STUDY THEORETICAL PHYSICS IN ZURICH, DID YOU OR DID YOU NOT COME UP WITH HIS THEORIES?
I DON'T KNOW WHAT YOU'RE GETTING AT. ALBERT USES HIS MIND, HIS MUSIC, AND HIS VISION TO COMBINE THINGS THAT NOBODY EVER COULD. ELECTRICITY AND MOTION. MASS AND ENERGY.

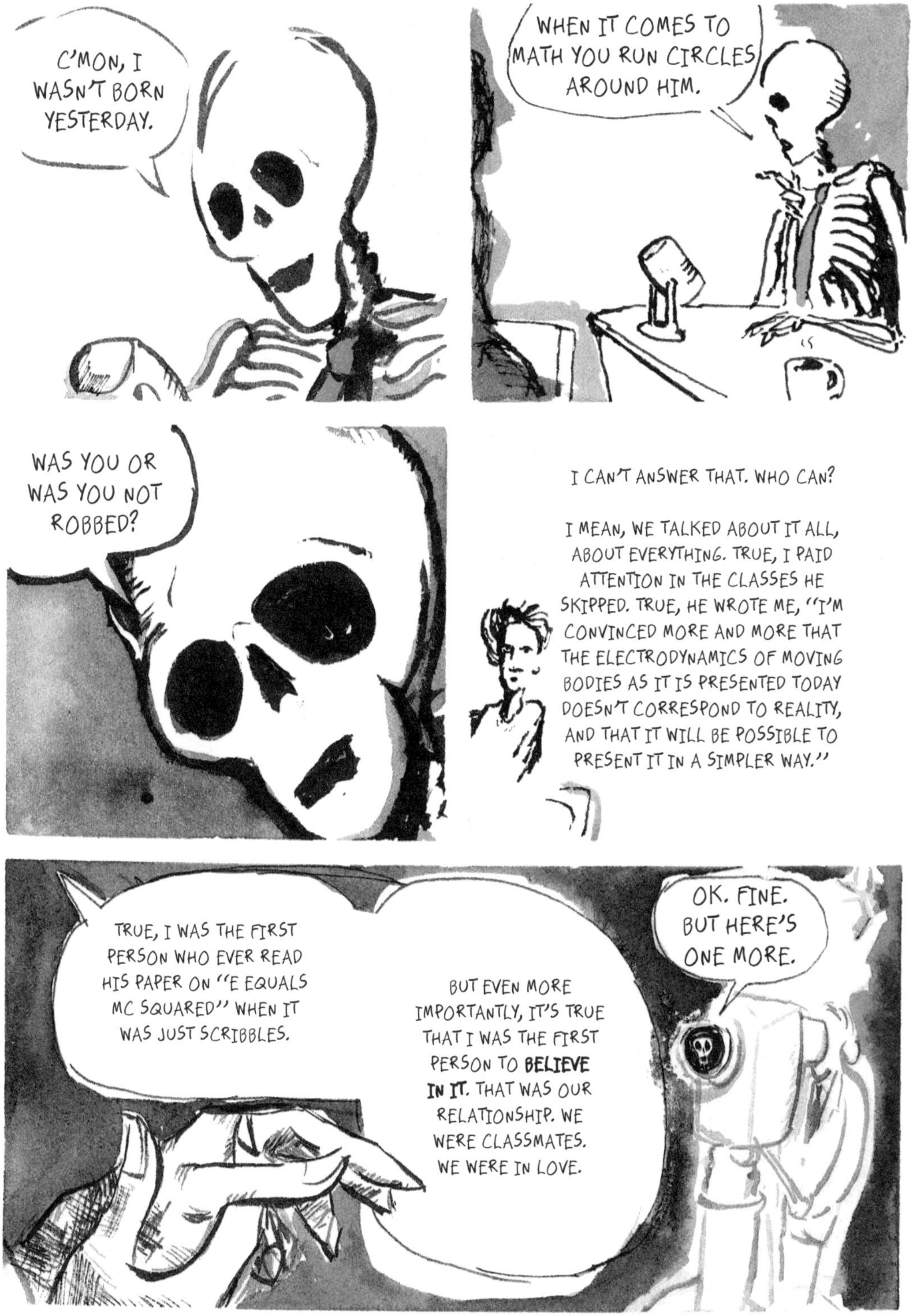
C'MON, I WASN'T BORN YESTERDAY.
WHEN IT COMES TO MATH YOU RUN CIRCLES AROUND HIM.
WAS YOU OR WAS YOU NOT ROBBED?
I CAN'T ANSWER THAT. WHO CAN?
I MEAN, WE TALKED ABOUT IT ALL, ABOUT EVERYTHING. TRUE, I PAID ATTENTION IN THE CLASSES HE SKIPPED. TRUE, HE WROTE ME, "I'M CONVINCED MORE AND MORE THAT THE ELECTRODYNAMICS OF MOVING BODIES AS IT IS PRESENTED TODAY DOESN'T CORRESPOND TO REALITY, AND THAT IT WILL BE POSSIBLE TO PRESENT IT IN A SIMPLER WAY."
TRUE, I WAS THE FIRST PERSON WHO EVER READ HIS PAPER ON "E EQUALS MC SQUARED" WHEN IT WAS JUST SCRIBBLES.
BUT EVEN MORE IMPORTANTLY, IT'S TRUE THAT I WAS THE FIRST PERSON TO **BELIEVE IN IT**. THAT WAS OUR RELATIONSHIP. WE WERE CLASSMATES. WE WERE IN LOVE.
OK. FINE. BUT HERE'S ONE MORE.

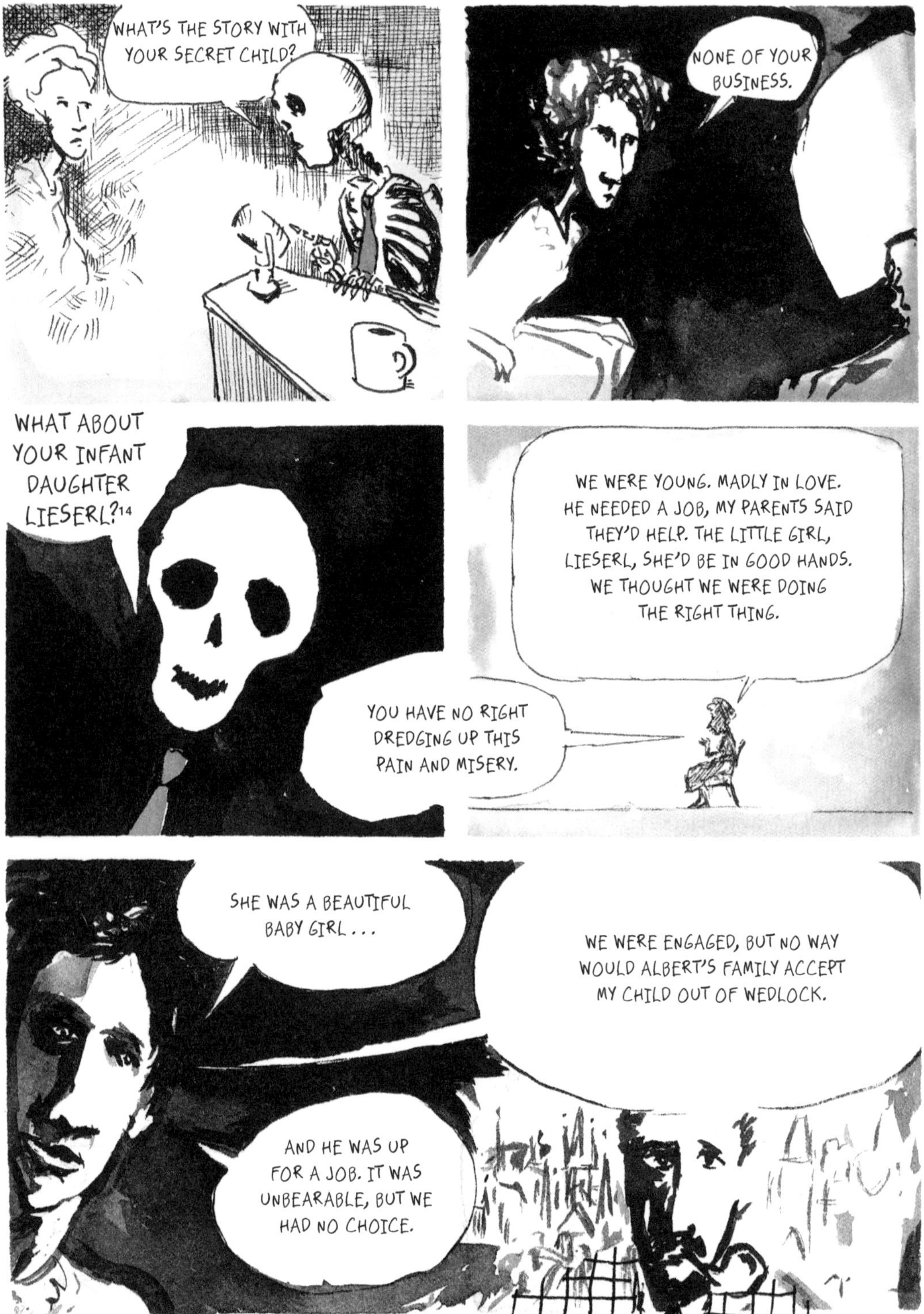
WHAT'S THE STORY WITH YOUR SECRET CHILD?
NONE OF YOUR BUSINESS.
WHAT ABOUT YOUR INFANT DAUGHTER LIESERL?[14]
WE WERE YOUNG. MADLY IN LOVE. HE NEEDED A JOB, MY PARENTS SAID THEY'D HELP. THE LITTLE GIRL, LIESERL, SHE'D BE IN GOOD HANDS. WE THOUGHT WE WERE DOING THE RIGHT THING.
YOU HAVE NO RIGHT DREDGING UP THIS PAIN AND MISERY.
SHE WAS A BEAUTIFUL BABY GIRL . . .
WE WERE ENGAGED, BUT NO WAY WOULD ALBERT'S FAMILY ACCEPT MY CHILD OUT OF WEDLOCK.
AND HE WAS UP FOR A JOB. IT WAS UNBEARABLE, BUT WE HAD NO CHOICE.

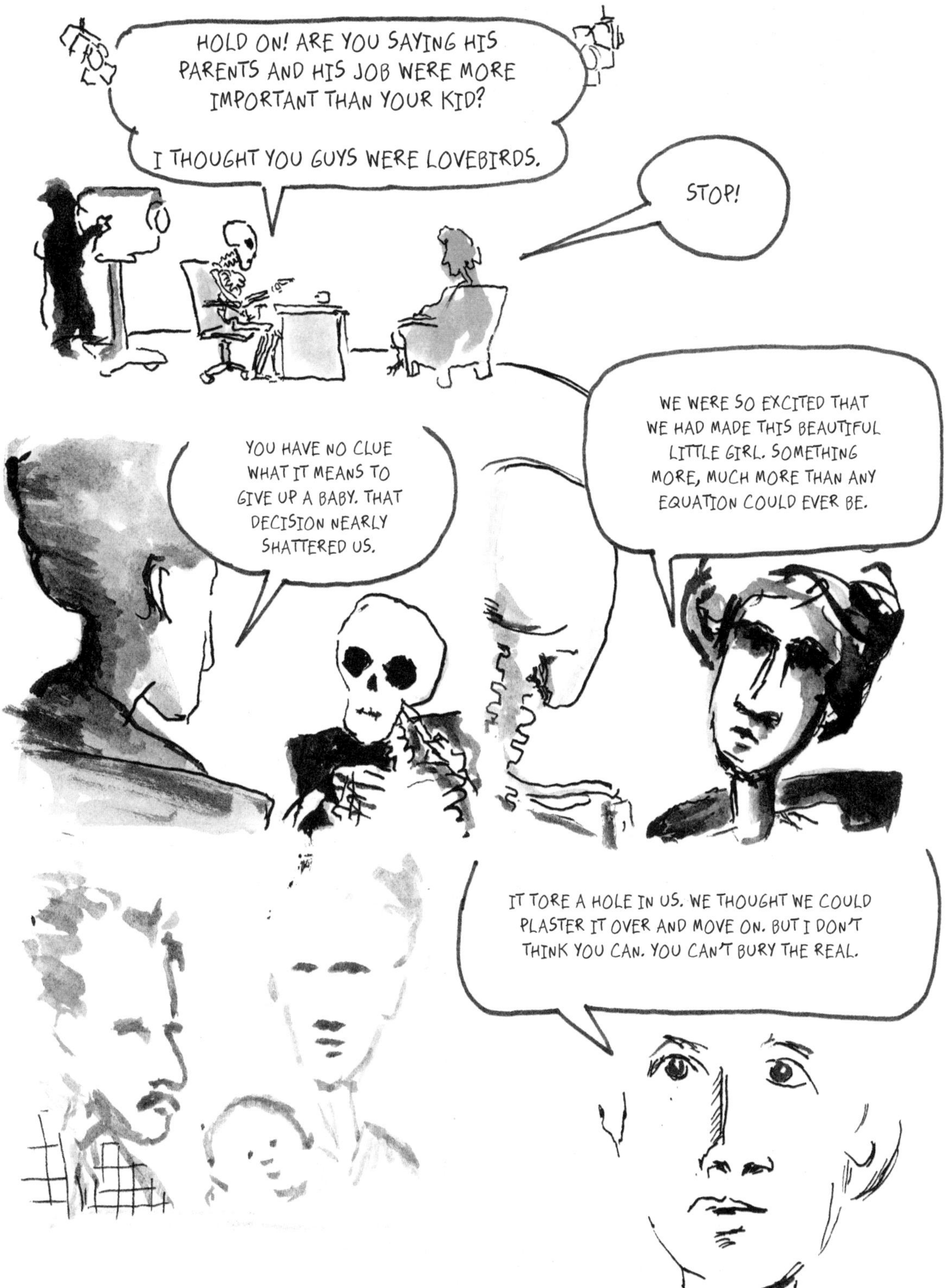
HOLD ON! ARE YOU SAYING HIS PARENTS AND HIS JOB WERE MORE IMPORTANT THAN YOUR KID?
I THOUGHT YOU GUYS WERE LOVEBIRDS.
STOP!
YOU HAVE NO CLUE WHAT IT MEANS TO GIVE UP A BABY. THAT DECISION NEARLY SHATTERED US.
WE WERE SO EXCITED THAT WE HAD MADE THIS BEAUTIFUL LITTLE GIRL. SOMETHING MORE, MUCH MORE THAN ANY EQUATION COULD EVER BE.
IT TORE A HOLE IN US. WE THOUGHT WE COULD PLASTER IT OVER AND MOVE ON. BUT I DON'T THINK YOU CAN. YOU CAN'T BURY THE REAL.

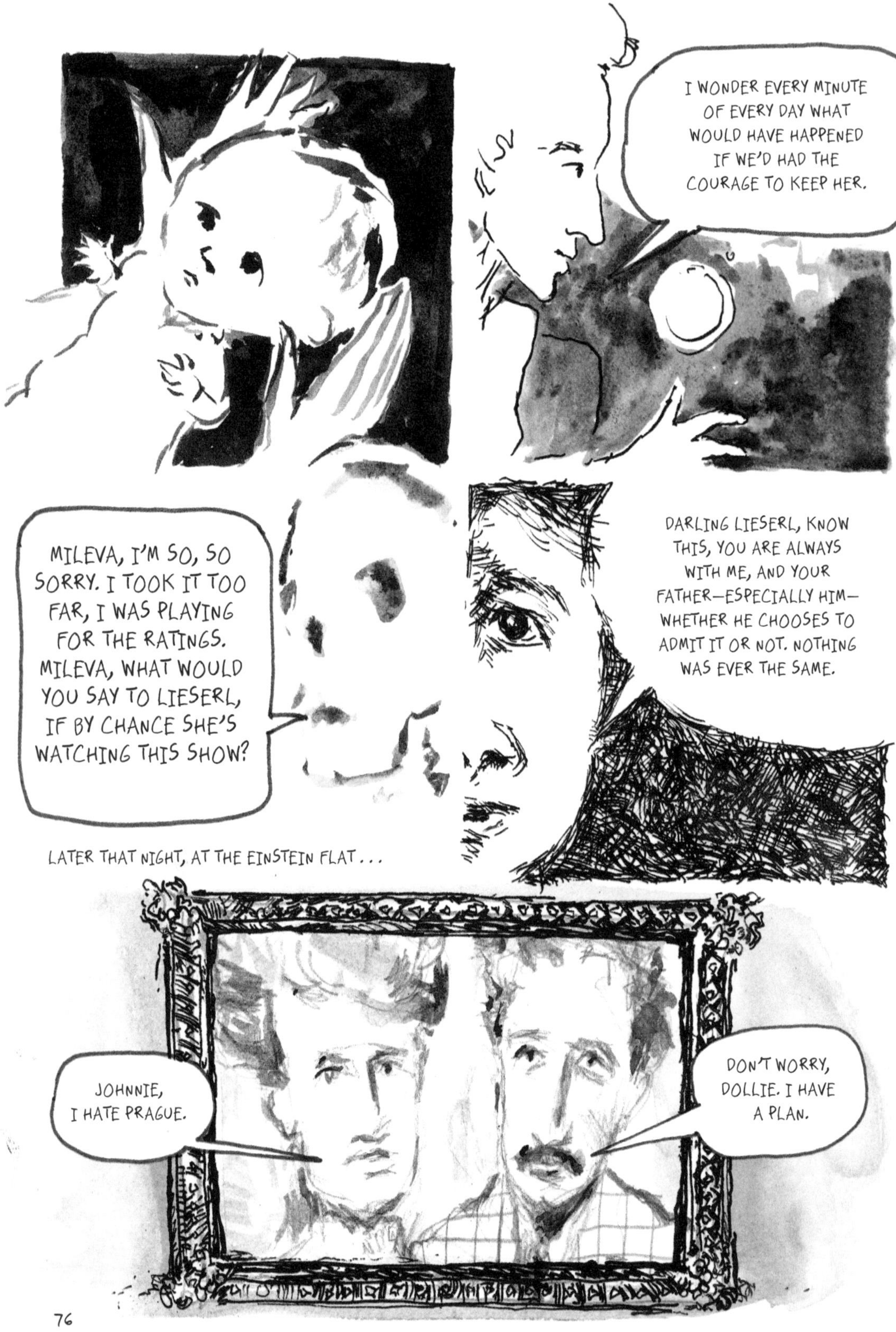
I WONDER EVERY MINUTE OF EVERY DAY WHAT WOULD HAVE HAPPENED IF WE'D HAD THE COURAGE TO KEEP HER.
MILEVA, I'M SO, SO SORRY. I TOOK IT TOO FAR, I WAS PLAYING FOR THE RATINGS. MILEVA, WHAT WOULD YOU SAY TO LIESERL, IF BY CHANCE SHE'S WATCHING THIS SHOW?
DARLING LIESERL, KNOW THIS, YOU ARE ALWAYS WITH ME, AND YOUR FATHER—ESPECIALLY HIM—WHETHER HE CHOOSES TO ADMIT IT OR NOT. NOTHING WAS EVER THE SAME.
LATER THAT NIGHT, AT THE EINSTEIN FLAT . . .
JOHNNIE, I HATE PRAGUE.
DON'T WORRY, DOLLIE. I HAVE A PLAN.

CHAPTER VII

Berta Fanta's Mad Tea Party (After)
May 24, 1911

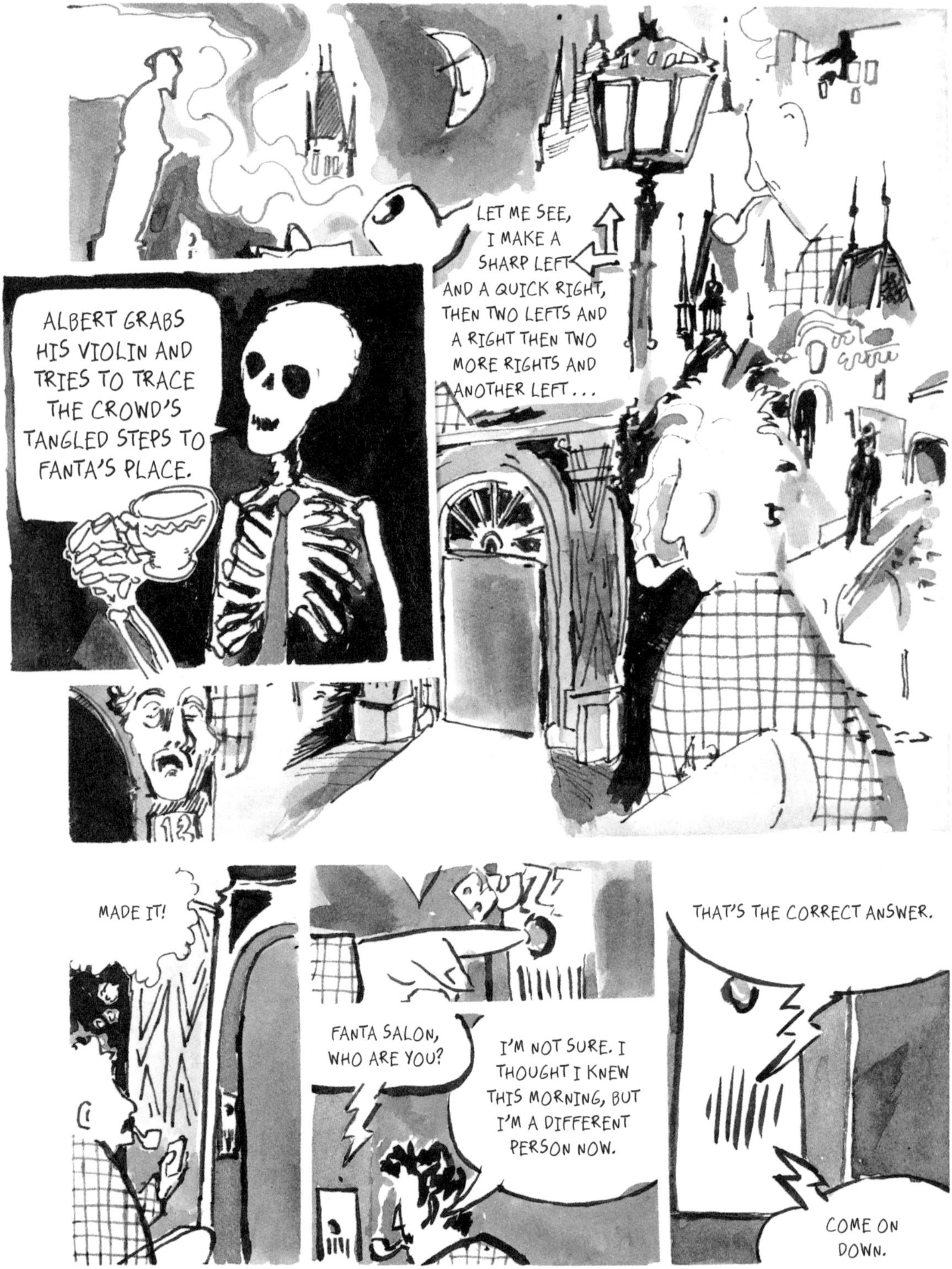
ALBERT GRABS HIS VIOLIN AND TRIES TO TRACE THE CROWD'S TANGLED STEPS TO FANTA'S PLACE.
LET ME SEE, I MAKE A SHARP LEFT AND A QUICK RIGHT, THEN TWO LEFTS AND A RIGHT THEN TWO MORE RIGHTS AND ANOTHER LEFT . . .
MADE IT!
FANTA SALON, WHO ARE YOU?
I'M NOT SURE. I THOUGHT I KNEW THIS MORNING, BUT I'M A DIFFERENT PERSON NOW.
THAT'S THE CORRECT ANSWER.
COME ON DOWN.

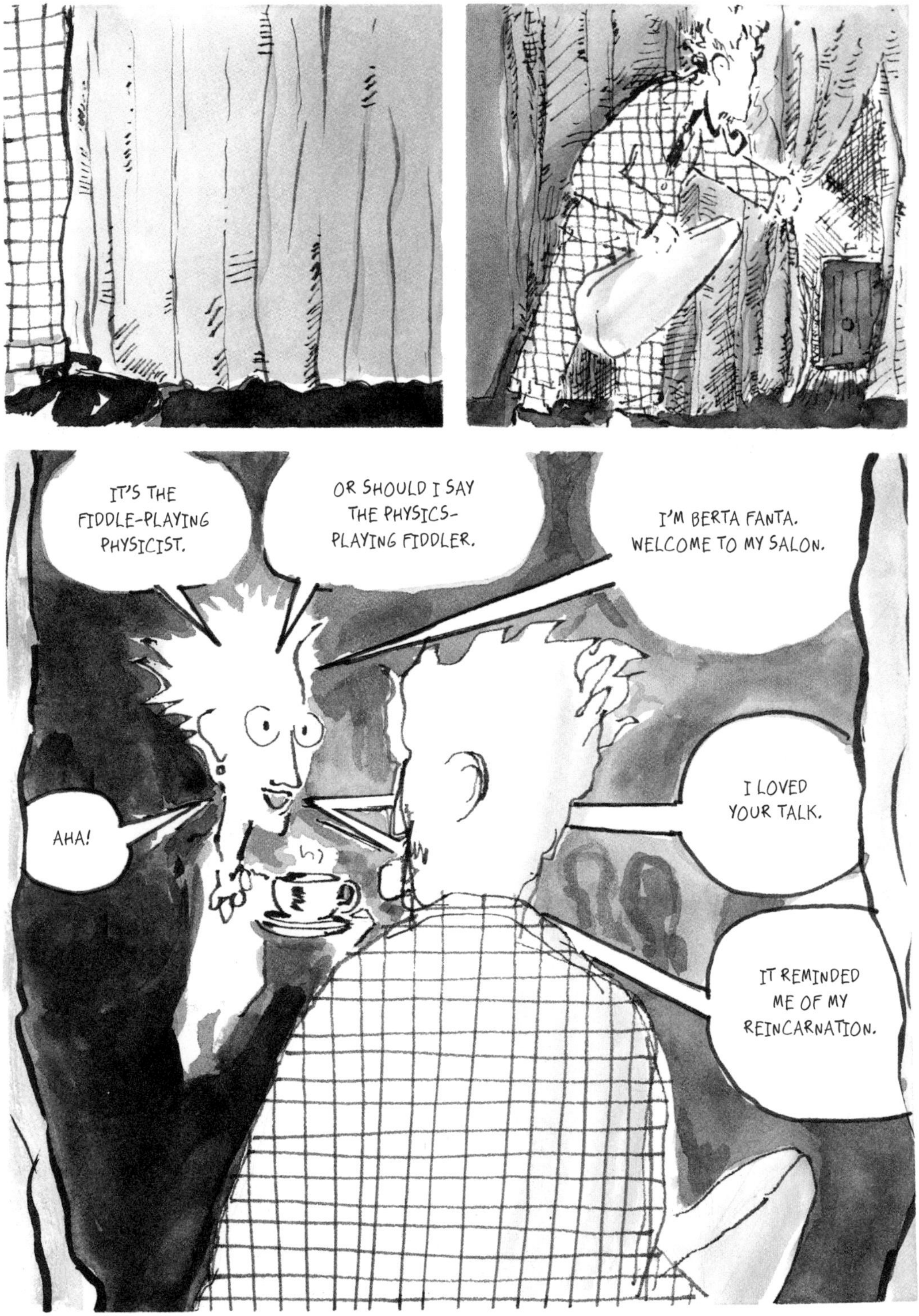
IT'S THE FIDDLE-PLAYING PHYSICIST.
OR SHOULD I SAY THE PHYSICS-PLAYING FIDDLER.
I'M BERTA FANTA. WELCOME TO MY SALON.
AHA!
I LOVED YOUR TALK.
IT REMINDED ME OF MY REINCARNATION.

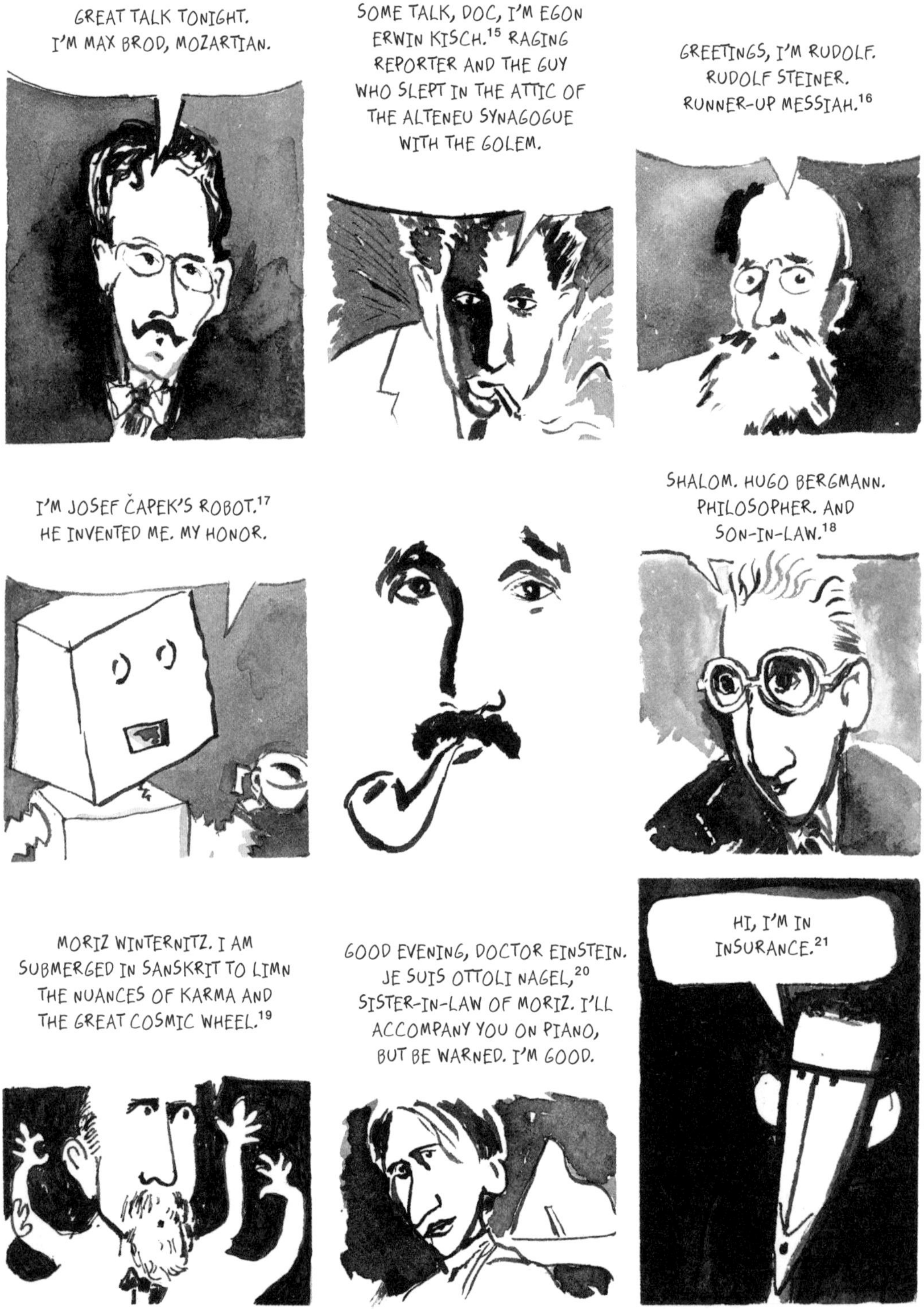
GREAT TALK TONIGHT. I'M MAX BROD, MOZARTIAN.
SOME TALK, DOC, I'M EGON ERWIN KISCH.[15] RAGING REPORTER AND THE GUY WHO SLEPT IN THE ATTIC OF THE ALTENEU SYNAGOGUE WITH THE GOLEM.
GREETINGS, I'M RUDOLF. RUDOLF STEINER. RUNNER-UP MESSIAH.[16]
I'M JOSEF ČAPEK'S ROBOT.[17] HE INVENTED ME. MY HONOR.
SHALOM. HUGO BERGMANN. PHILOSOPHER. AND SON-IN-LAW.[18]
MORIZ WINTERNITZ. I AM SUBMERGED IN SANSKRIT TO LIMN THE NUANCES OF KARMA AND THE GREAT COSMIC WHEEL.[19]
GOOD EVENING, DOCTOR EINSTEIN. JE SUIS OTTOLI NAGEL,[20] SISTER-IN-LAW OF MORIZ. I'LL ACCOMPANY YOU ON PIANO, BUT BE WARNED. I'M GOOD.
HI, I'M IN INSURANCE.[21]

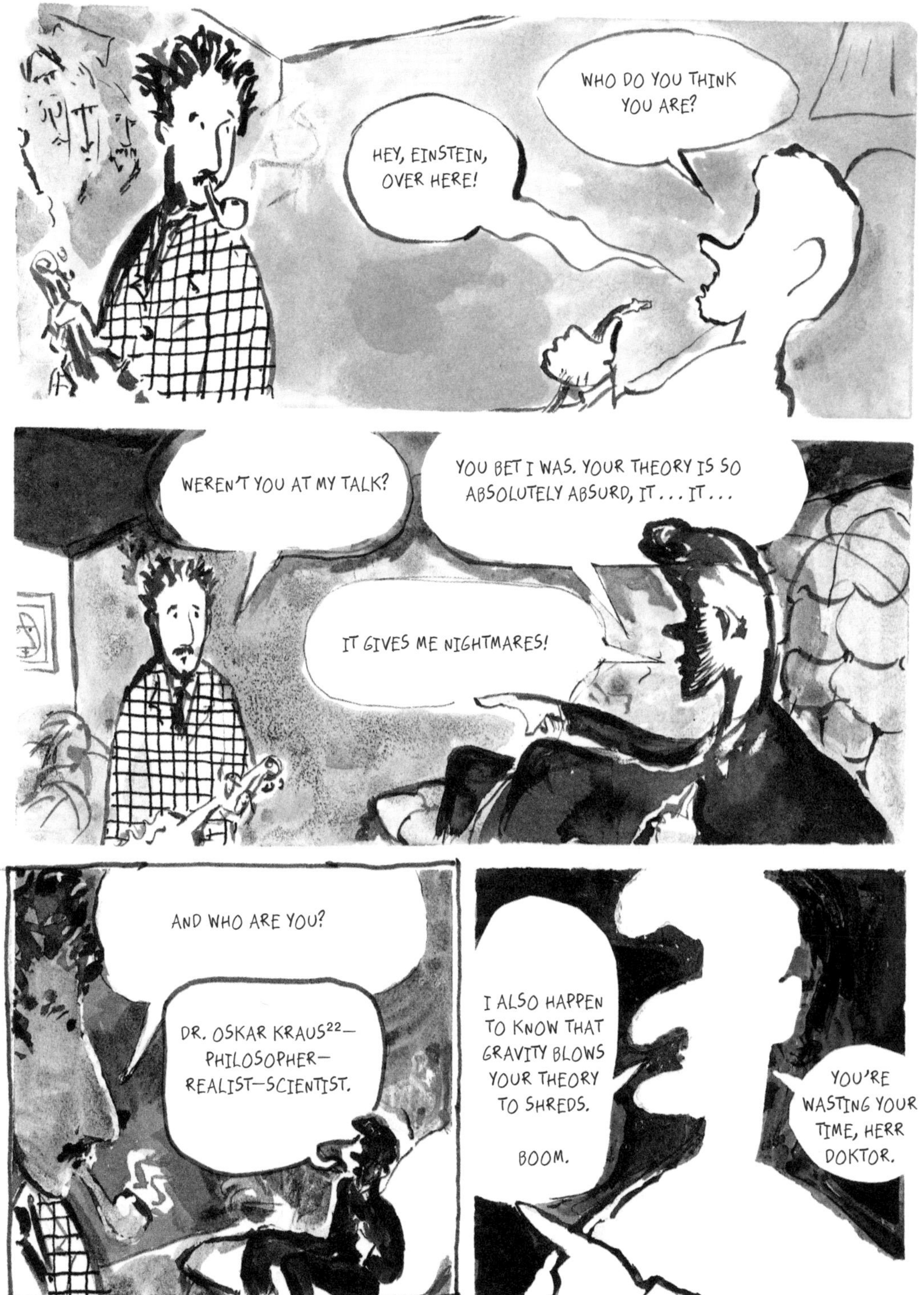
HEY, EINSTEIN, OVER HERE!
WHO DO YOU THINK YOU ARE?
WEREN'T YOU AT MY TALK?
YOU BET I WAS. YOUR THEORY IS SO ABSOLUTELY ABSURD, IT . . . IT . . .
IT GIVES ME NIGHTMARES!
AND WHO ARE YOU?
DR. OSKAR KRAUS22—PHILOSOPHER—REALIST—SCIENTIST.
I ALSO HAPPEN TO KNOW THAT GRAVITY BLOWS YOUR THEORY TO SHREDS.
BOOM.
YOU'RE WASTING YOUR TIME, HERR DOKTOR.

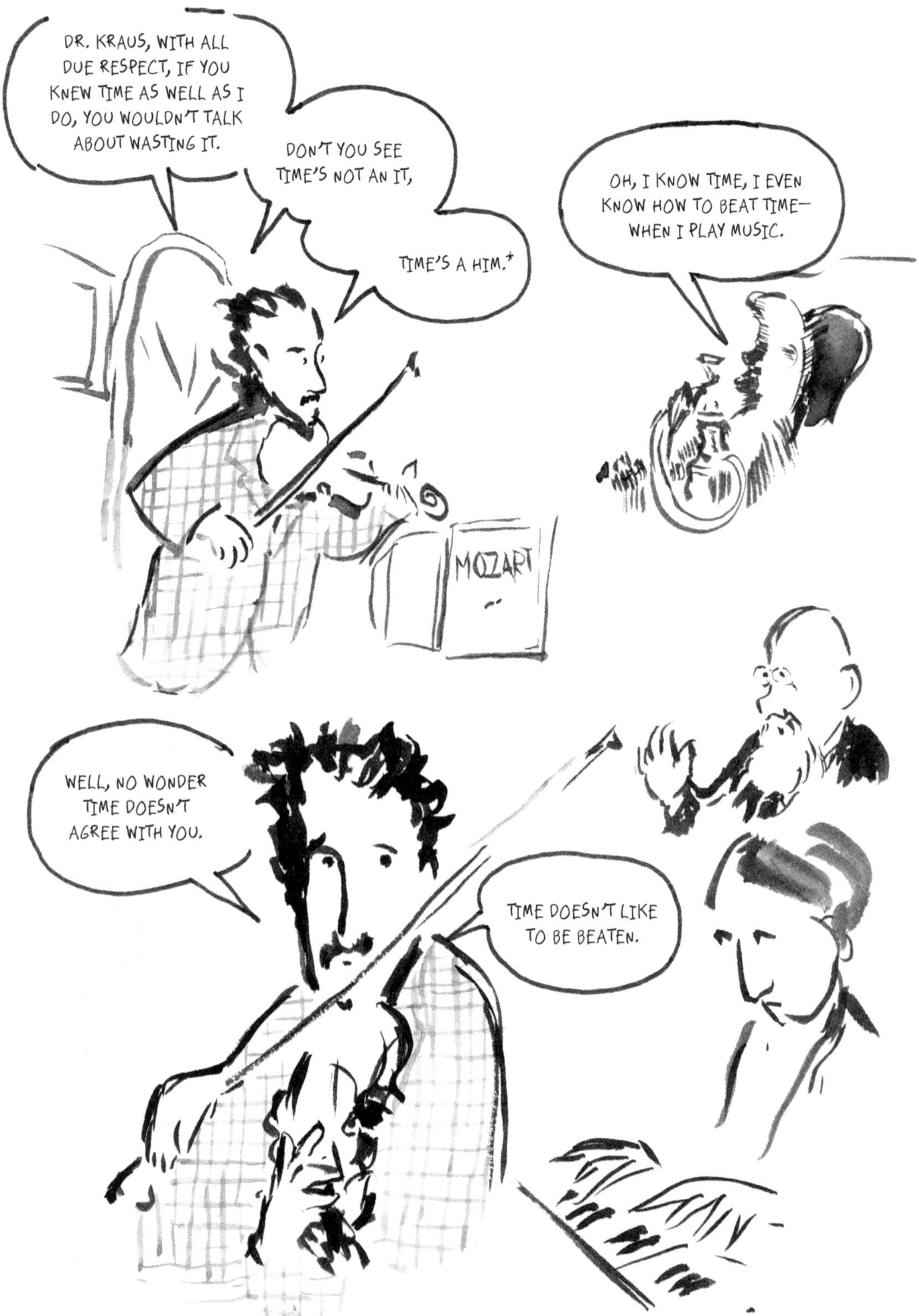
DR. KRAUS, WITH ALL DUE RESPECT, IF YOU KNEW TIME AS WELL AS I DO, YOU WOULDN'T TALK ABOUT WASTING IT.
DON'T YOU SEE TIME'S NOT AN IT,
TIME'S A HIM.*
OH, I KNOW TIME, I EVEN KNOW HOW TO BEAT TIME— WHEN I PLAY MUSIC.
MOZART
WELL, NO WONDER TIME DOESN'T AGREE WITH YOU.
TIME DOESN'T LIKE TO BE BEATEN.

**Alice's Adventures In Wonderland*, Chapter VII (London: R. Clay, Son and Taylor, 1874), p. 101.

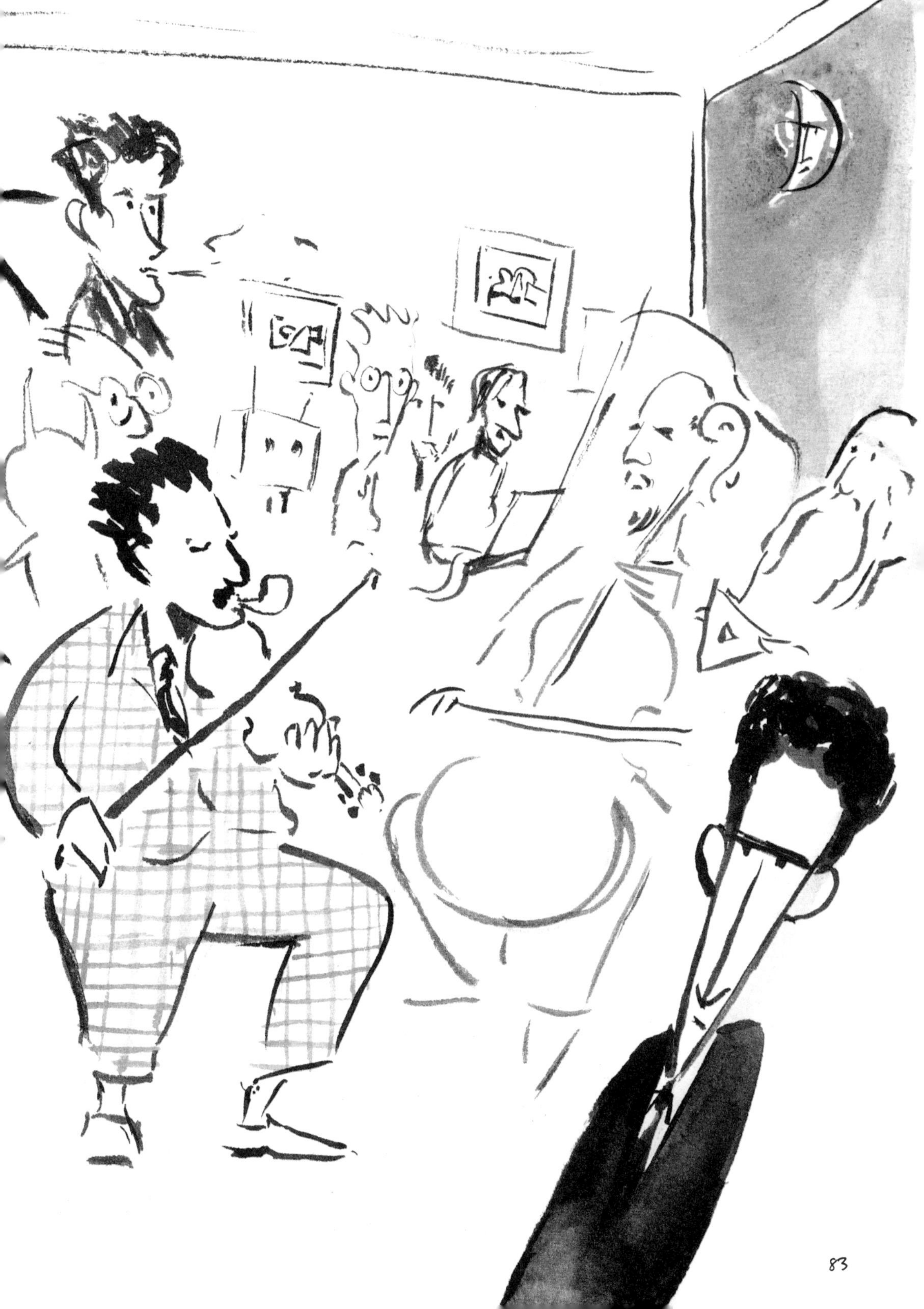

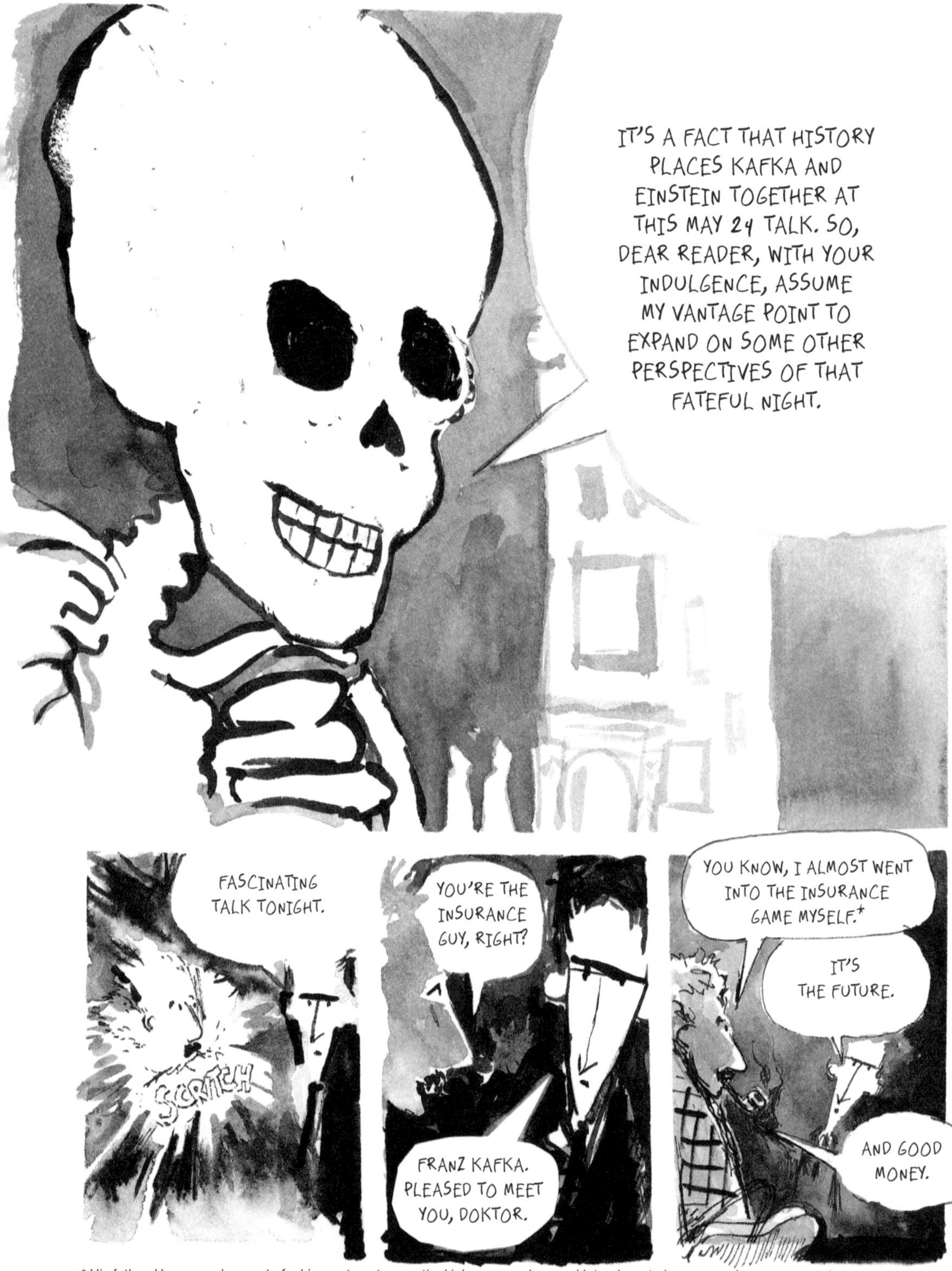
IT'S A FACT THAT HISTORY PLACES KAFKA AND EINSTEIN TOGETHER AT THIS MAY 24 TALK. SO, DEAR READER, WITH YOUR INDULGENCE, ASSUME MY VANTAGE POINT TO EXPAND ON SOME OTHER PERSPECTIVES OF THAT FATEFUL NIGHT.
FASCINATING TALK TONIGHT.
SCRATCH
YOU'RE THE INSURANCE GUY, RIGHT?
FRANZ KAFKA. PLEASED TO MEET YOU, DOKTOR.
YOU KNOW, I ALMOST WENT INTO THE INSURANCE GAME MYSELF.*
IT'S THE FUTURE.
AND GOOD MONEY.

* His father, Hermann, desperate for his son to get a practical job, arranged several interviews to become an insurance agent and it's said that the frustrated fledgling physicist even tried his hand at selling policies for a few months.

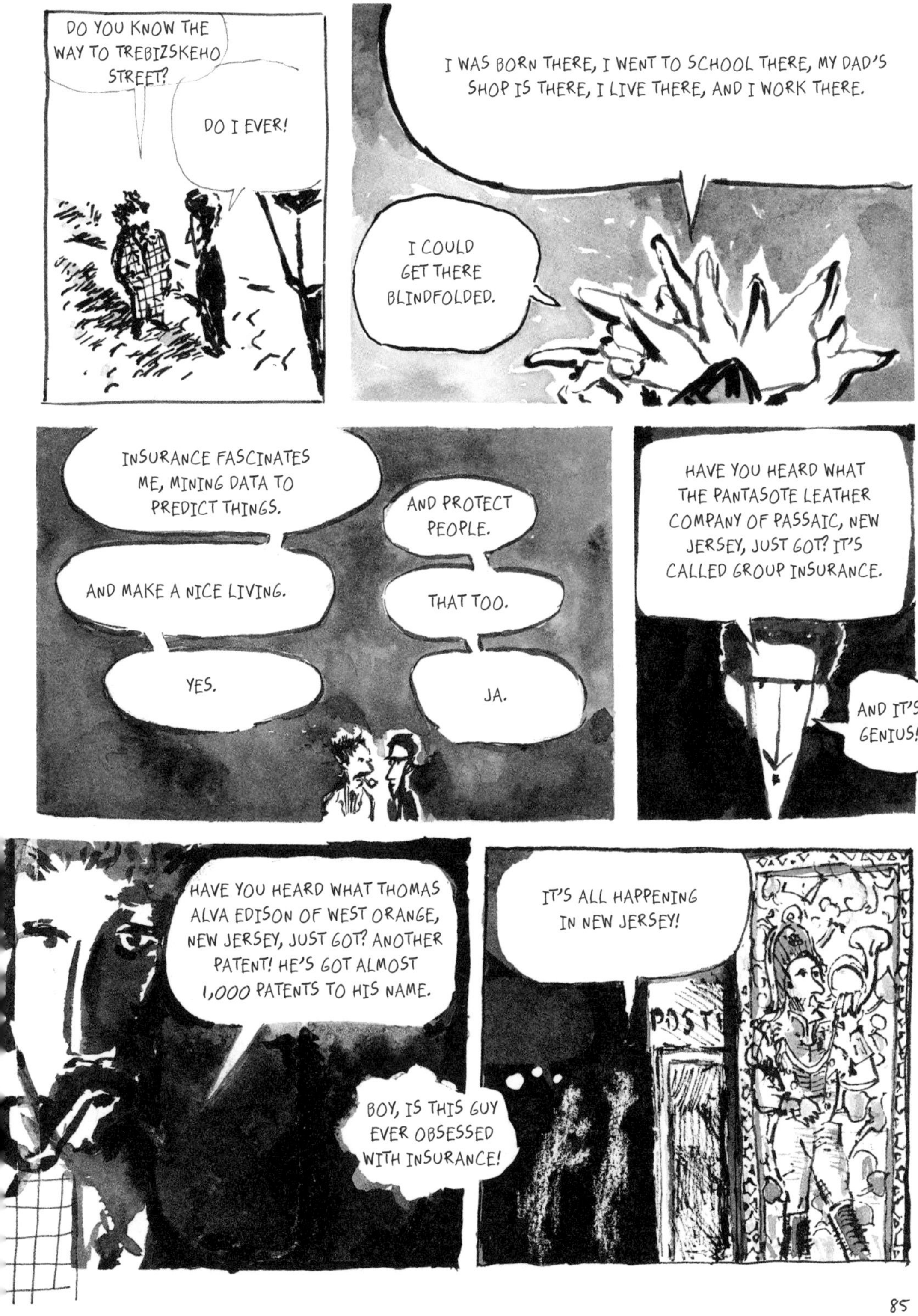
DO YOU KNOW THE WAY TO TREBIZSKEHO STREET?
DO I EVER!
I WAS BORN THERE, I WENT TO SCHOOL THERE, MY DAD'S SHOP IS THERE, I LIVE THERE, AND I WORK THERE.
I COULD GET THERE BLINDFOLDED.
INSURANCE FASCINATES ME, MINING DATA TO PREDICT THINGS.
AND PROTECT PEOPLE.
AND MAKE A NICE LIVING.
THAT TOO.
YES.
JA.
HAVE YOU HEARD WHAT THE PANTASOTE LEATHER COMPANY OF PASSAIC, NEW JERSEY, JUST GOT? IT'S CALLED GROUP INSURANCE.
AND IT'S GENIUS!
HAVE YOU HEARD WHAT THOMAS ALVA EDISON OF WEST ORANGE, NEW JERSEY, JUST GOT? ANOTHER PATENT! HE'S GOT ALMOST 1,000 PATENTS TO HIS NAME.
BOY, IS THIS GUY EVER OBSESSED WITH INSURANCE!
IT'S ALL HAPPENING IN NEW JERSEY!

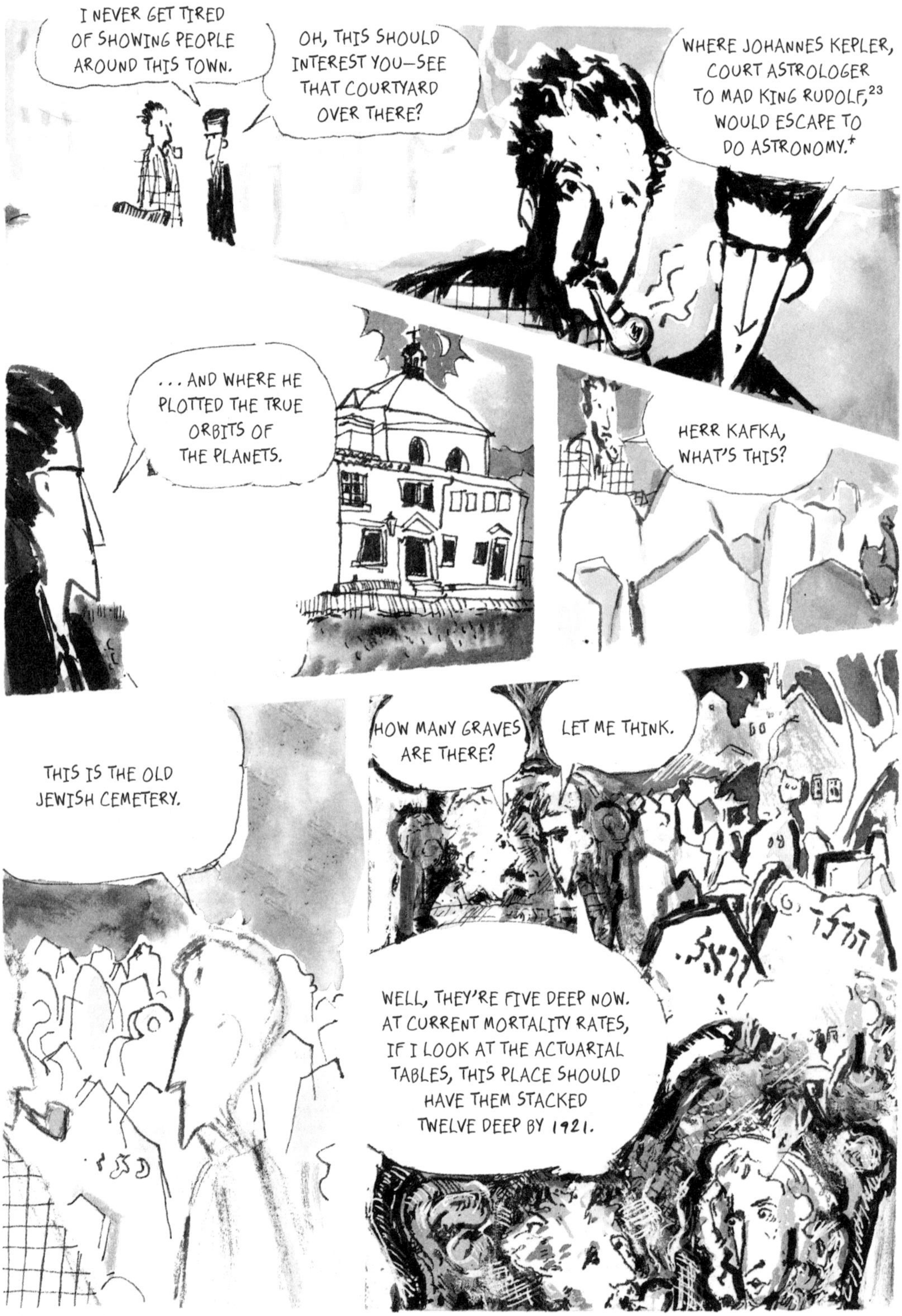
I NEVER GET TIRED OF SHOWING PEOPLE AROUND THIS TOWN.
OH, THIS SHOULD INTEREST YOU—SEE THAT COURTYARD OVER THERE?
WHERE JOHANNES KEPLER, COURT ASTROLOGER TO MAD KING RUDOLF,[23] WOULD ESCAPE TO DO ASTRONOMY.*
. . . AND WHERE HE PLOTTED THE TRUE ORBITS OF THE PLANETS.
HERR KAFKA, WHAT'S THIS?
THIS IS THE OLD JEWISH CEMETERY.
HOW MANY GRAVES ARE THERE?
LET ME THINK.
WELL, THEY'RE FIVE DEEP NOW. AT CURRENT MORTALITY RATES, IF I LOOK AT THE ACTUARIAL TABLES, THIS PLACE SHOULD HAVE THEM STACKED TWELVE DEEP BY 1921.

* And write the first ever sci-fi novel!

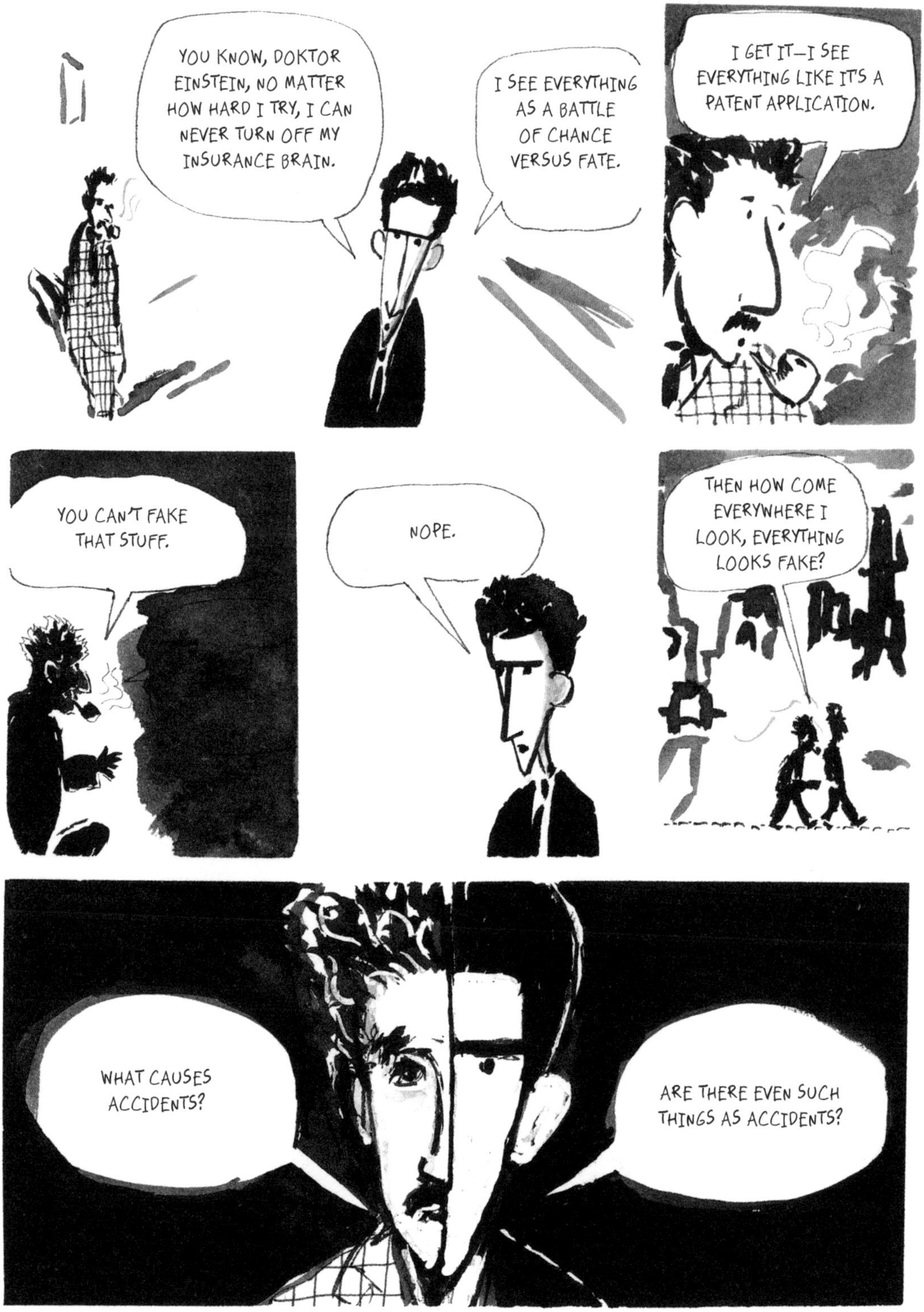
YOU KNOW, DOKTOR EINSTEIN, NO MATTER HOW HARD I TRY, I CAN NEVER TURN OFF MY INSURANCE BRAIN.
I SEE EVERYTHING AS A BATTLE OF CHANCE VERSUS FATE.
I GET IT—I SEE EVERYTHING LIKE IT'S A PATENT APPLICATION.
YOU CAN'T FAKE THAT STUFF.
NOPE.
THEN HOW COME EVERYWHERE I LOOK, EVERYTHING LOOKS FAKE?
WHAT CAUSES ACCIDENTS?
ARE THERE EVEN SUCH THINGS AS ACCIDENTS?

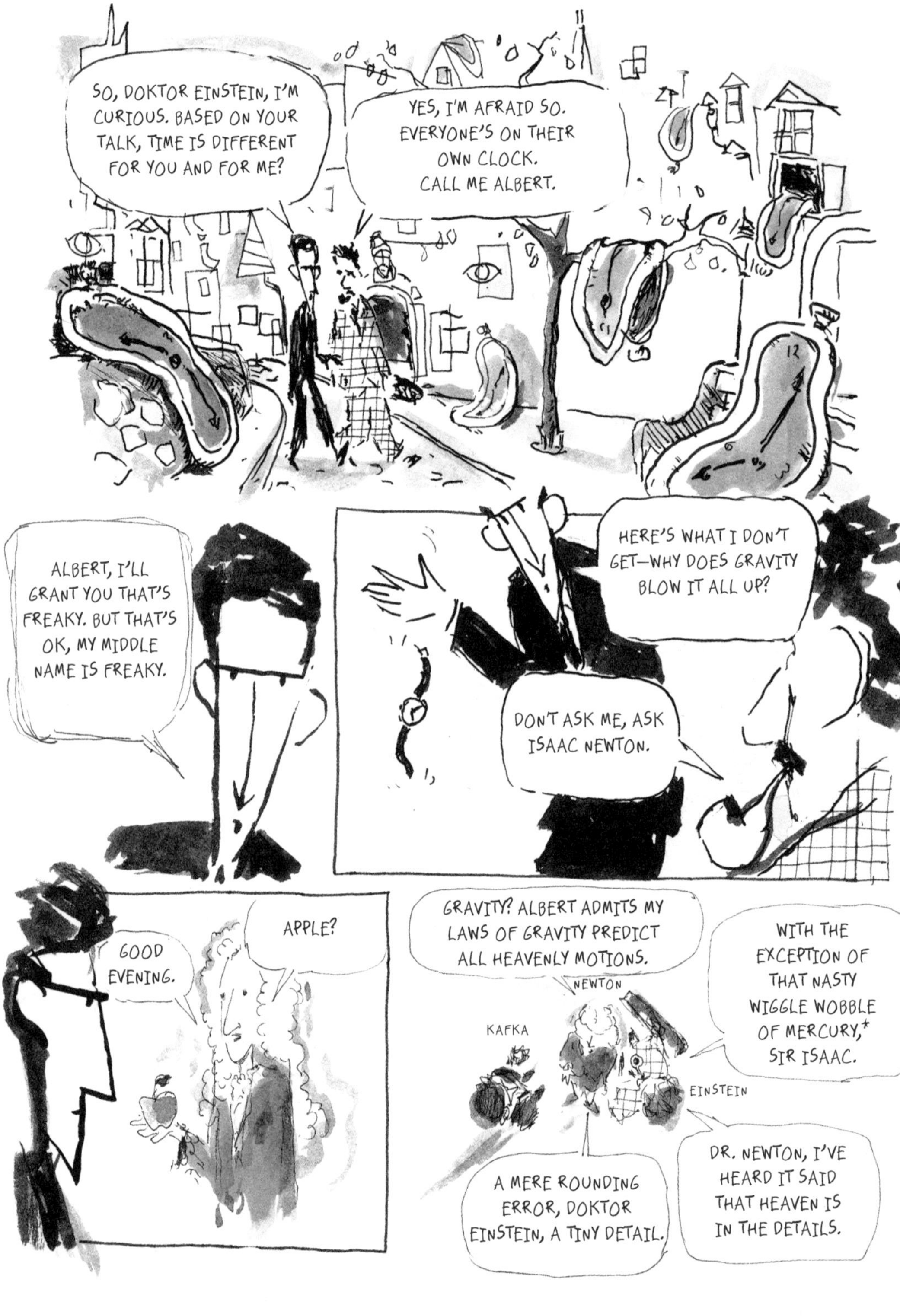
SO, DOKTOR EINSTEIN, I'M CURIOUS. BASED ON YOUR TALK, TIME IS DIFFERENT FOR YOU AND FOR ME?
YES, I'M AFRAID SO. EVERYONE'S ON THEIR OWN CLOCK. CALL ME ALBERT.
ALBERT, I'LL GRANT YOU THAT'S FREAKY. BUT THAT'S OK, MY MIDDLE NAME IS FREAKY.
HERE'S WHAT I DON'T GET—WHY DOES GRAVITY BLOW IT ALL UP?
DON'T ASK ME, ASK ISAAC NEWTON.
GOOD EVENING.
APPLE?
GRAVITY? ALBERT ADMITS MY LAWS OF GRAVITY PREDICT ALL HEAVENLY MOTIONS.
NEWTON
KAFKA
EINSTEIN
WITH THE EXCEPTION OF THAT NASTY WIGGLE WOBBLE OF MERCURY,* SIR ISAAC.
A MERE ROUNDING ERROR, DOKTOR EINSTEIN, A TINY DETAIL.
DR. NEWTON, I'VE HEARD IT SAID THAT HEAVEN IS IN THE DETAILS.

* The perihelion of Mercury, an anomaly of Mercury's orbit that didn't square with Newton's math, that vexed physicists for more than 200 years, and that was perfectly explained by Einstein's eventual insights.

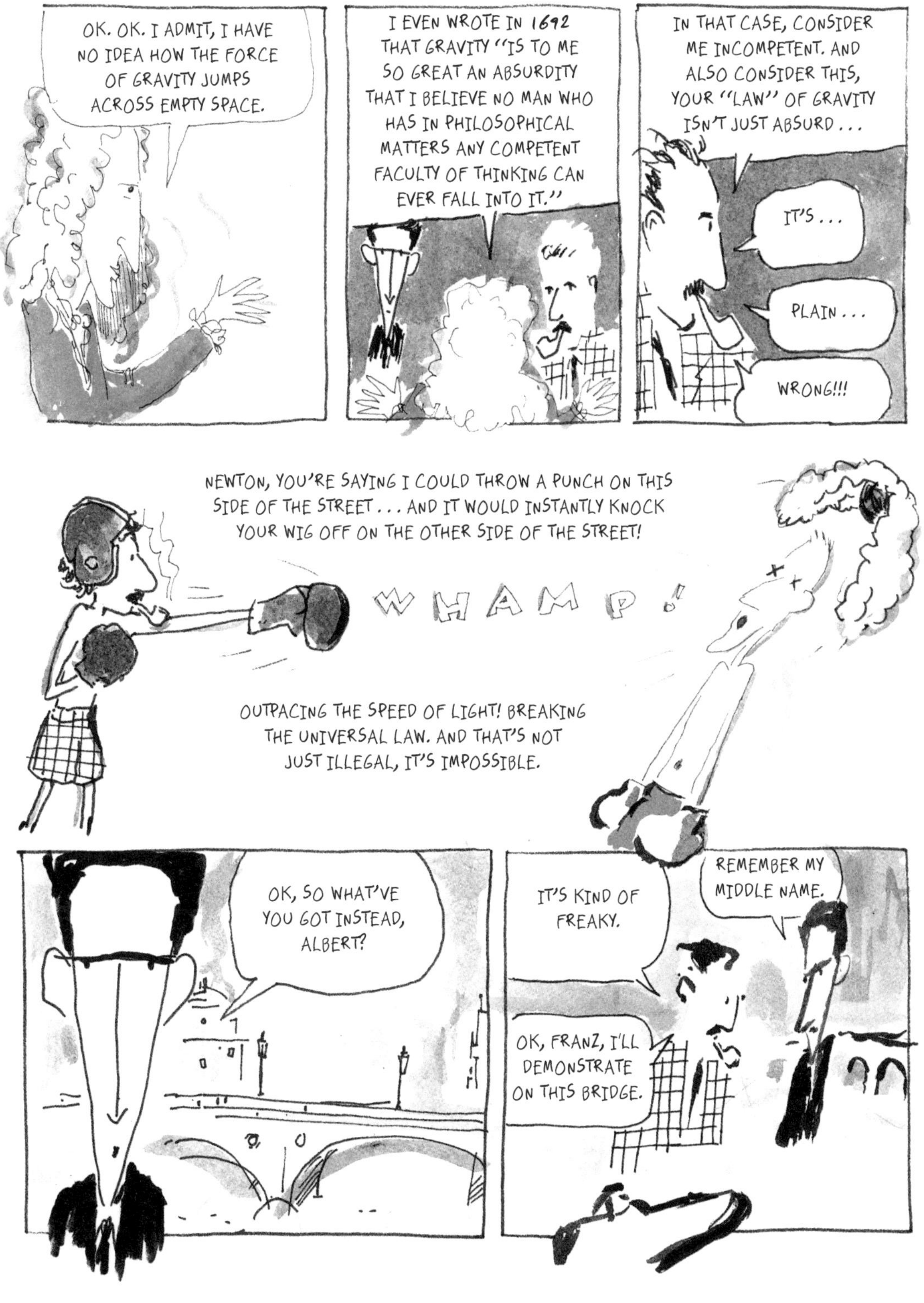

OK. OK. I ADMIT, I HAVE NO IDEA HOW THE FORCE OF GRAVITY JUMPS ACROSS EMPTY SPACE.
I EVEN WROTE IN 1692 THAT GRAVITY "IS TO ME SO GREAT AN ABSURDITY THAT I BELIEVE NO MAN WHO HAS IN PHILOSOPHICAL MATTERS ANY COMPETENT FACULTY OF THINKING CAN EVER FALL INTO IT."
IN THAT CASE, CONSIDER ME INCOMPETENT. AND ALSO CONSIDER THIS, YOUR "LAW" OF GRAVITY ISN'T JUST ABSURD . . .
IT'S . . .
PLAIN . . .
WRONG!!!
NEWTON, YOU'RE SAYING I COULD THROW A PUNCH ON THIS SIDE OF THE STREET . . . AND IT WOULD INSTANTLY KNOCK YOUR WIG OFF ON THE OTHER SIDE OF THE STREET!
WHAMP!
OUTPACING THE SPEED OF LIGHT! BREAKING THE UNIVERSAL LAW. AND THAT'S NOT JUST ILLEGAL, IT'S IMPOSSIBLE.
OK, SO WHAT'VE YOU GOT INSTEAD, ALBERT?
IT'S KIND OF FREAKY.
REMEMBER MY MIDDLE NAME.
OK, FRANZ, I'LL DEMONSTRATE ON THIS BRIDGE.

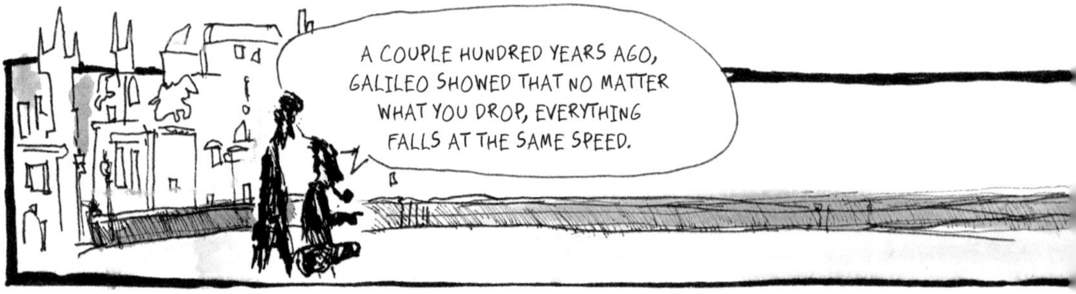
A COUPLE HUNDRED YEARS AGO, GALILEO SHOWED THAT NO MATTER WHAT YOU DROP, EVERYTHING FALLS AT THE SAME SPEED.

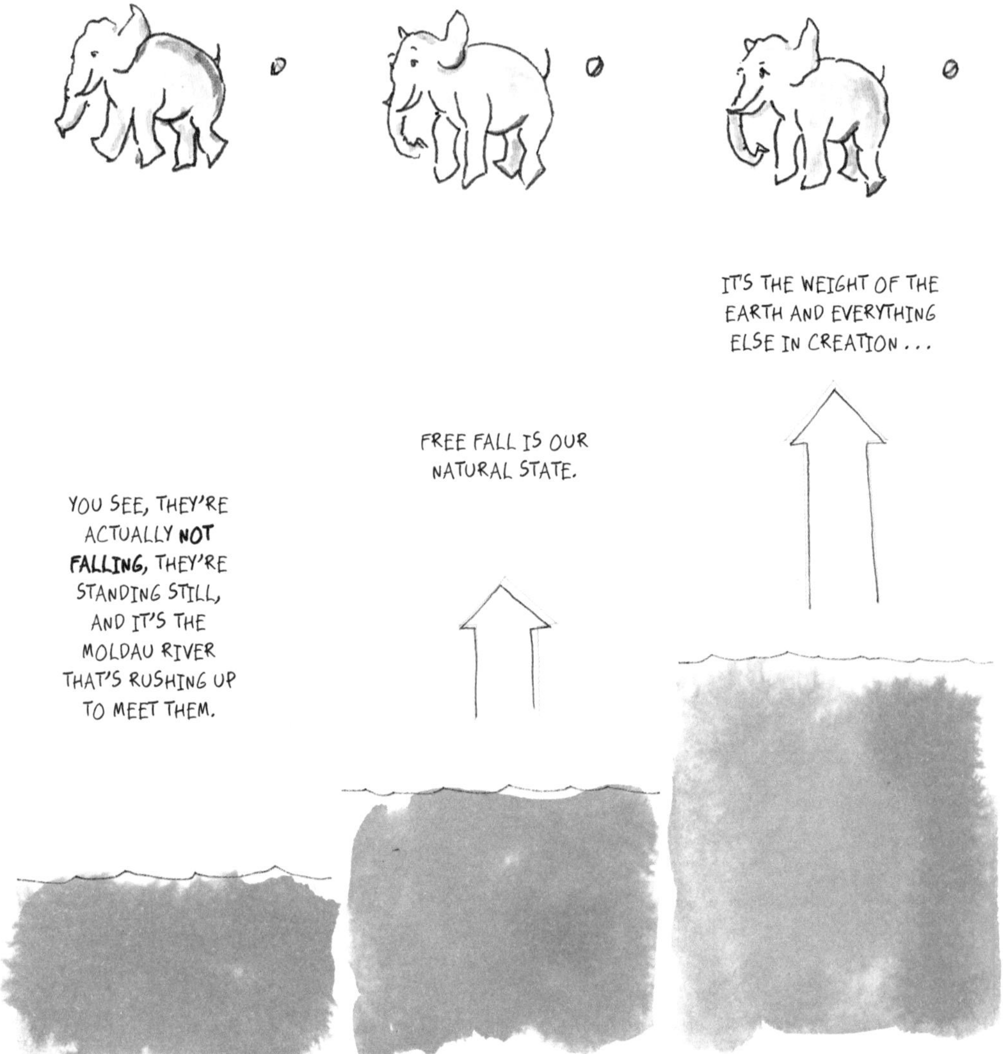
YOU SEE, THEY'RE ACTUALLY **NOT FALLING**, THEY'RE STANDING STILL, AND IT'S THE MOLDAU RIVER THAT'S RUSHING UP TO MEET THEM.
FREE FALL IS OUR NATURAL STATE.
IT'S THE WEIGHT OF THE EARTH AND EVERYTHING ELSE IN CREATION . . .

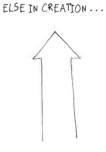

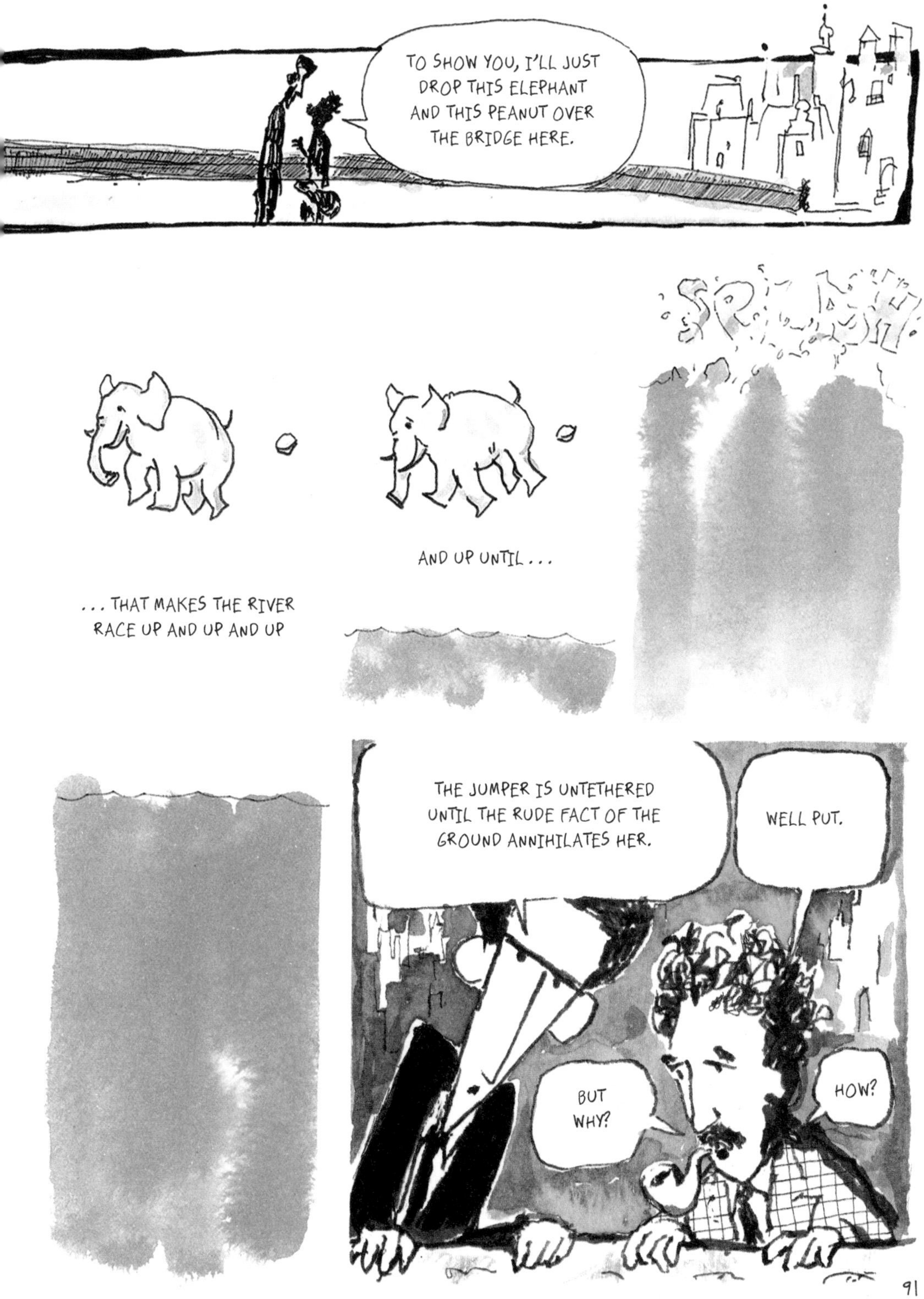
TO SHOW YOU, I'LL JUST DROP THIS ELEPHANT AND THIS PEANUT OVER THE BRIDGE HERE.
... THAT MAKES THE RIVER RACE UP AND UP AND UP
AND UP UNTIL ...
SPLASH
THE JUMPER IS UNTETHERED UNTIL THE RUDE FACT OF THE GROUND ANNIHILATES HER.
WELL PUT.
BUT WHY?
HOW?

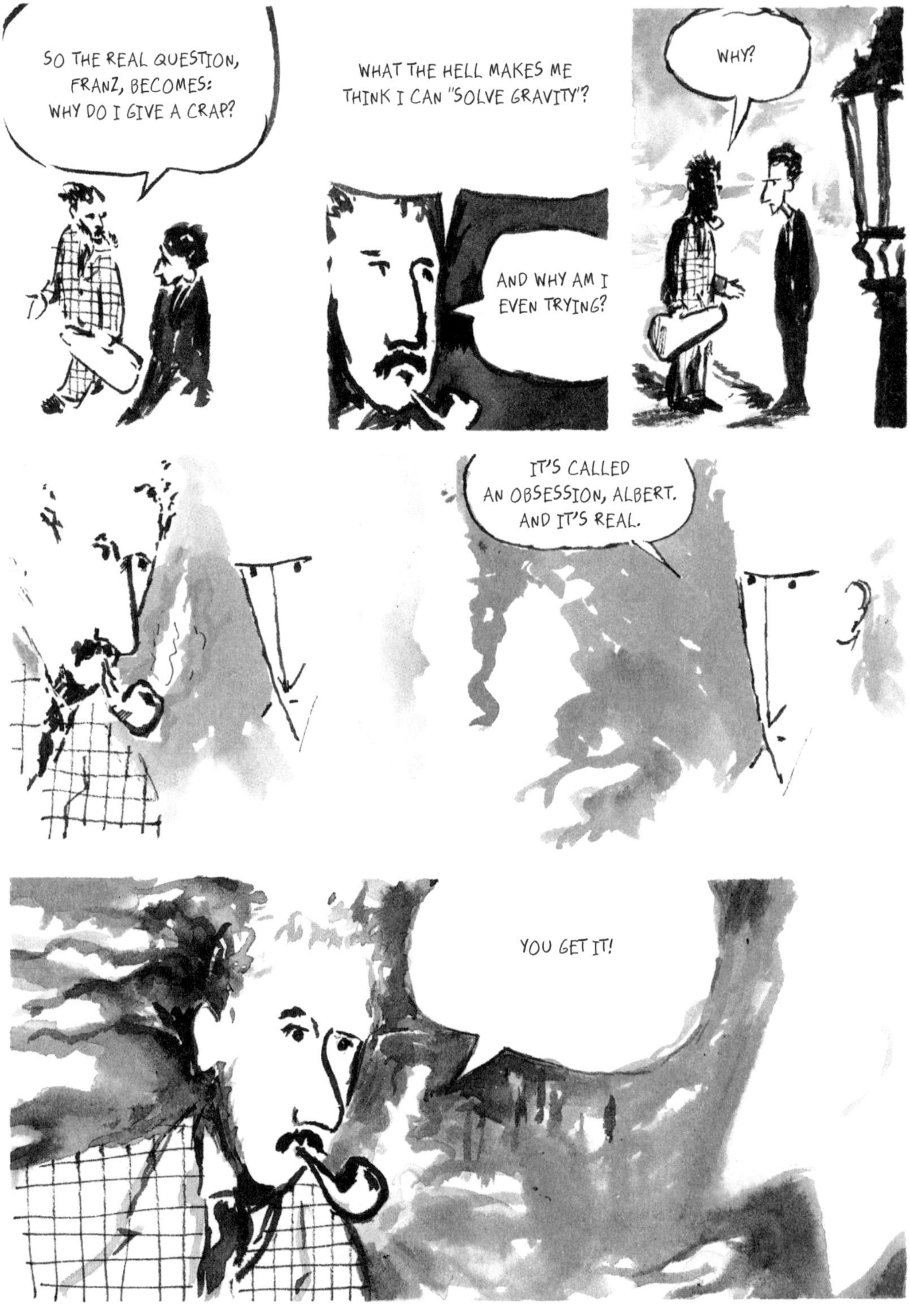
SO THE REAL QUESTION, FRANZ, BECOMES: WHY DO I GIVE A CRAP?
WHAT THE HELL MAKES ME THINK I CAN "SOLVE GRAVITY"?
AND WHY AM I EVEN TRYING?
WHY?
IT'S CALLED AN OBSESSION, ALBERT. AND IT'S REAL.
YOU GET IT!

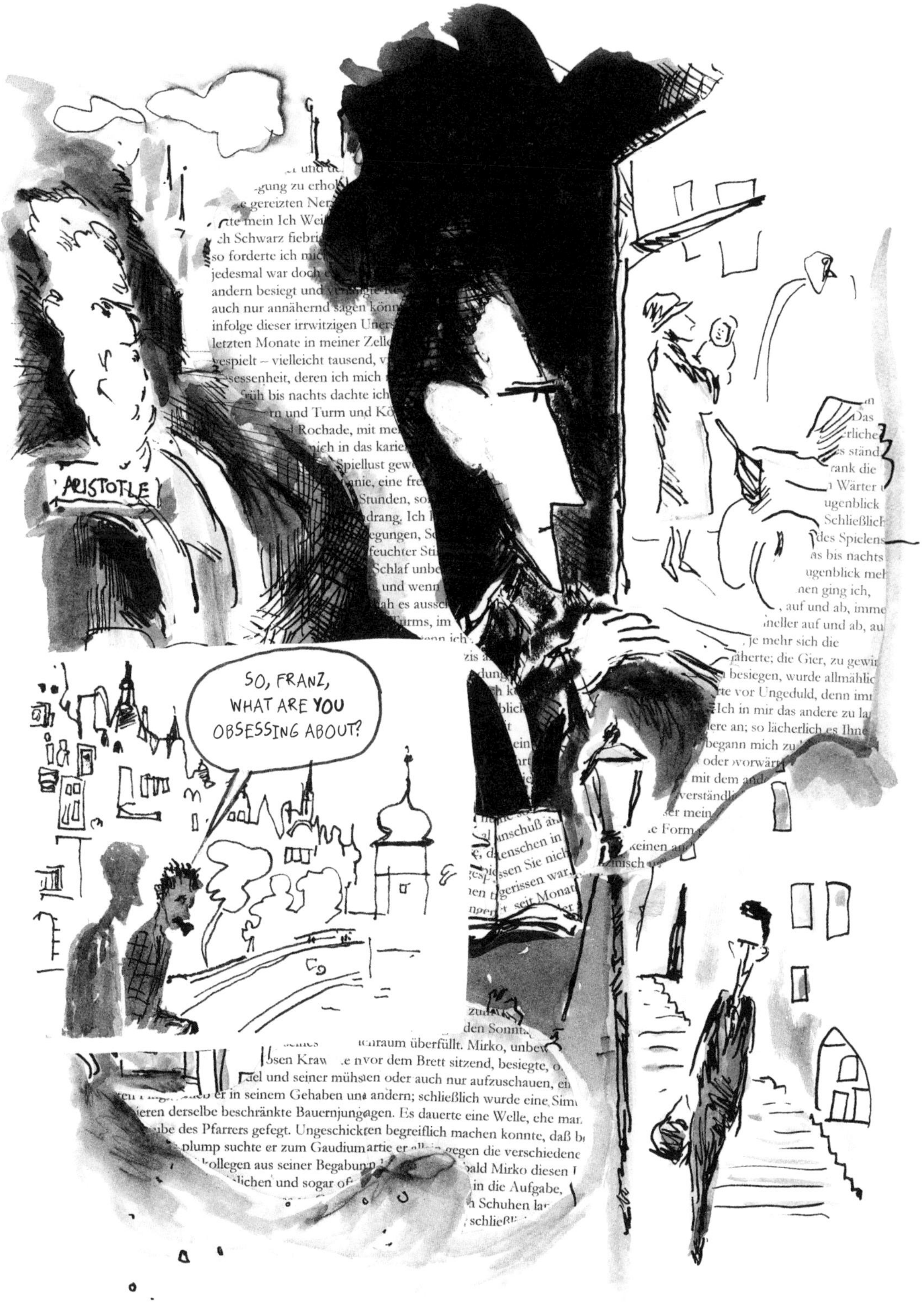
ARISTOTLE
SO, FRANZ, WHAT ARE YOU OBSESSING ABOUT?

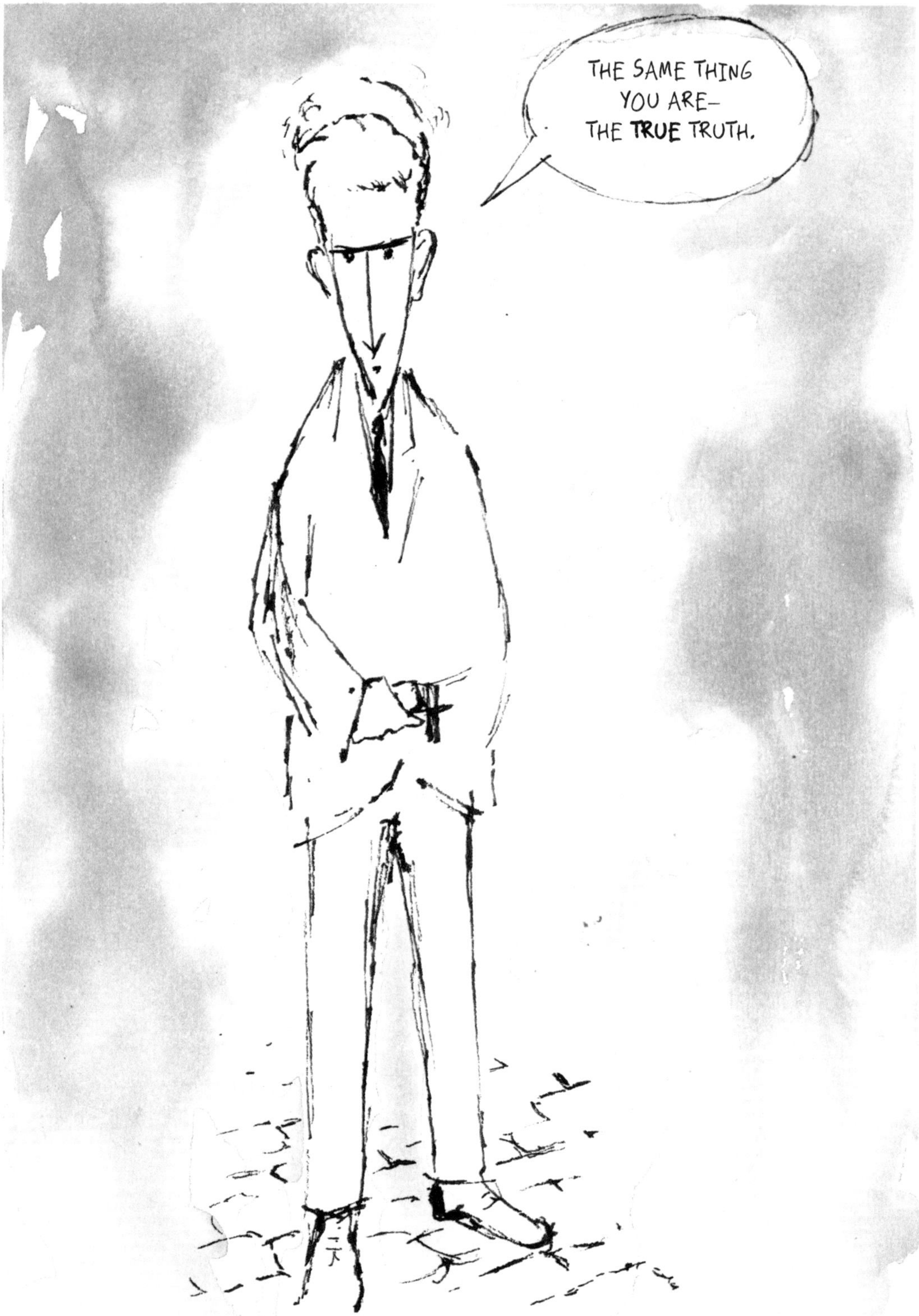
THE SAME THING YOU ARE– THE **TRUE** TRUTH.

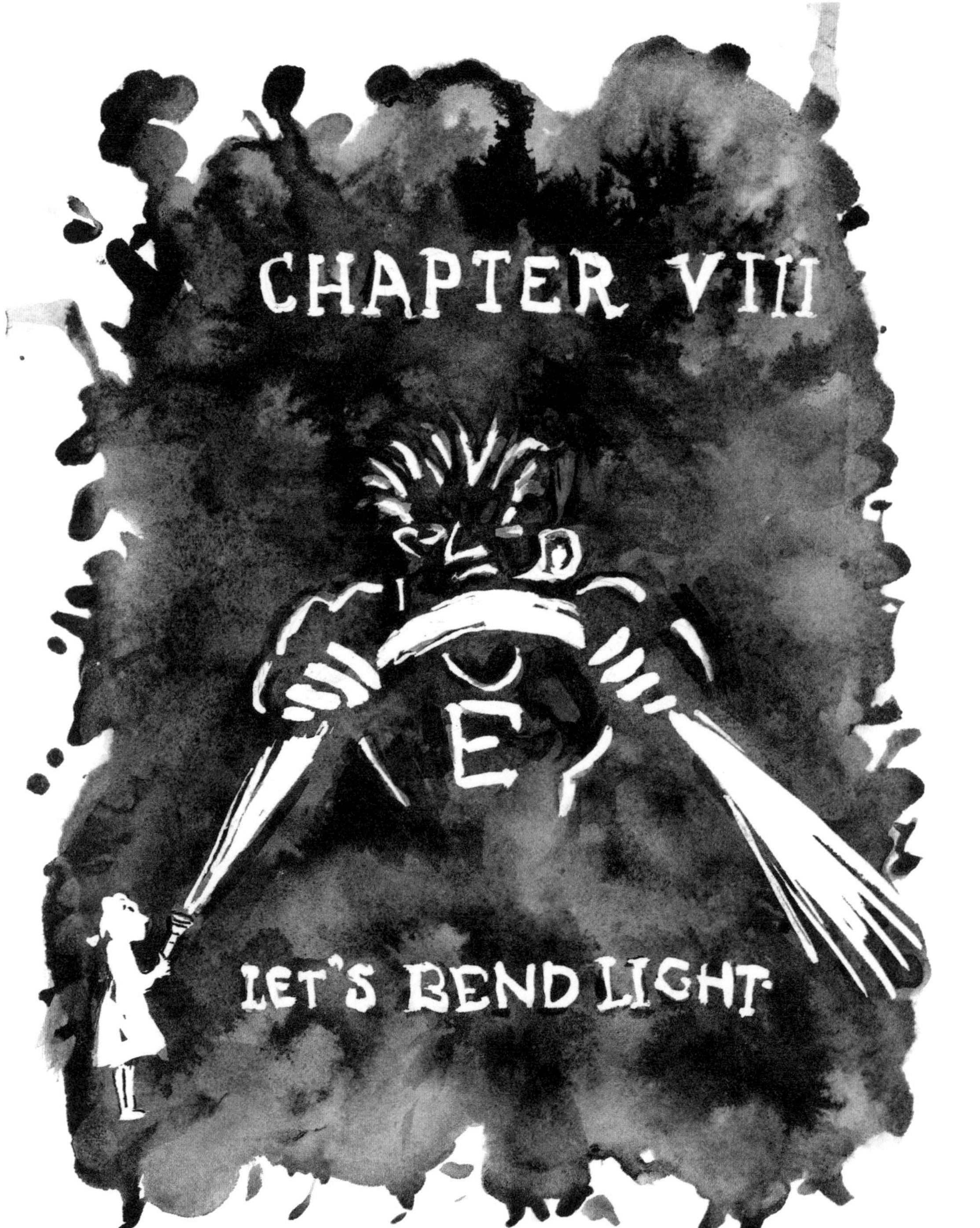
CHAPTER VIII
LET'S BEND LIGHT

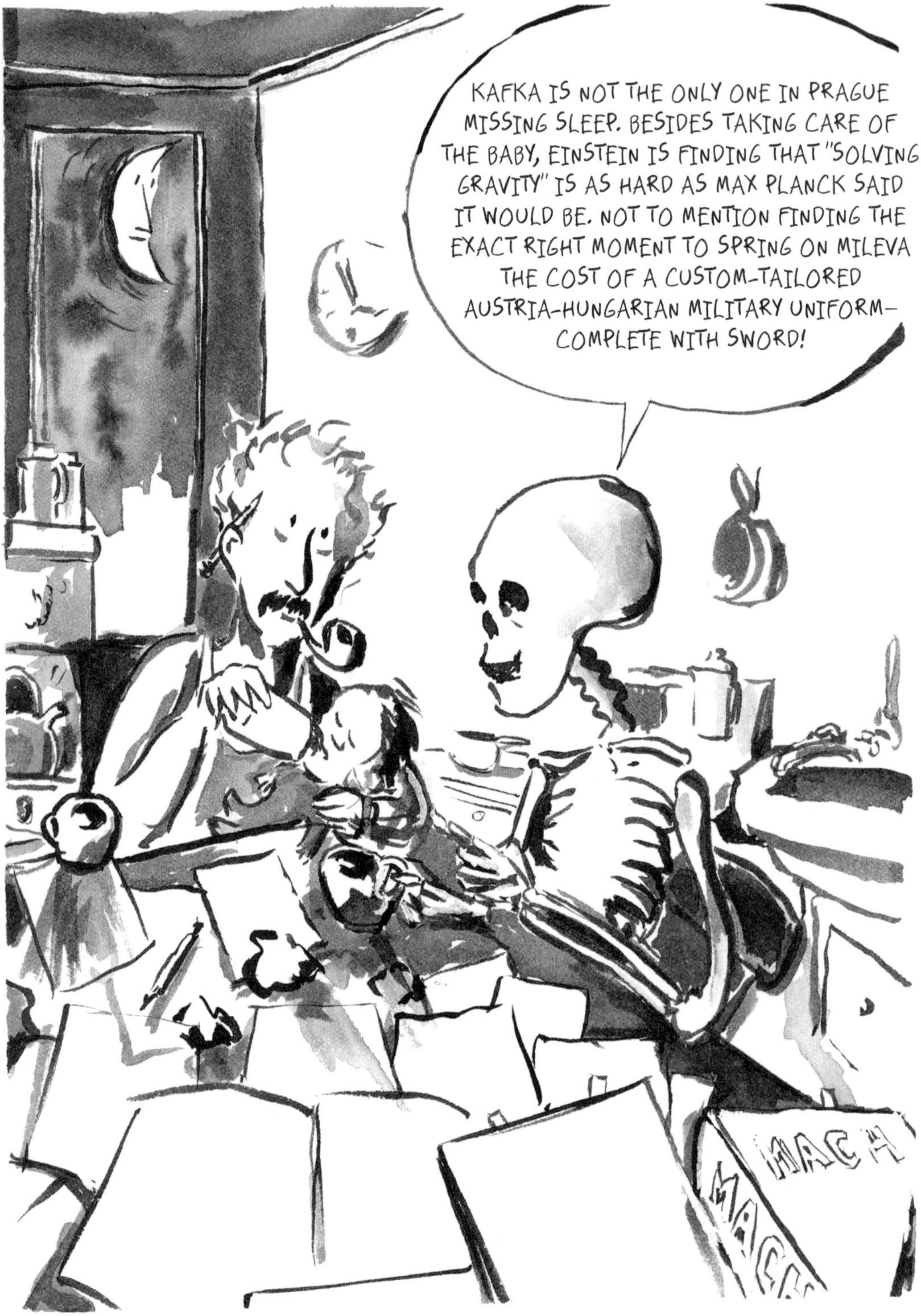
KAFKA IS NOT THE ONLY ONE IN PRAGUE MISSING SLEEP. BESIDES TAKING CARE OF THE BABY, EINSTEIN IS FINDING THAT "SOLVING GRAVITY" IS AS HARD AS MAX PLANCK SAID IT WOULD BE. NOT TO MENTION FINDING THE EXACT RIGHT MOMENT TO SPRING ON MILEVA THE COST OF A CUSTOM-TAILORED AUSTRIA-HUNGARIAN MILITARY UNIFORM—COMPLETE WITH SWORD!
MACH

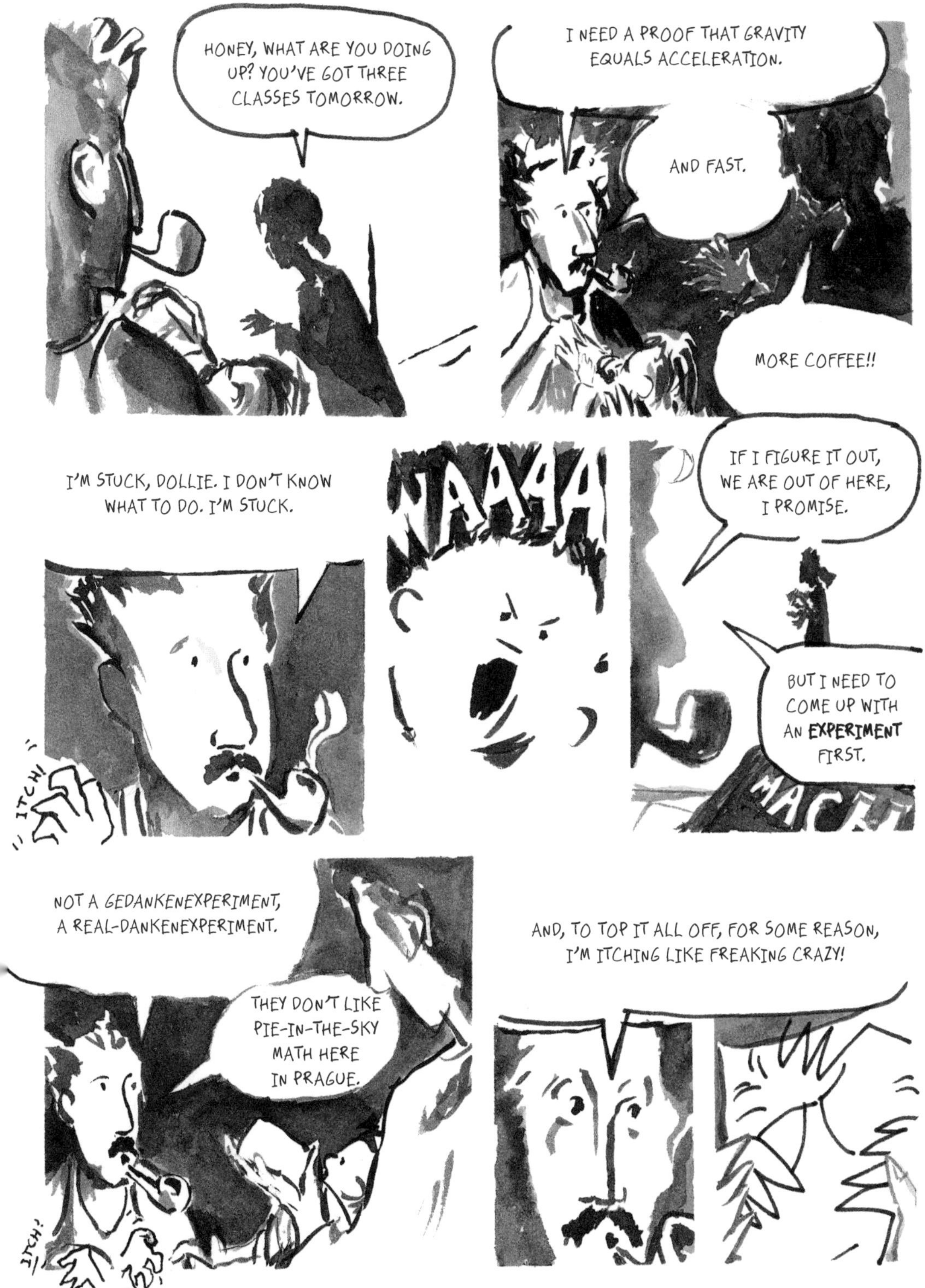
HONEY, WHAT ARE YOU DOING UP? YOU'VE GOT THREE CLASSES TOMORROW.
I NEED A PROOF THAT GRAVITY EQUALS ACCELERATION.
AND FAST.
MORE COFFEE!!
I'M STUCK, DOLLIE. I DON'T KNOW WHAT TO DO. I'M STUCK.
"ITCH!"
WAAAA
IF I FIGURE IT OUT, WE ARE OUT OF HERE, I PROMISE.
BUT I NEED TO COME UP WITH AN **EXPERIMENT** FIRST.
NOT A GEDANKENEXPERIMENT, A REAL-DANKENEXPERIMENT.
THEY DON'T LIKE PIE-IN-THE-SKY MATH HERE IN PRAGUE.
ITCH!
ITCH!
AND, TO TOP IT ALL OFF, FOR SOME REASON, I'M ITCHING LIKE FREAKING CRAZY!

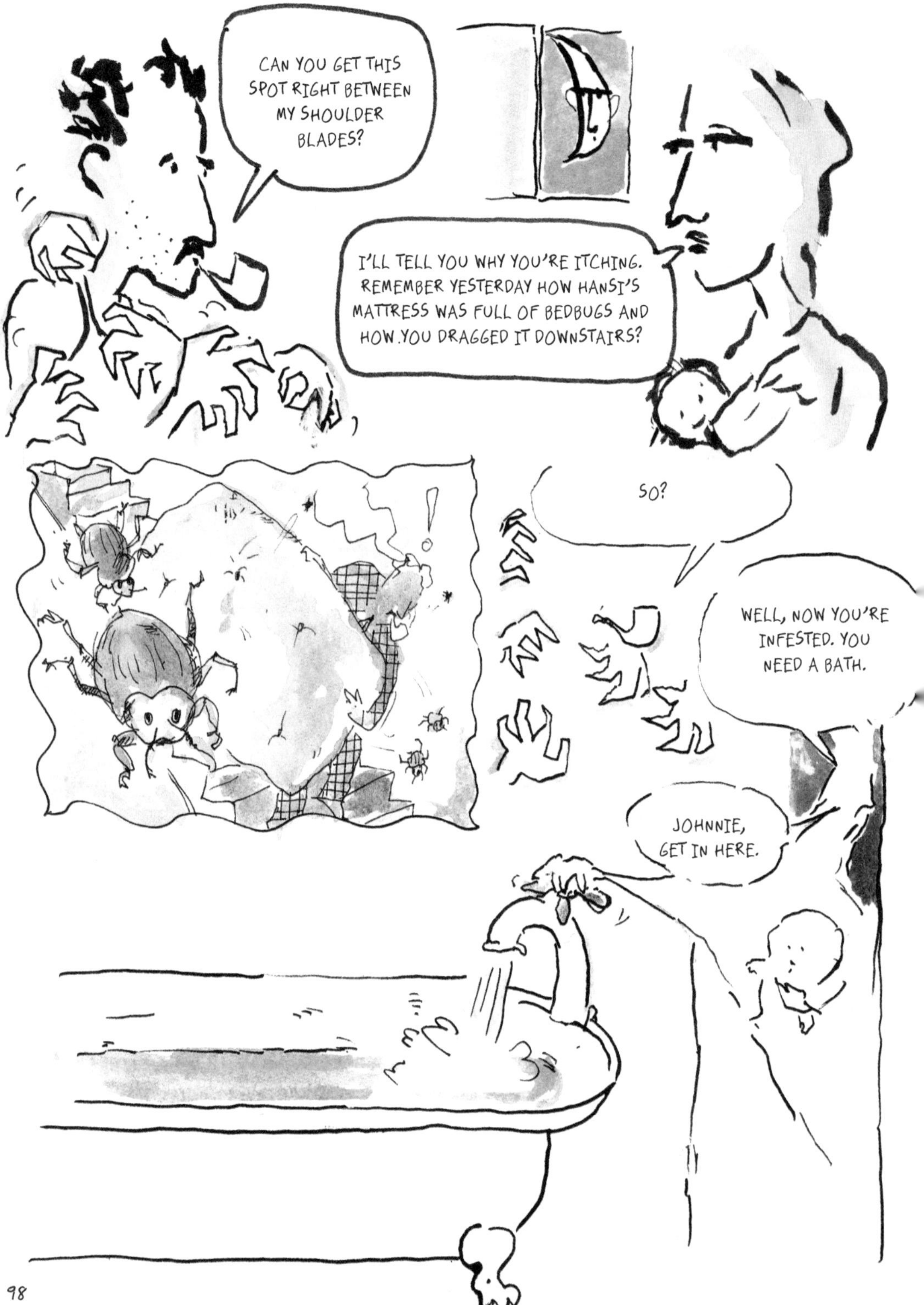
CAN YOU GET THIS SPOT RIGHT BETWEEN MY SHOULDER BLADES?
I'LL TELL YOU WHY YOU'RE ITCHING. REMEMBER YESTERDAY HOW HANSI'S MATTRESS WAS FULL OF BEDBUGS AND HOW YOU DRAGGED IT DOWNSTAIRS?
SO?
WELL, NOW YOU'RE INFESTED. YOU NEED A BATH.
JOHNNIE, GET IN HERE.

EUREKA! I'VE GOT IT! HOW ABOUT INSTEAD OF PIE-IN-THE-SKY, I SERVE THESE PRAGUERS UP A BIG FAT SLICE OF STARLIGHT IN THE SKY? DOLLIE, **LET'S BEND LIGHT!** WE JUST TAKE A PICTURE OF, SAY, THE NORTH STAR ONE NIGHT. AND THEN, WHEN IT'S CLOSE TO THE SUN IN THE SKY, WE TAKE ANOTHER PICTURE OF IT. AND VOILA, THE NORTH STAR, SHE'LL BECOME THE SOUTH STAR! THE NORTH STAR'S POSITION WILL APPEAR TO ADJUST AND BEND AROUND THE SUN'S GRAVITY. THEN, WHEN THE SUN MOVES, PRESTO CHANGO, IT POPS BACK INTO PLACE, IT'S THE NORTH STAR AGAIN! PHOTO-FREAKING-GRAPHIC PRAGUE PROOF. IF MASS BENDS LIGHT, GRAVITY IS THE SAME THING AS ACCELERATION, AND BING BANG BOOM WE ARE OUT OF HERE.

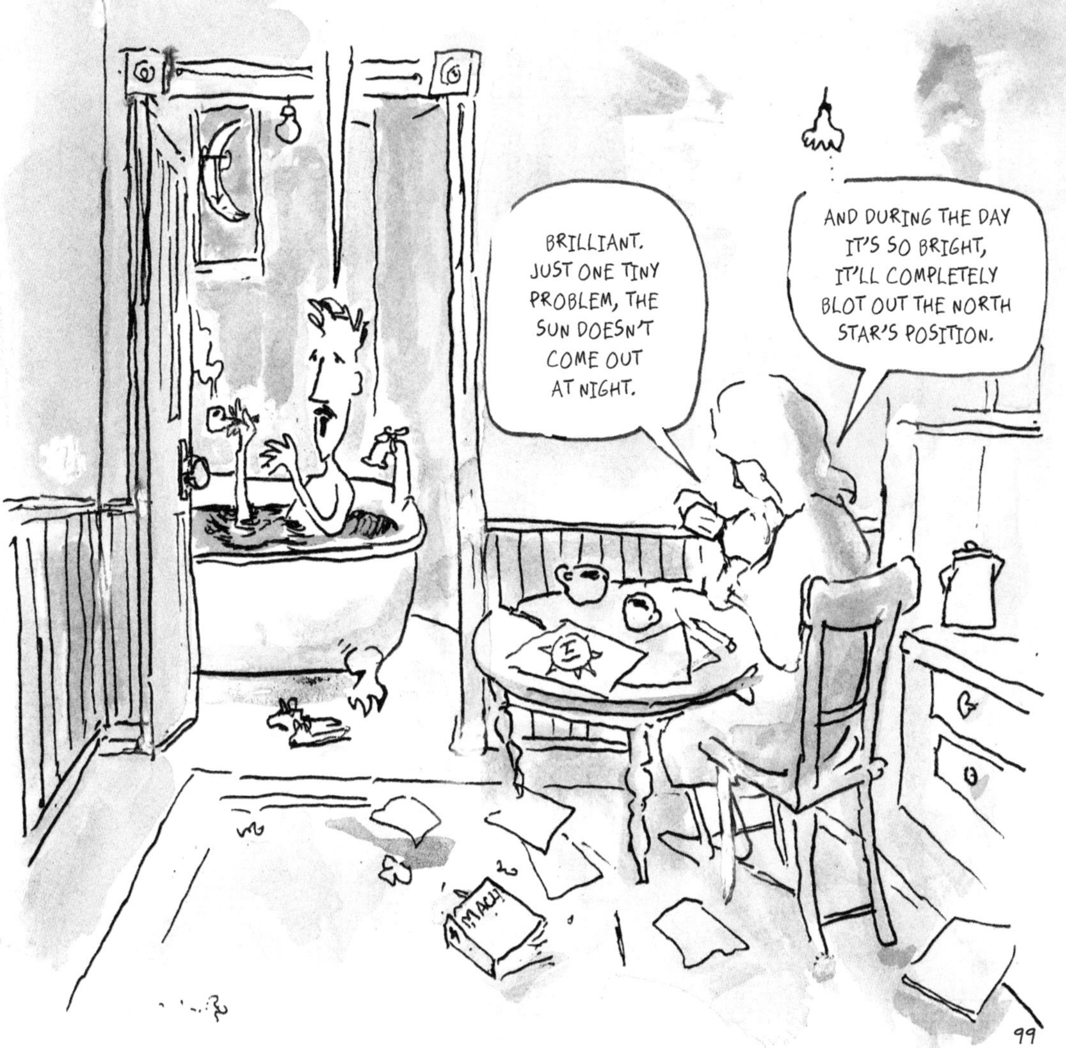
BRILLIANT. JUST ONE TINY PROBLEM, THE SUN DOESN'T COME OUT AT NIGHT.
AND DURING THE DAY IT'S SO BRIGHT, IT'LL COMPLETELY BLOT OUT THE NORTH STAR'S POSITION.
MACH

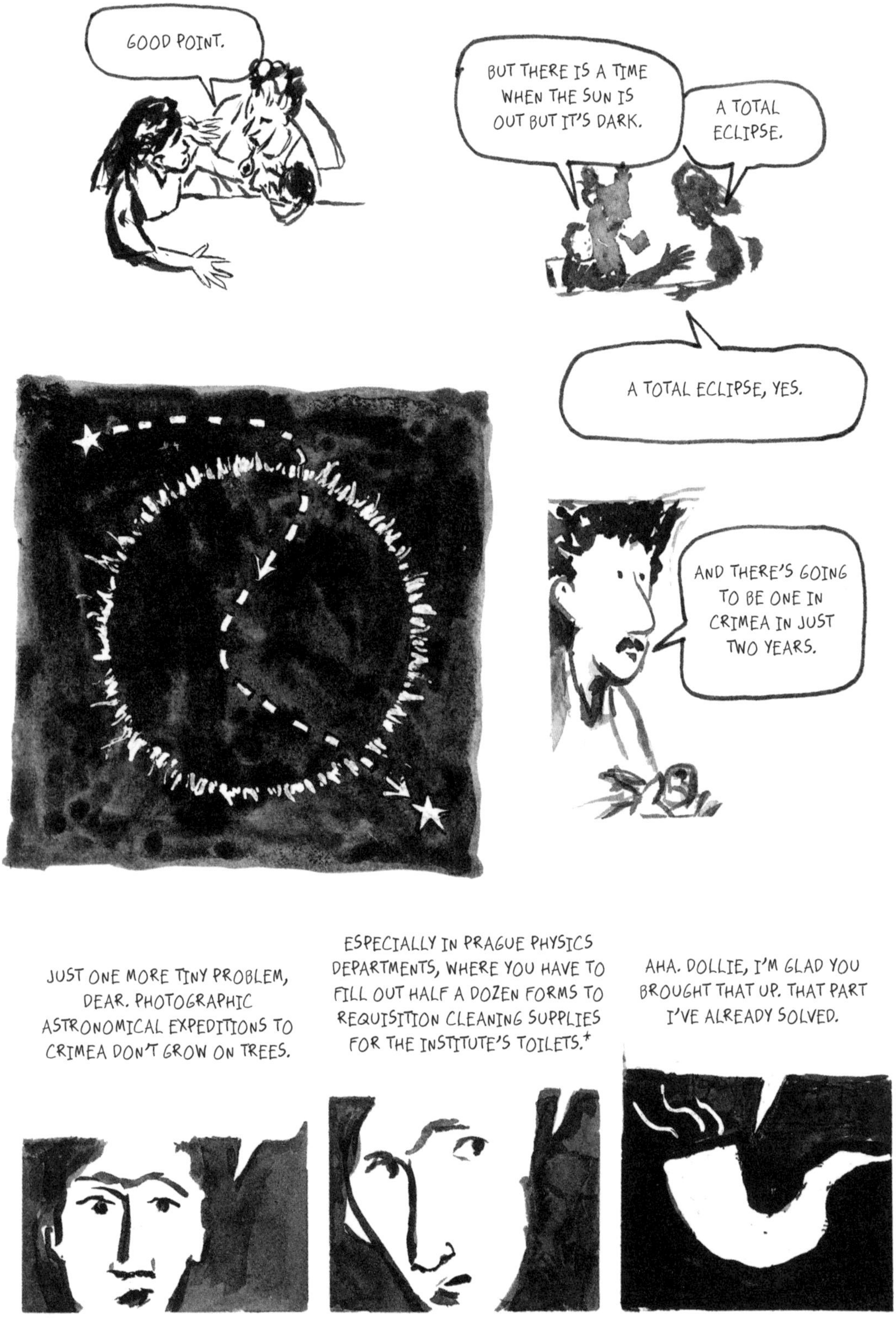
GOOD POINT.
BUT THERE IS A TIME WHEN THE SUN IS OUT BUT IT'S DARK.
A TOTAL ECLIPSE.
A TOTAL ECLIPSE, YES.
AND THERE'S GOING TO BE ONE IN CRIMEA IN JUST TWO YEARS.
JUST ONE MORE TINY PROBLEM, DEAR. PHOTOGRAPHIC ASTRONOMICAL EXPEDITIONS TO CRIMEA DON'T GROW ON TREES.
ESPECIALLY IN PRAGUE PHYSICS DEPARTMENTS, WHERE YOU HAVE TO FILL OUT HALF A DOZEN FORMS TO REQUISITION CLEANING SUPPLIES FOR THE INSTITUTE'S TOILETS.*
AHA. DOLLIE, I'M GLAD YOU BROUGHT THAT UP. THAT PART I'VE ALREADY SOLVED.

* Referenced by Einstein directly, calling it "bureaucratic ink-shitting" in a letter to family friends Alfred and Clara Stern, March 17, 1912.

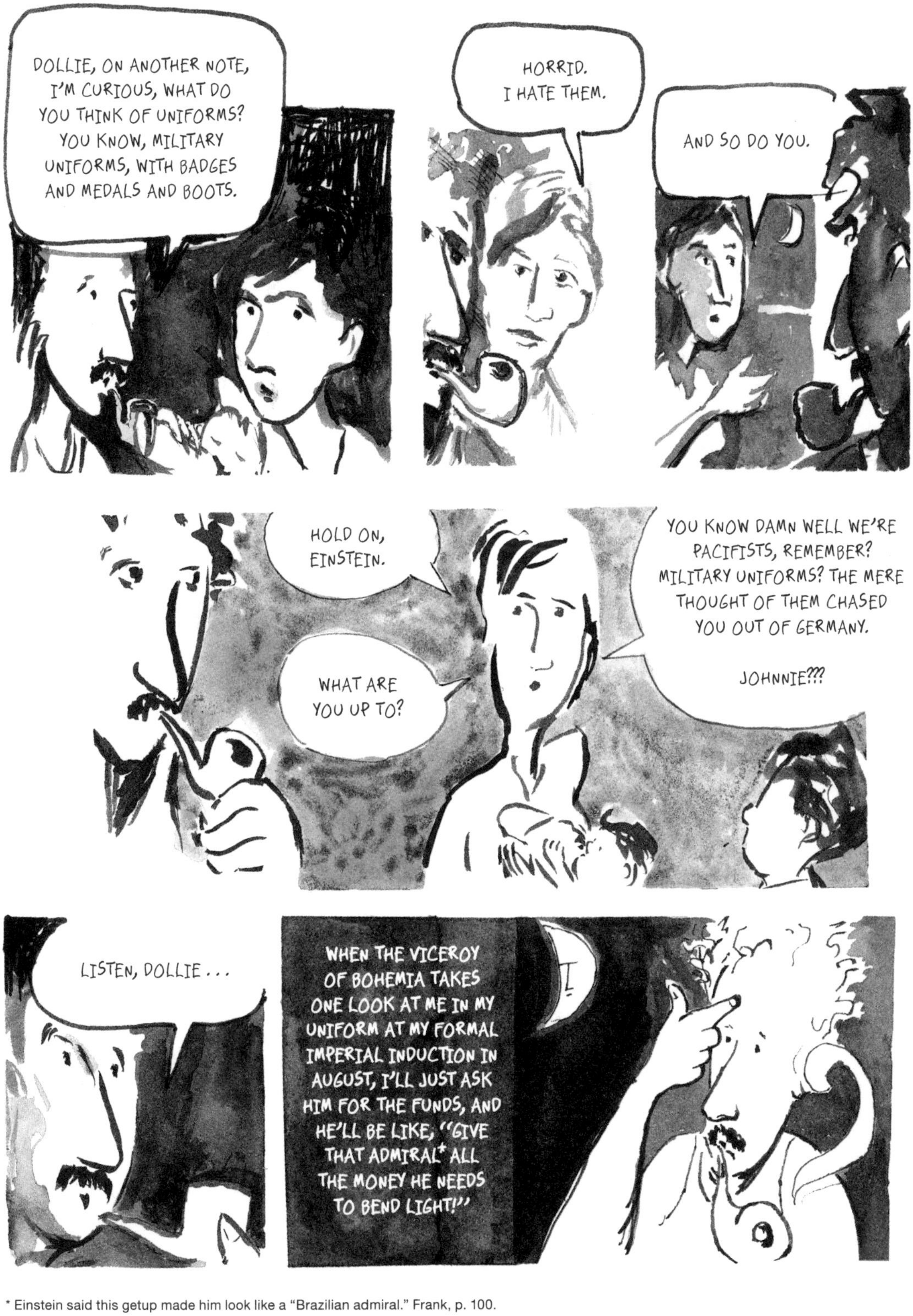

DOLLIE, ON ANOTHER NOTE, I'M CURIOUS, WHAT DO YOU THINK OF UNIFORMS? YOU KNOW, MILITARY UNIFORMS, WITH BADGES AND MEDALS AND BOOTS.
HORRID. I HATE THEM.
AND SO DO YOU.
HOLD ON, EINSTEIN.
WHAT ARE YOU UP TO?
YOU KNOW DAMN WELL WE'RE PACIFISTS, REMEMBER? MILITARY UNIFORMS? THE MERE THOUGHT OF THEM CHASED YOU OUT OF GERMANY.
JOHNNIE???
LISTEN, DOLLIE . . .
WHEN THE VICEROY OF BOHEMIA TAKES ONE LOOK AT ME IN MY UNIFORM AT MY FORMAL IMPERIAL INDUCTION IN AUGUST, I'LL JUST ASK HIM FOR THE FUNDS, AND HE'LL BE LIKE, "GIVE THAT ADMIRAL* ALL THE MONEY HE NEEDS TO BEND LIGHT!"

* Einstein said this getup made him look like a "Brazilian admiral." Frank, p. 100.

WAIT A MINUTE. IS THIS INDUCTION ANOTHER ONE OF THOSE INSANE PRAGUE THINGS INVOLVING LOTS OF MEDALS AND WEAPONS AND ANTISEMITIC ANTISLAVIC PRO-AUSTRIA-HUNGARY EMPIRE BS?

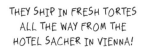

THEY SHIP IN FRESH TORTES ALL THE WAY FROM THE HOTEL SACHER IN VIENNA!

HMMM . . . WELL, AT LEAST THE UNIVERSITY IS PICKING UP THE COST OF THE UNIFORM.

AREN'T THEY?

YUM. TORTS!

PICK SAID HE'D GET ME A GREAT DEAL ON IT. HE KNOWS A GUY.

DOLLIE, IT'S JUST A SILLY FORMALITY. AND AS SOON AS WE PUBLISH OUR BEND LIGHT PAPER, WE ARE OUT OF HERE.

AT THE VERY LEAST, I'LL HAVE SOMETHING THAT'LL SLICE THAT PESKY MAX ABRAHAM TO RIBBONS.

TOUCHÉ!

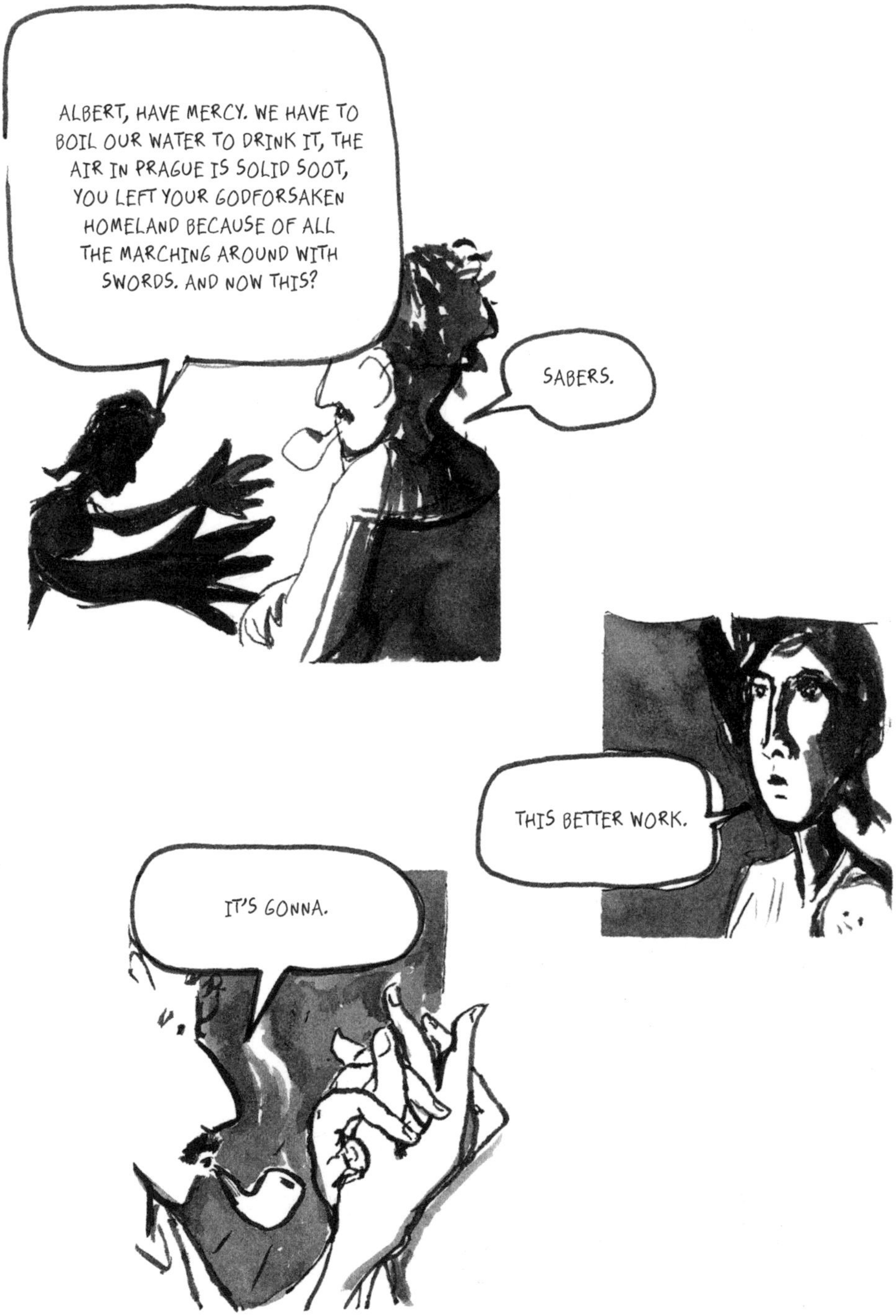
ALBERT, HAVE MERCY. WE HAVE TO BOIL OUR WATER TO DRINK IT, THE AIR IN PRAGUE IS SOLID SOOT, YOU LEFT YOUR GODFORSAKEN HOMELAND BECAUSE OF ALL THE MARCHING AROUND WITH SWORDS. AND NOW THIS?
SABERS.
THIS BETTER WORK.
IT'S GONNA.

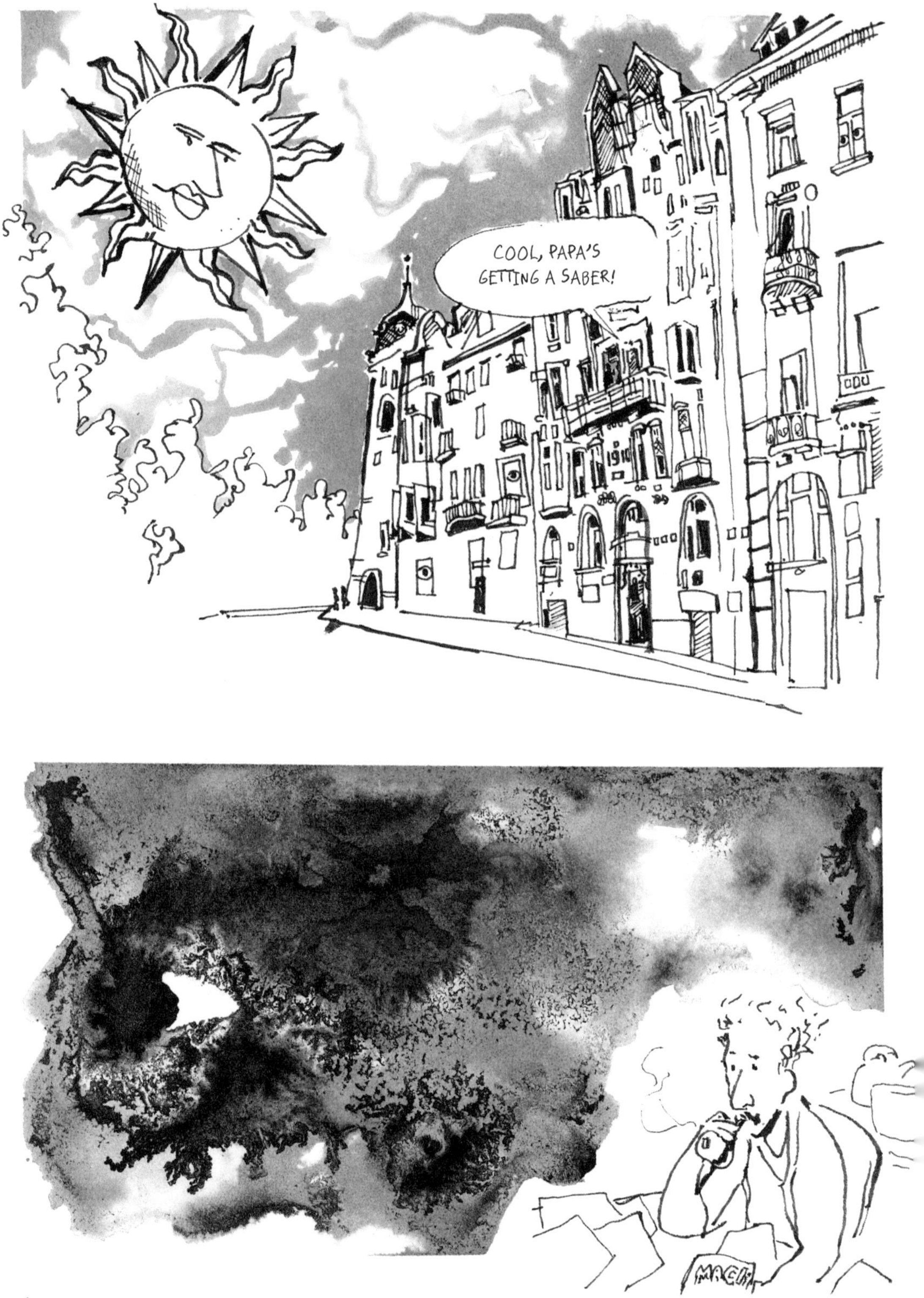
COOL, PAPA'S GETTING A SABER!
1910

CHAPTER IX

The Crime of the Century
August 21, 1911

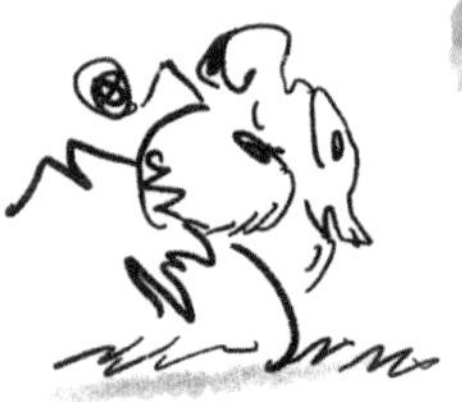

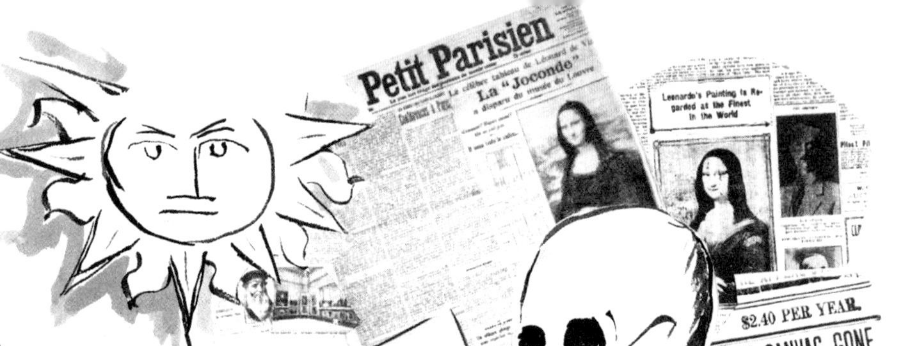

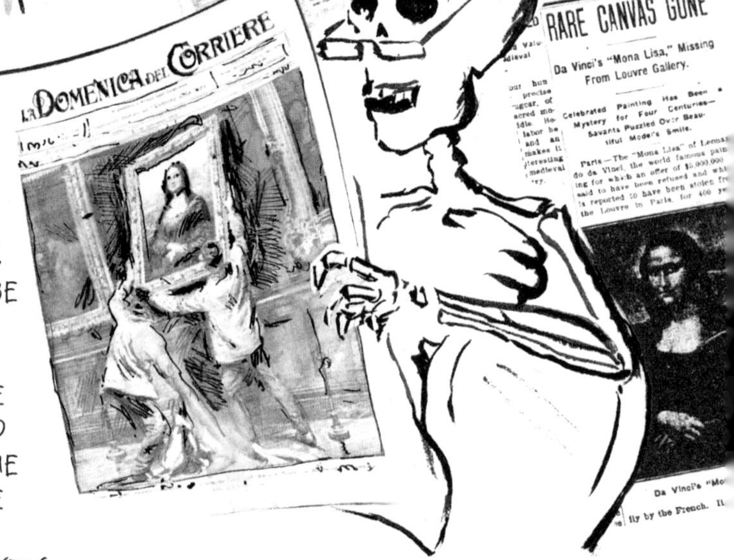
Petit Parisien
La "Joconde"
Leonardo's Painting Is Regarded as the Finest in the World
$2.40 PER YEAR.
RARE CANVAS GONE
Da Vinci's "Mona Lisa," Missing From Louvre Gallery.
Celebrated Painting Has Been a Mystery for Four Centuries—Savants Puzzled Over Beautiful Model's Smile.
LA DOMENICA DEL CORRIERE

IN THE LATE SUMMER OF 1911, EINSTEIN'S WORLD ISN'T THE ONLY ONE THAT'S GOING CRAZY. THE WHOLE WORLD SEEMS TO BE SPIRALING—WITH NEWSPAPERS AROUND THE WORLD WRITING ABOUT THE RECORD HEAT WAVE THAT WAS DRIVING PEOPLE MAD FROM BRITAIN TO BALI, AND THE FIRST APPEARANCE OF GEORGE HERRIMAN'S KRAZY KAT IN THE FUNNY PAGES, AND, MOST SHOCKING OF ALL, THAT THE *MONA LISA* HAD BEEN STOLEN RIGHT OFF THE WALL OF THE LOUVRE IN BROAD DAYLIGHT.

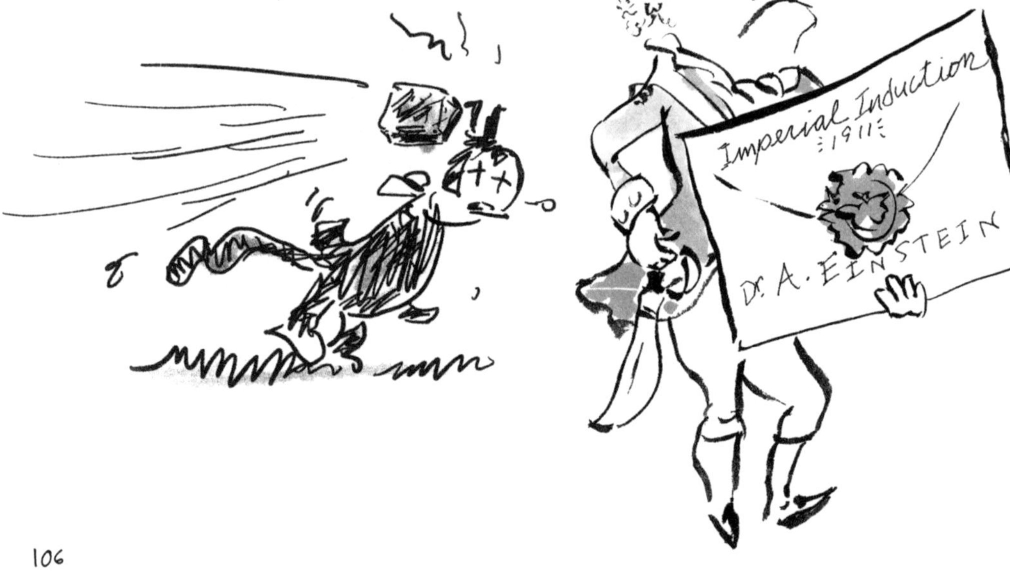
Imperial Induction
1911
Dr. A. EINSTEIN

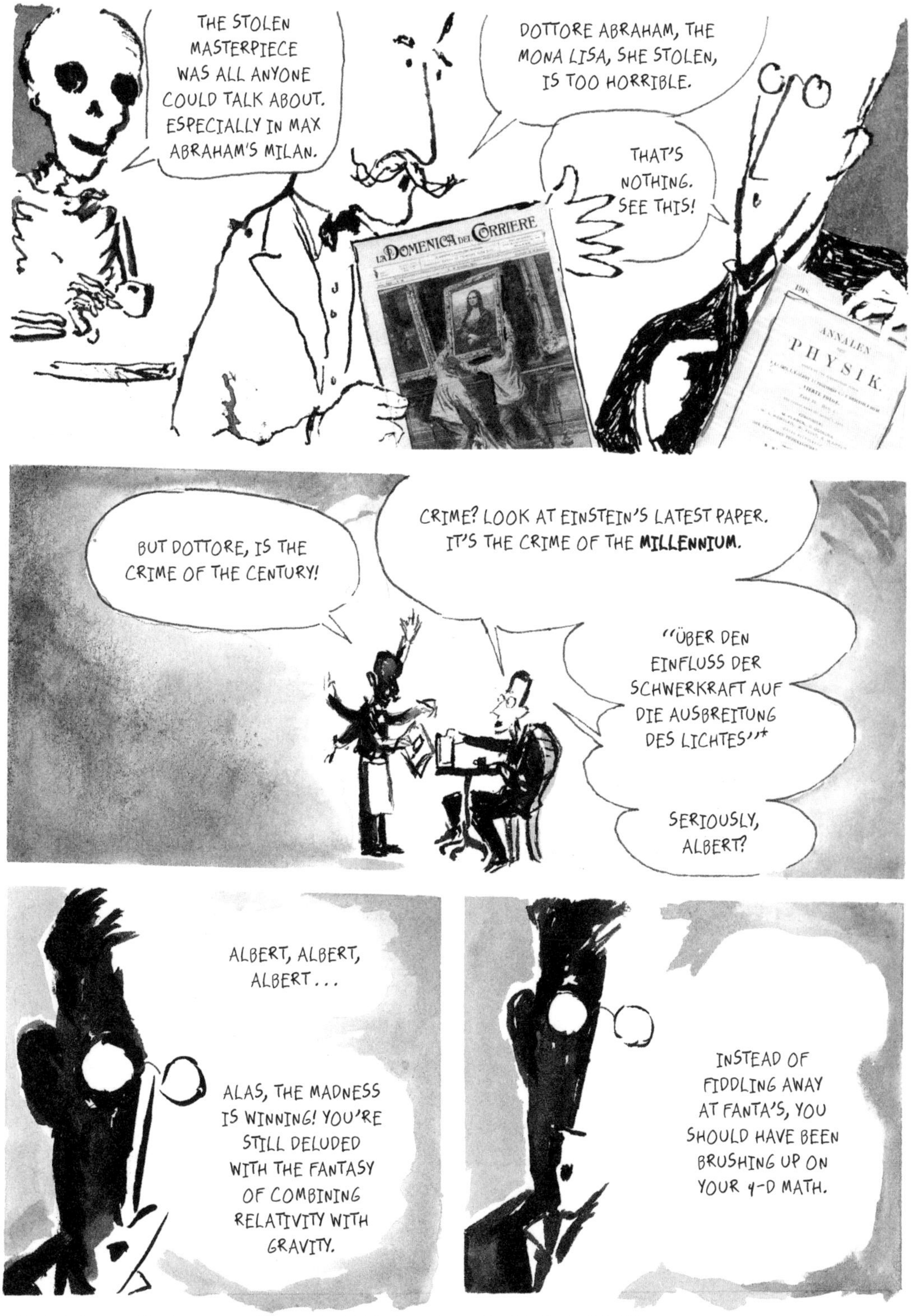
THE STOLEN MASTERPIECE WAS ALL ANYONE COULD TALK ABOUT. ESPECIALLY IN MAX ABRAHAM'S MILAN.
DOTTORE ABRAHAM, THE MONA LISA, SHE STOLEN, IS TOO HORRIBLE.
THAT'S NOTHING. SEE THIS!
LA DOMENICA DEL CORRIERE
ANNALEN
PHYSIK
BUT DOTTORE, IS THE CRIME OF THE CENTURY!
CRIME? LOOK AT EINSTEIN'S LATEST PAPER. IT'S THE CRIME OF THE **MILLENNIUM**.
"ÜBER DEN EINFLUSS DER SCHWERKRAFT AUF DIE AUSBREITUNG DES LICHTES"*
SERIOUSLY, ALBERT?
ALBERT, ALBERT, ALBERT . . .
ALAS, THE MADNESS IS WINNING! YOU'RE STILL DELUDED WITH THE FANTASY OF COMBINING RELATIVITY WITH GRAVITY.
INSTEAD OF FIDDLING AWAY AT FANTA'S, YOU SHOULD HAVE BEEN BRUSHING UP ON YOUR 4-D MATH.

* "On the Influence of Gravitation on the Propagation of Light," *Annalen der Physik* 35 (1911): 898–908.

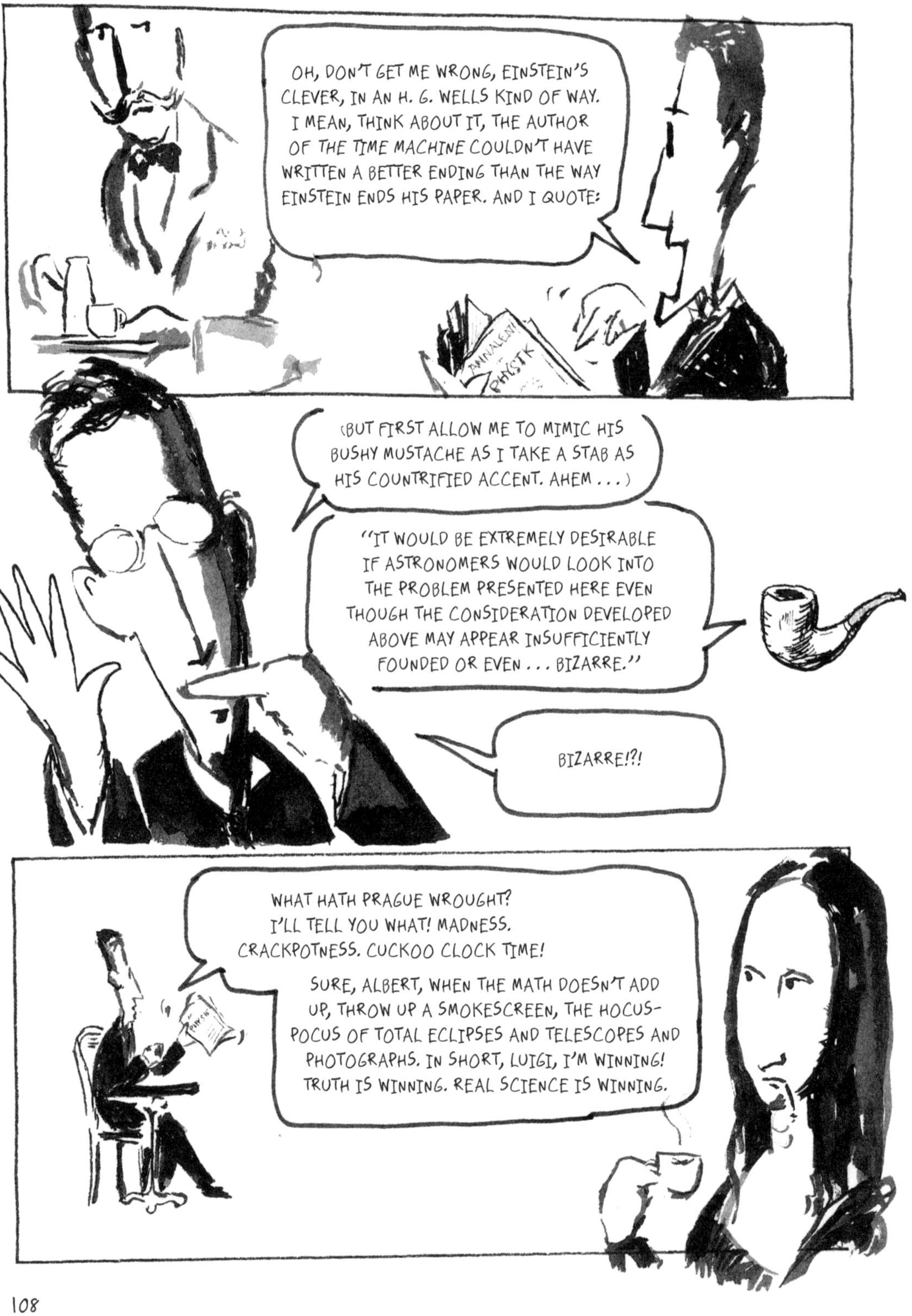
OH, DON'T GET ME WRONG, EINSTEIN'S CLEVER, IN AN H. G. WELLS KIND OF WAY. I MEAN, THINK ABOUT IT, THE AUTHOR OF THE TIME MACHINE COULDN'T HAVE WRITTEN A BETTER ENDING THAN THE WAY EINSTEIN ENDS HIS PAPER. AND I QUOTE:
(BUT FIRST ALLOW ME TO MIMIC HIS BUSHY MUSTACHE AS I TAKE A STAB AS HIS COUNTRIFIED ACCENT. AHEM . . .)
"IT WOULD BE EXTREMELY DESIRABLE IF ASTRONOMERS WOULD LOOK INTO THE PROBLEM PRESENTED HERE EVEN THOUGH THE CONSIDERATION DEVELOPED ABOVE MAY APPEAR INSUFFICIENTLY FOUNDED OR EVEN . . . BIZARRE."
BIZARRE!?!
WHAT HATH PRAGUE WROUGHT? I'LL TELL YOU WHAT! MADNESS. CRACKPOTNESS. CUCKOO CLOCK TIME!
SURE, ALBERT, WHEN THE MATH DOESN'T ADD UP, THROW UP A SMOKESCREEN, THE HOCUS-POCUS OF TOTAL ECLIPSES AND TELESCOPES AND PHOTOGRAPHS. IN SHORT, LUIGI, I'M WINNING! TRUTH IS WINNING. REAL SCIENCE IS WINNING.

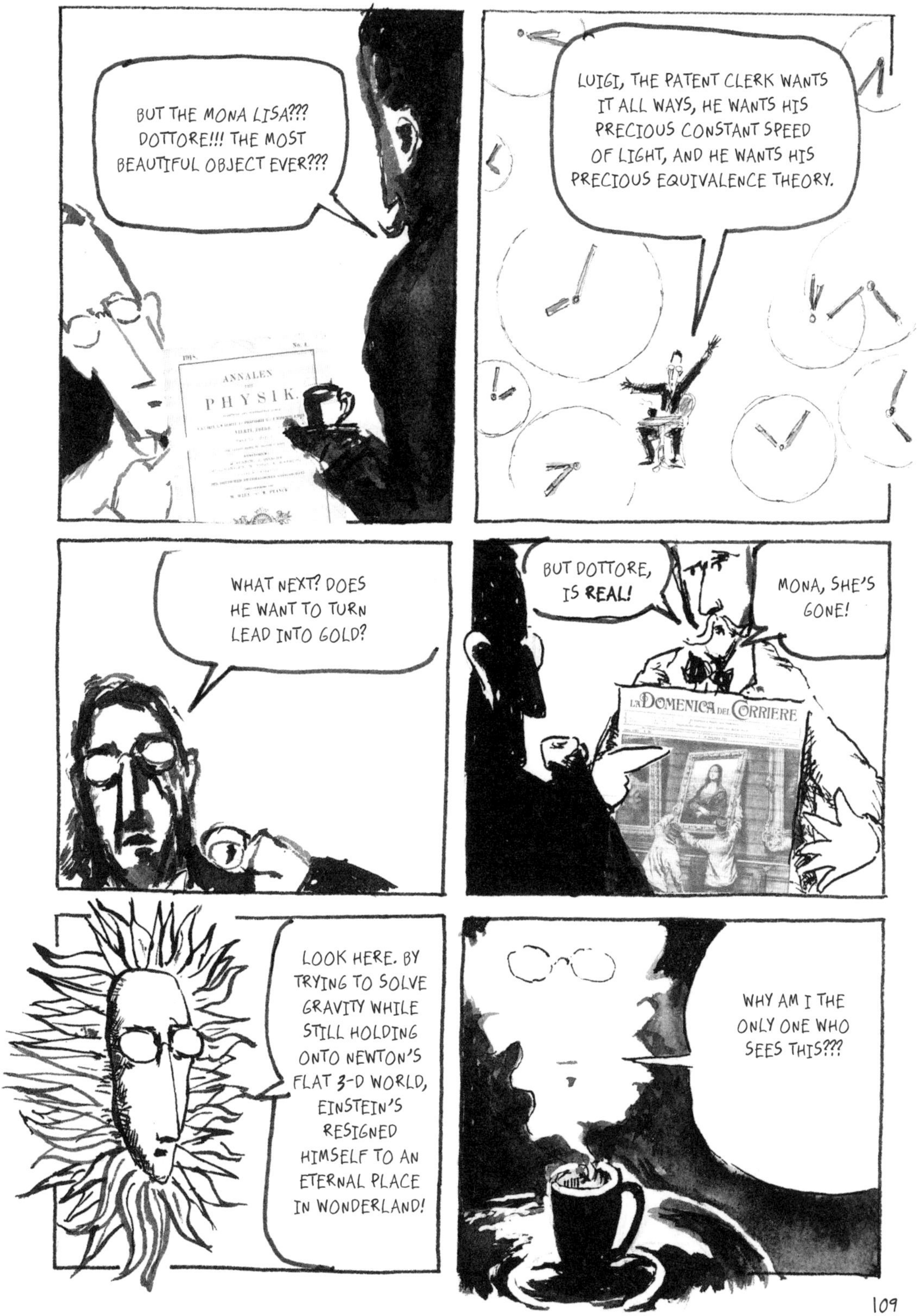
BUT THE MONA LISA??? DOTTORE!!! THE MOST BEAUTIFUL OBJECT EVER???
ANNALEN
PHYSIK.
LUIGI, THE PATENT CLERK WANTS IT ALL WAYS, HE WANTS HIS PRECIOUS CONSTANT SPEED OF LIGHT, AND HE WANTS HIS PRECIOUS EQUIVALENCE THEORY.
WHAT NEXT? DOES HE WANT TO TURN LEAD INTO GOLD?
BUT DOTTORE, IS **REAL!**
MONA, SHE'S GONE!
LA DOMENICA DEL CORRIERE
LOOK HERE. BY TRYING TO SOLVE GRAVITY WHILE STILL HOLDING ONTO NEWTON'S FLAT 3-D WORLD, EINSTEIN'S RESIGNED HIMSELF TO AN ETERNAL PLACE IN WONDERLAND!
WHY AM I THE ONLY ONE WHO SEES THIS???

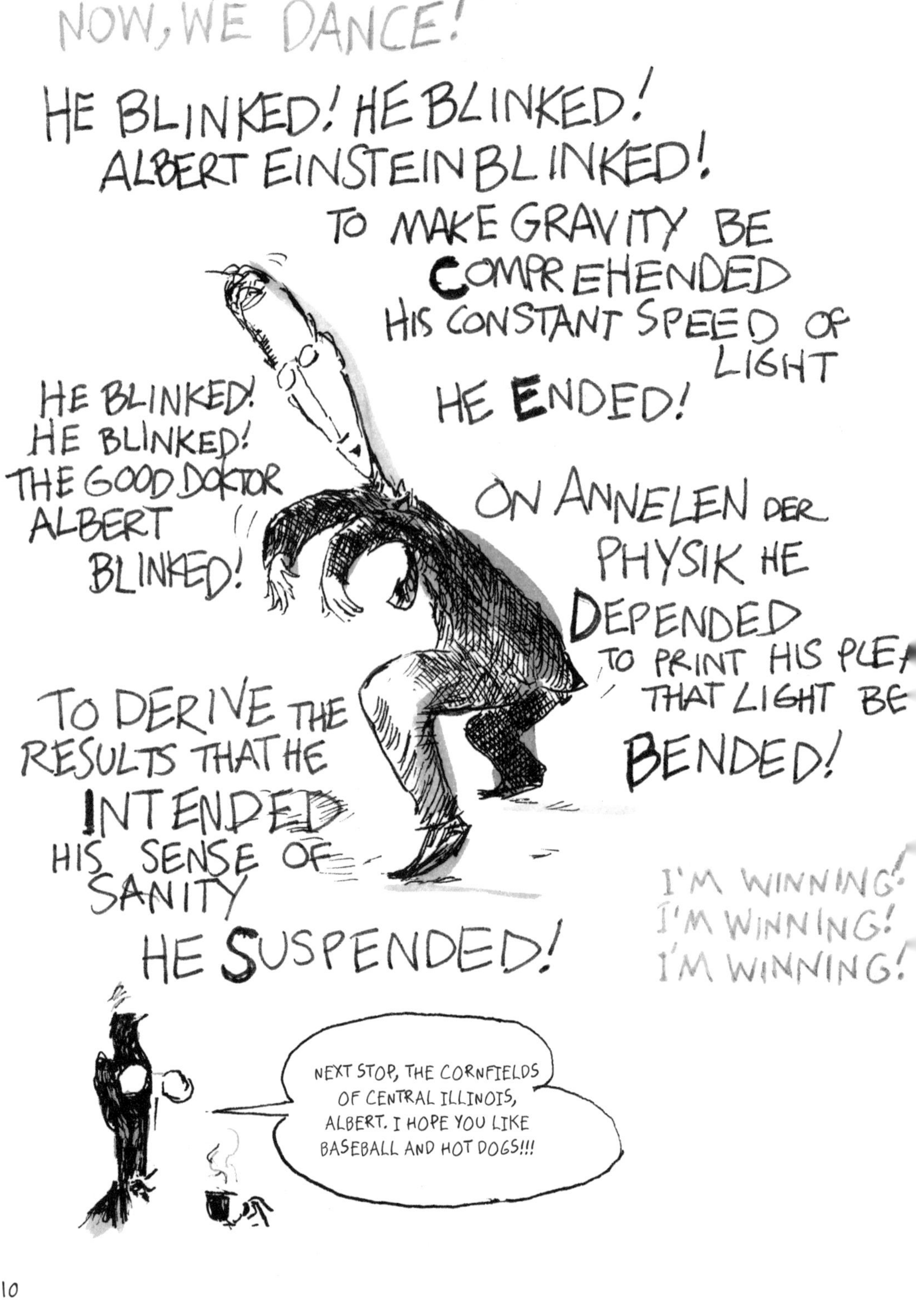
NOW, WE DANCE!
HE BLINKED! HE BLINKED! ALBERT EINSTEIN BLINKED!
TO MAKE GRAVITY BE COMPREHENDED
HIS CONSTANT SPEED OF LIGHT
HE ENDED!
HE BLINKED! HE BLINKED! THE GOOD DOKTOR ALBERT BLINKED!
ON ANNELEN DER PHYSIK HE DEPENDED
TO PRINT HIS PLE
THAT LIGHT BE BENDED!
TO DERIVE THE RESULTS THAT HE INTENDED
HIS SENSE OF SANITY
HE SUSPENDED!
I'M WINNING! I'M WINNING! I'M WINNING!
NEXT STOP, THE CORNFIELDS OF CENTRAL ILLINOIS, ALBERT. I HOPE YOU LIKE BASEBALL AND HOT DOGS!!!

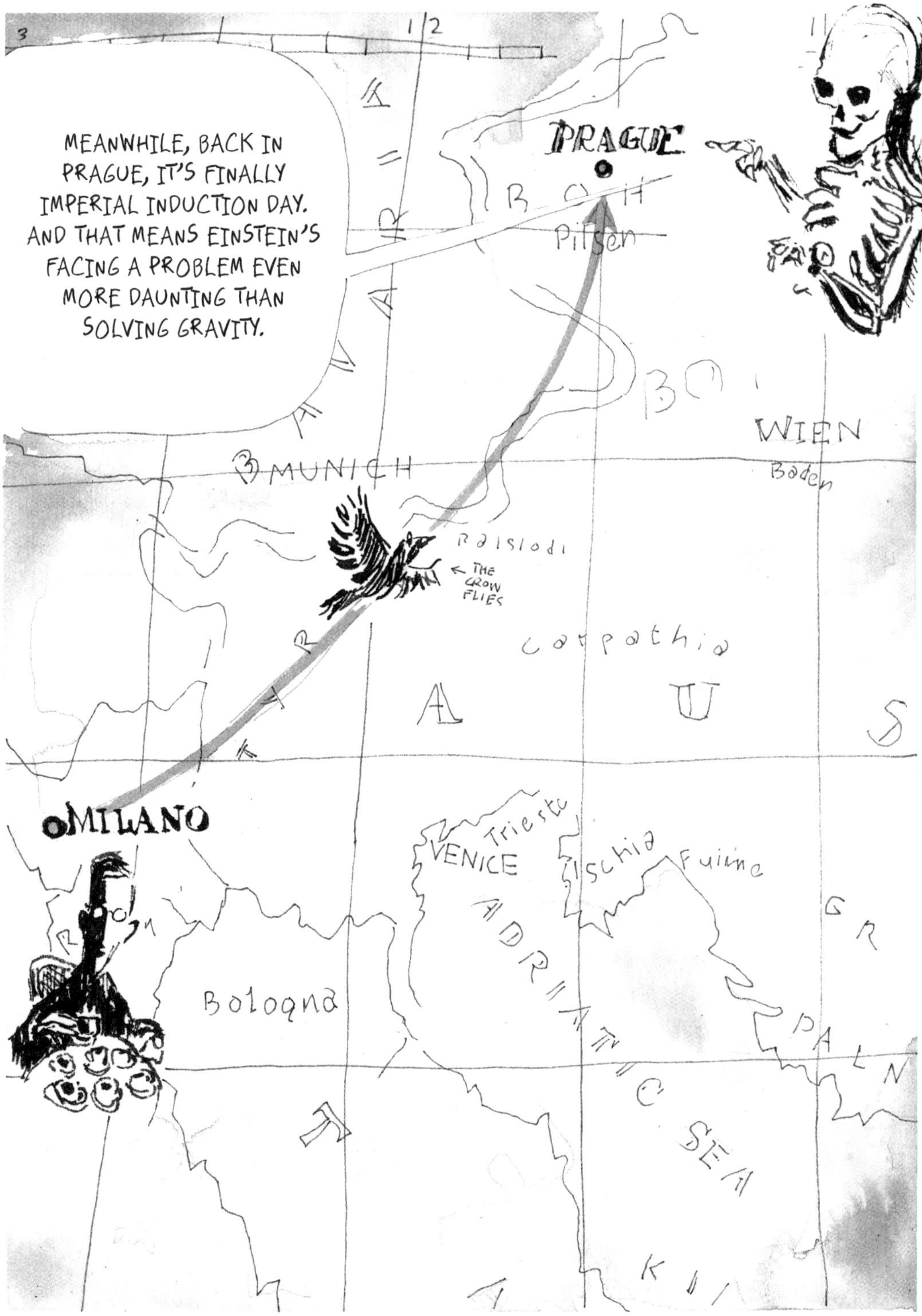
MEANWHILE, BACK IN PRAGUE, IT'S FINALLY IMPERIAL INDUCTION DAY. AND THAT MEANS EINSTEIN'S FACING A PROBLEM EVEN MORE DAUNTING THAN SOLVING GRAVITY.
PRAGUE
Pilsen
WIEN
Baden
MUNICH
THE CROW FLIES
Carpathia
MILANO
VENICE
Trieste
Ischia
Fuime
Bologna
ADRIATIC SEA

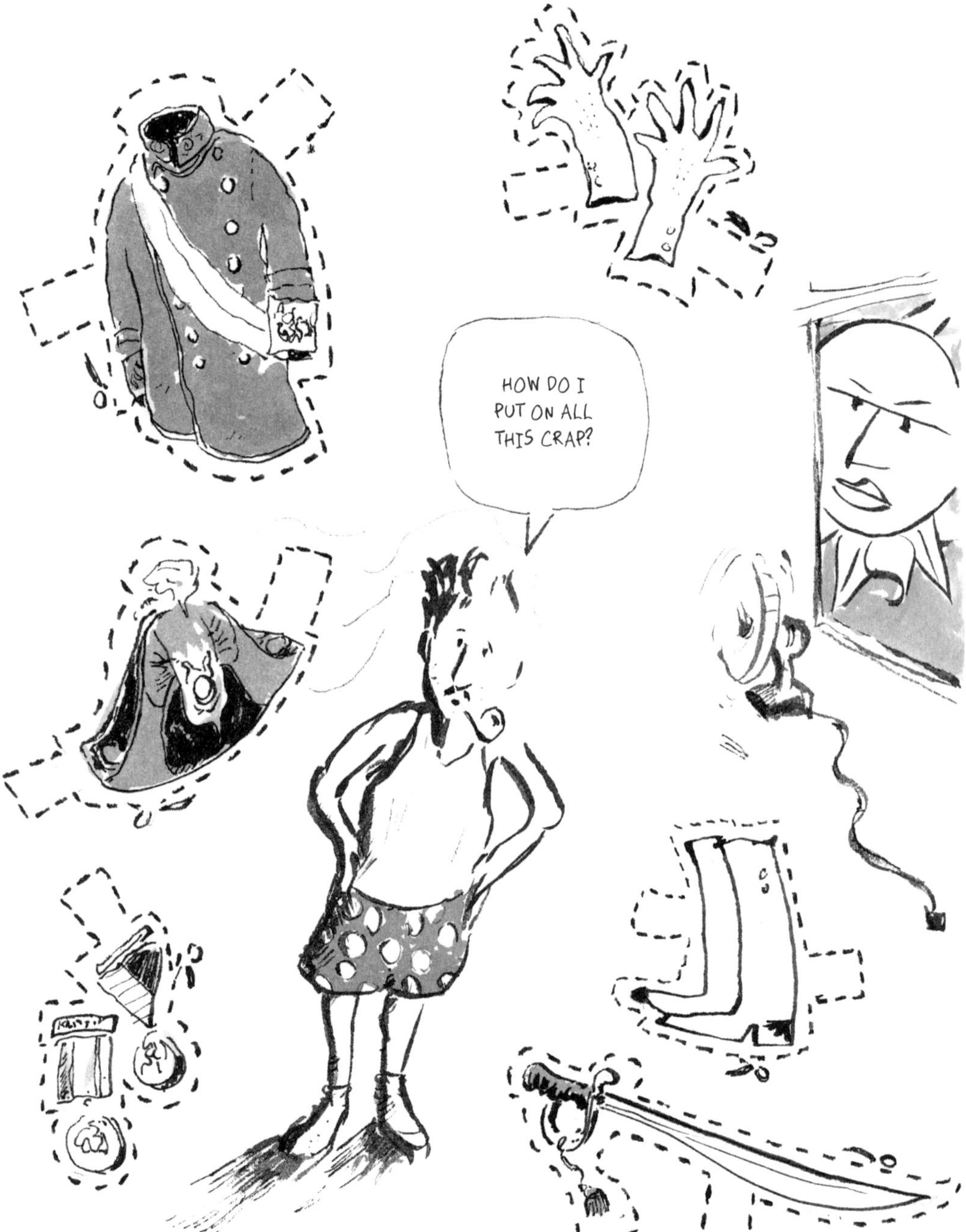
HOW DO I PUT ON ALL THIS CRAP?

SPEAK.

IS THIS THE VICEROY'S[24] INDUCTION CEREMONY?

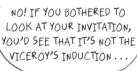

NO! IF YOU BOTHERED TO LOOK AT YOUR INVITATION, YOU'D SEE THAT IT'S NOT THE VICEROY'S INDUCTION . . .

IT'S YOURS.

HE WAS INDUCTED YEARS AGO!

ENTER!

AND NO SMOKING.

PRIVATE HASAK, DID I DETECT A FRENCH ACCENT?

COULD **HE** HAVE STOLEN THE MONA LISA?

I, COUNT FRANZ THUN UND HOHENSTEIN, THE GOVERNOR OF BOHEMIA ACTING AS VICEROY ON BEHALF OF HIS IMPERIAL APOSTOLIC MAJESTY FRANZ JOSEPH, AS A RESULT OF HIS DENUNCIATION OF BEING AN UNBELIEVER AND INSTEAD BEING OF THE MOSAIC FAITH, FORMALLY INDUCT ALBERT EINSTEIN TO THE POSITION OF FULL PROFESSOR AT KARL-FERDINAND UNIVERSITY AND DO NOW PROCLAIM THE BAR OPEN!

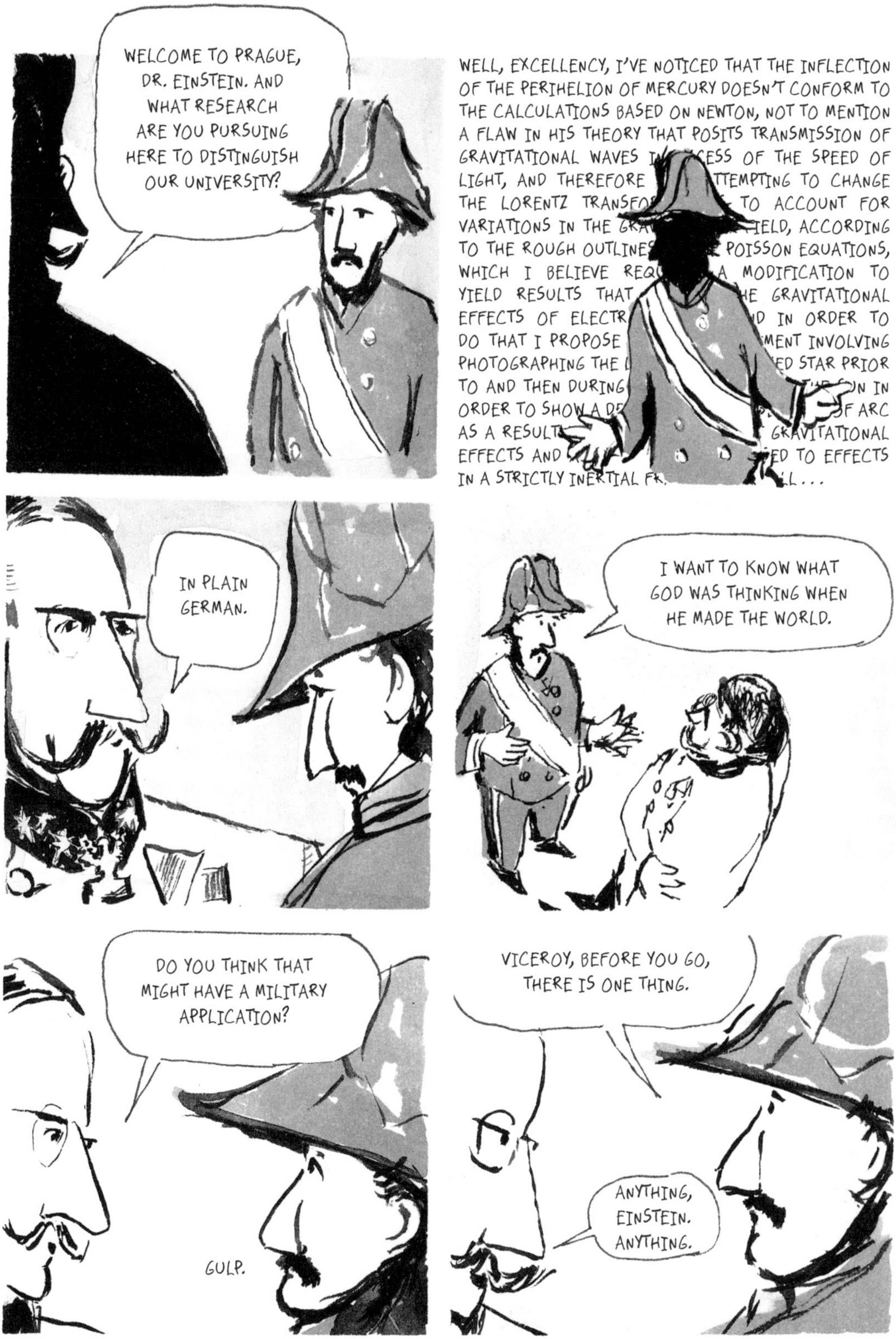

WELCOME TO PRAGUE, DR. EINSTEIN. AND WHAT RESEARCH ARE YOU PURSUING HERE TO DISTINGUISH OUR UNIVERSITY?
WELL, EXCELLENCY, I'VE NOTICED THAT THE INFLECTION OF THE PERIHELION OF MERCURY DOESN'T CONFORM TO THE CALCULATIONS BASED ON NEWTON, NOT TO MENTION A FLAW IN HIS THEORY THAT POSITS TRANSMISSION OF GRAVITATIONAL WAVES I CESS OF THE SPEED OF LIGHT, AND THEREFORE TTEMPTING TO CHANGE THE LORENTZ TRANSFO TO ACCOUNT FOR VARIATIONS IN THE GRA IELD, ACCORDING TO THE ROUGH OUTLINES POISSON EQUATIONS, WHICH I BELIEVE REQ A MODIFICATION TO YIELD RESULTS THAT HE GRAVITATIONAL EFFECTS OF ELECTR D IN ORDER TO DO THAT I PROPOSE MENT INVOLVING PHOTOGRAPHING THE ED STAR PRIOR TO AND THEN DURING N IN ORDER TO SHOW A D F ARC AS A RESULT GRAVITATIONAL EFFECTS AND ED TO EFFECTS IN A STRICTLY INERTIAL F L...
IN PLAIN GERMAN.
I WANT TO KNOW WHAT GOD WAS THINKING WHEN HE MADE THE WORLD.
DO YOU THINK THAT MIGHT HAVE A MILITARY APPLICATION?
GULP.
VICEROY, BEFORE YOU GO, THERE IS ONE THING.
ANYTHING, EINSTEIN. ANYTHING.

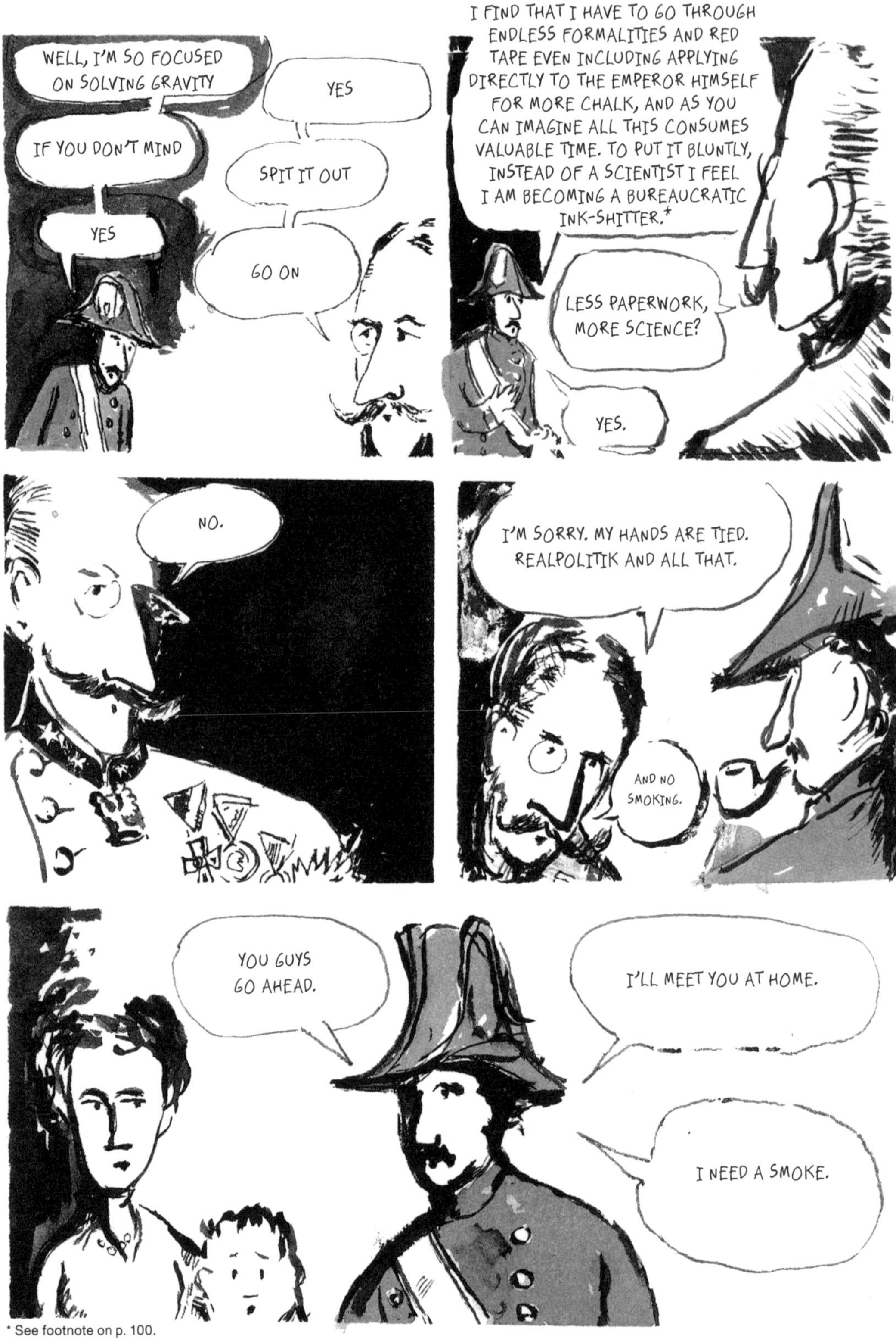
WELL, I'M SO FOCUSED ON SOLVING GRAVITY
YES
IF YOU DON'T MIND
SPIT IT OUT
YES
GO ON
I FIND THAT I HAVE TO GO THROUGH ENDLESS FORMALITIES AND RED TAPE EVEN INCLUDING APPLYING DIRECTLY TO THE EMPEROR HIMSELF FOR MORE CHALK, AND AS YOU CAN IMAGINE ALL THIS CONSUMES VALUABLE TIME. TO PUT IT BLUNTLY, INSTEAD OF A SCIENTIST I FEEL I AM BECOMING A BUREAUCRATIC INK-SHITTER.*
LESS PAPERWORK, MORE SCIENCE?
YES.
NO.
I'M SORRY. MY HANDS ARE TIED. REALPOLITIK AND ALL THAT.
AND NO SMOKING.
YOU GUYS GO AHEAD.
I'LL MEET YOU AT HOME.
I NEED A SMOKE.

* See footnote on p. 100.

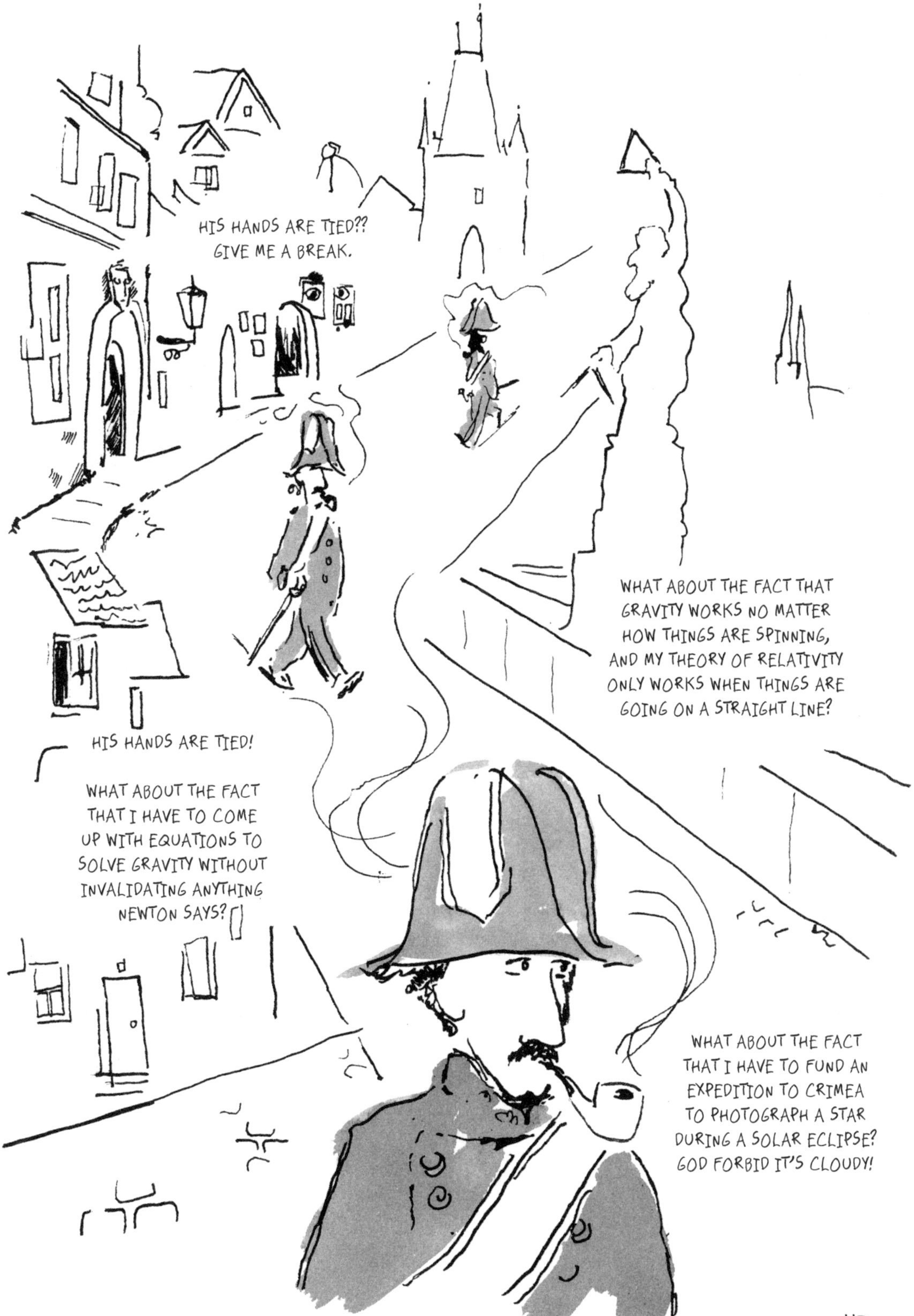
HIS HANDS ARE TIED?? GIVE ME A BREAK.
WHAT ABOUT THE FACT THAT GRAVITY WORKS NO MATTER HOW THINGS ARE SPINNING, AND MY THEORY OF RELATIVITY ONLY WORKS WHEN THINGS ARE GOING ON A STRAIGHT LINE?
HIS HANDS ARE TIED!
WHAT ABOUT THE FACT THAT I HAVE TO COME UP WITH EQUATIONS TO SOLVE GRAVITY WITHOUT INVALIDATING ANYTHING NEWTON SAYS?
WHAT ABOUT THE FACT THAT I HAVE TO FUND AN EXPEDITION TO CRIMEA TO PHOTOGRAPH A STAR DURING A SOLAR ECLIPSE? GOD FORBID IT'S CLOUDY!

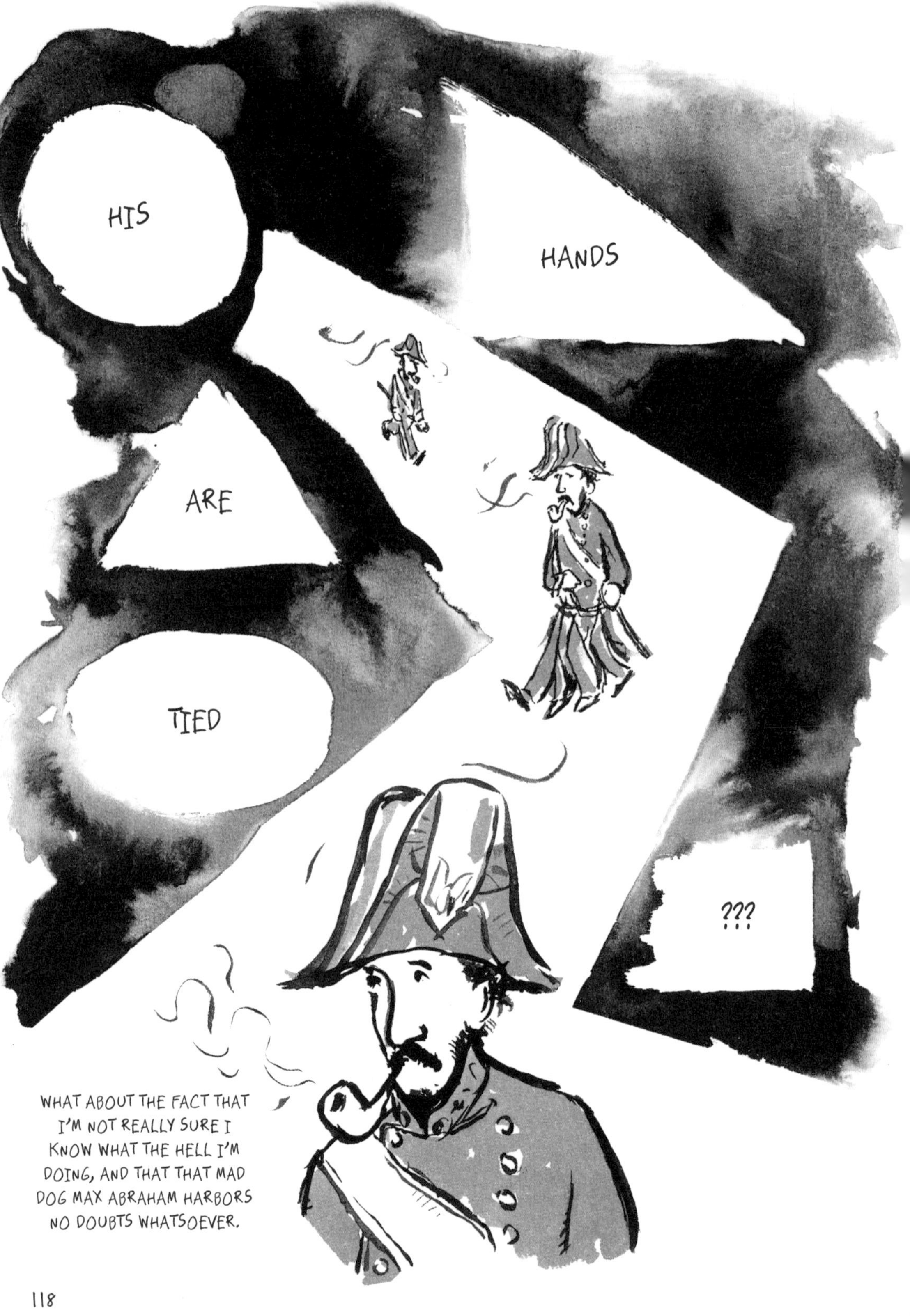
HIS
HANDS
ARE
TIED
???
WHAT ABOUT THE FACT THAT I'M NOT REALLY SURE I KNOW WHAT THE HELL I'M DOING, AND THAT THAT MAD DOG MAX ABRAHAM HARBORS NO DOUBTS WHATSOEVER.

WHOA, THAT IS ONE STRANGE-LOOKING CAT. BUT ALSO STRANGELY FAMILIAR.

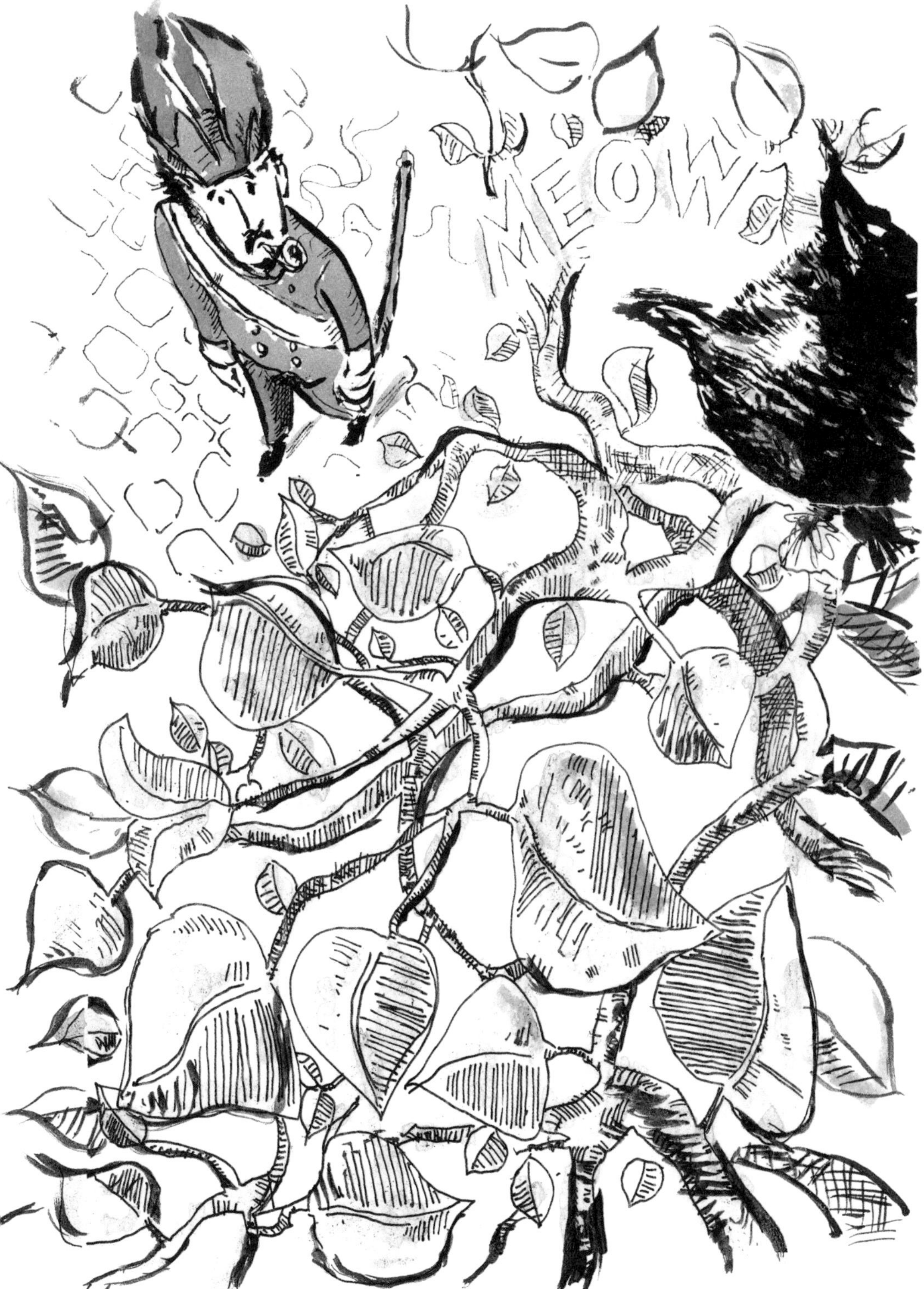
MEOW

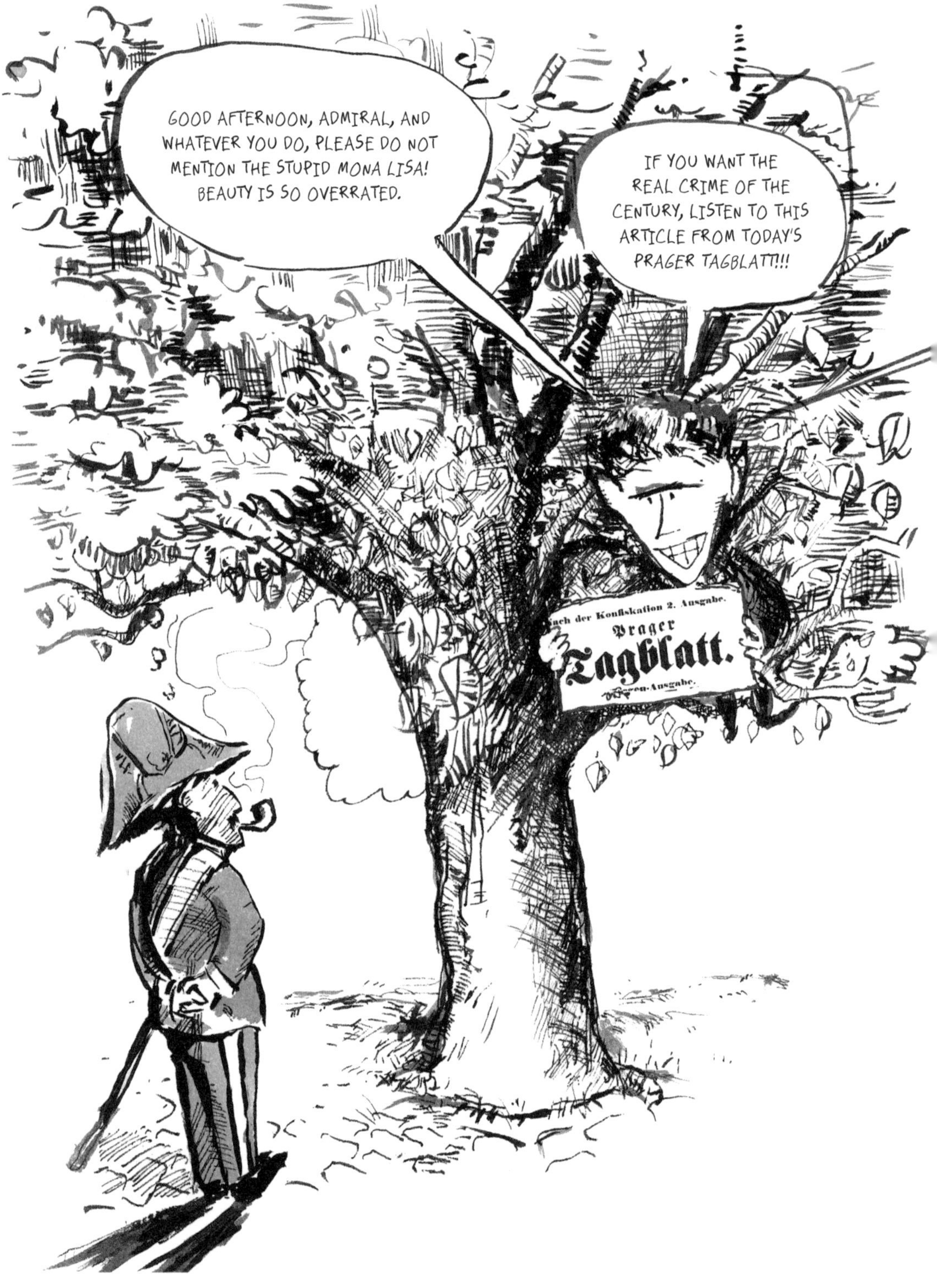
GOOD AFTERNOON, ADMIRAL, AND WHATEVER YOU DO, PLEASE DO NOT MENTION THE STUPID MONA LISA! BEAUTY IS SO OVERRATED.
IF YOU WANT THE REAL CRIME OF THE CENTURY, LISTEN TO THIS ARTICLE FROM TODAY'S PRAGER TAGBLATT!!!
Nach der Konfiskation 2. Ausgabe.
Prager
Tagblatt.

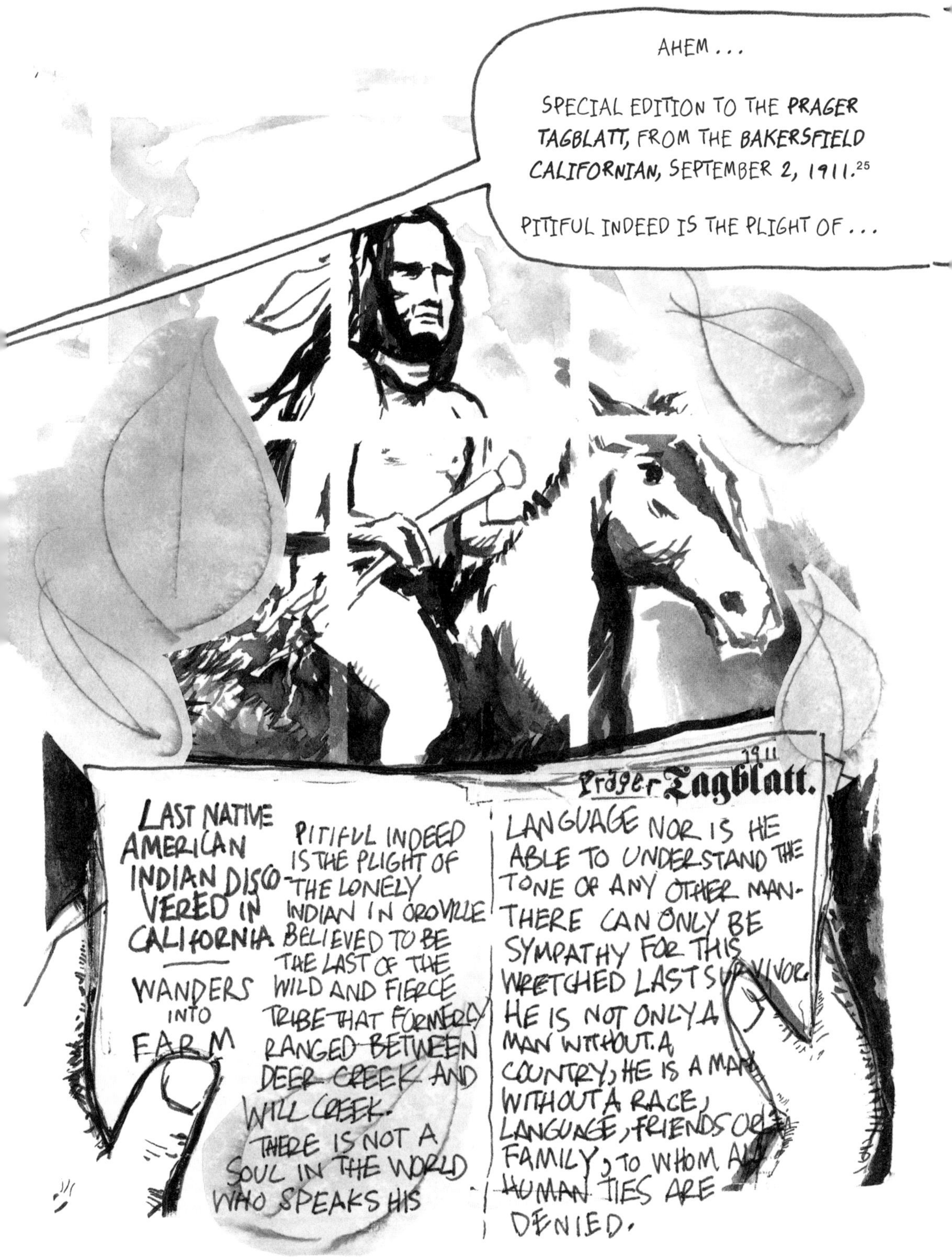
AHEM . . .
SPECIAL EDITION TO THE PRAGER TAGBLATT, FROM THE BAKERSFIELD CALIFORNIAN, SEPTEMBER 2, 1911.[25]
PITIFUL INDEED IS THE PLIGHT OF . . .
1911
Prager Tagblatt.
LAST NATIVE AMERICAN INDIAN DISCOVERED IN CALIFORNIA
WANDERS INTO FARM
PITIFUL INDEED IS THE PLIGHT OF THE LONELY INDIAN IN OROVILLE BELIEVED TO BE THE LAST OF THE WILD AND FIERCE TRIBE THAT FORMERLY RANGED BETWEEN DEER CREEK AND WILL CREEK.
THERE IS NOT A SOUL IN THE WORLD WHO SPEAKS HIS LANGUAGE NOR IS HE ABLE TO UNDERSTAND THE TONE OF ANY OTHER MAN. THERE CAN ONLY BE SYMPATHY FOR THIS WRETCHED LAST SURVIVOR. HE IS NOT ONLY A MAN WITHOUT A COUNTRY, HE IS A MAN WITHOUT A RACE, LANGUAGE, FRIENDS OR FAMILY, TO WHOM ALL HUMAN TIES ARE DENIED.

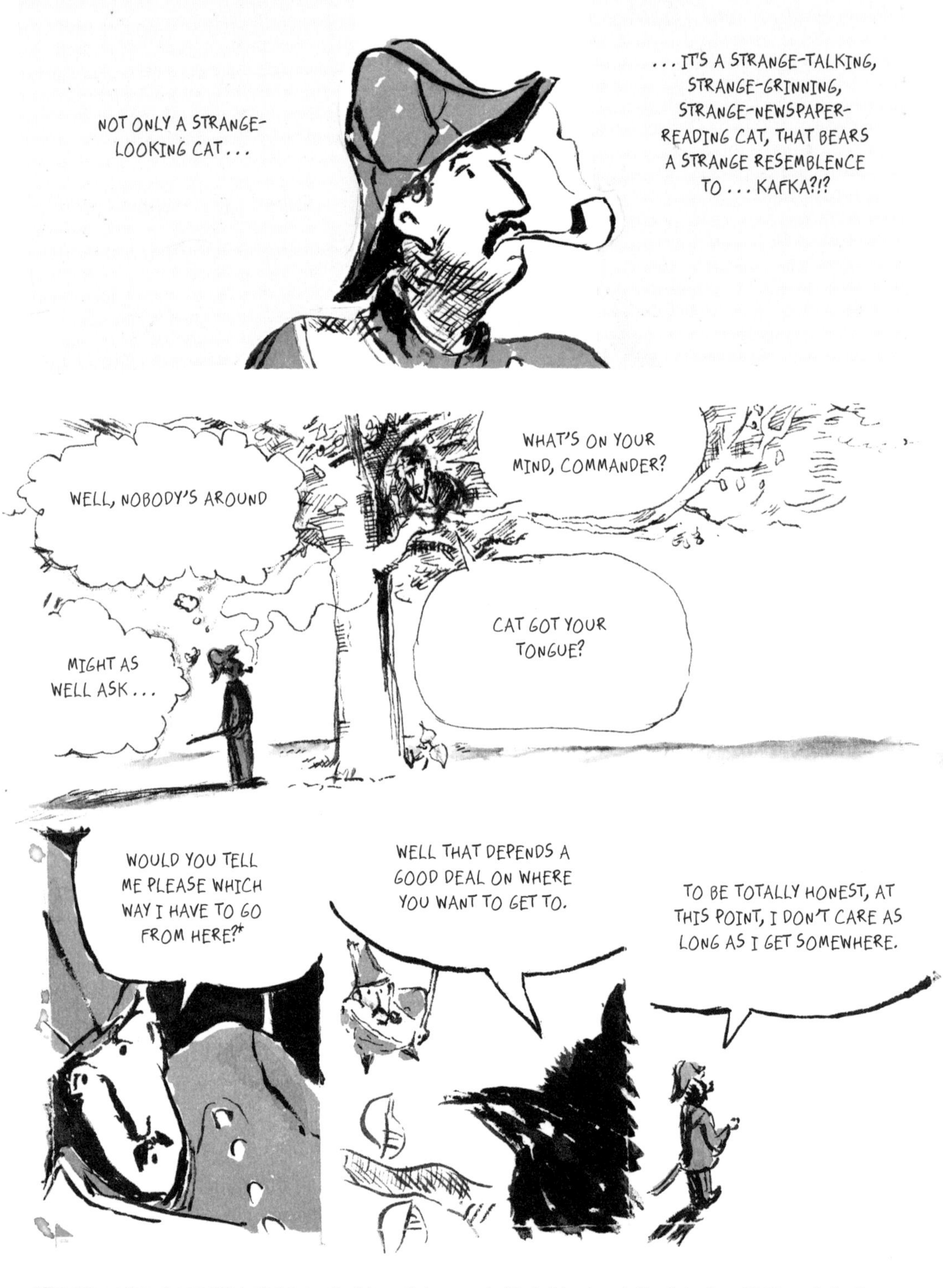
NOT ONLY A STRANGE-LOOKING CAT...
...IT'S A STRANGE-TALKING, STRANGE-GRINNING, STRANGE-NEWSPAPER-READING CAT, THAT BEARS A STRANGE RESEMBLENCE TO... KAFKA?!?
WELL, NOBODY'S AROUND
MIGHT AS WELL ASK...
WHAT'S ON YOUR MIND, COMMANDER?
CAT GOT YOUR TONGUE?
WOULD YOU TELL ME PLEASE WHICH WAY I HAVE TO GO FROM HERE?*
WELL THAT DEPENDS A GOOD DEAL ON WHERE YOU WANT TO GET TO.
TO BE TOTALLY HONEST, AT THIS POINT, I DON'T CARE AS LONG AS I GET SOMEWHERE.

* The dialogue through page 124 duplicates exactly dialogue that appears in *Alice's Adventures in Wonderland*, pp. 89–90, op cit. Don't worry, the madness is public domain.

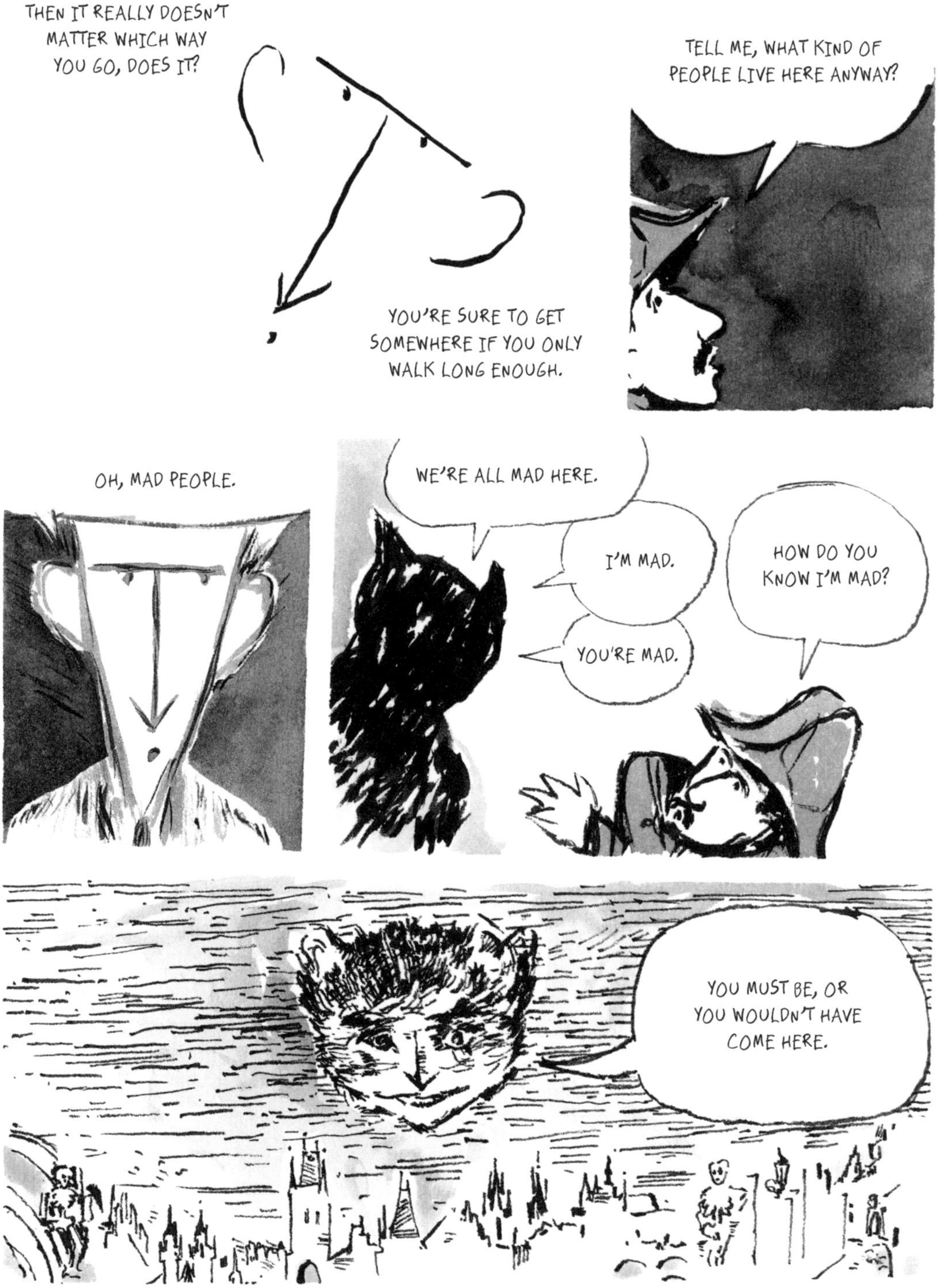

THEN IT REALLY DOESN'T MATTER WHICH WAY YOU GO, DOES IT?
YOU'RE SURE TO GET SOMEWHERE IF YOU ONLY WALK LONG ENOUGH.
TELL ME, WHAT KIND OF PEOPLE LIVE HERE ANYWAY?
OH, MAD PEOPLE.
WE'RE ALL MAD HERE.
I'M MAD.
YOU'RE MAD.
HOW DO YOU KNOW I'M MAD?
YOU MUST BE, OR YOU WOULDN'T HAVE COME HERE.

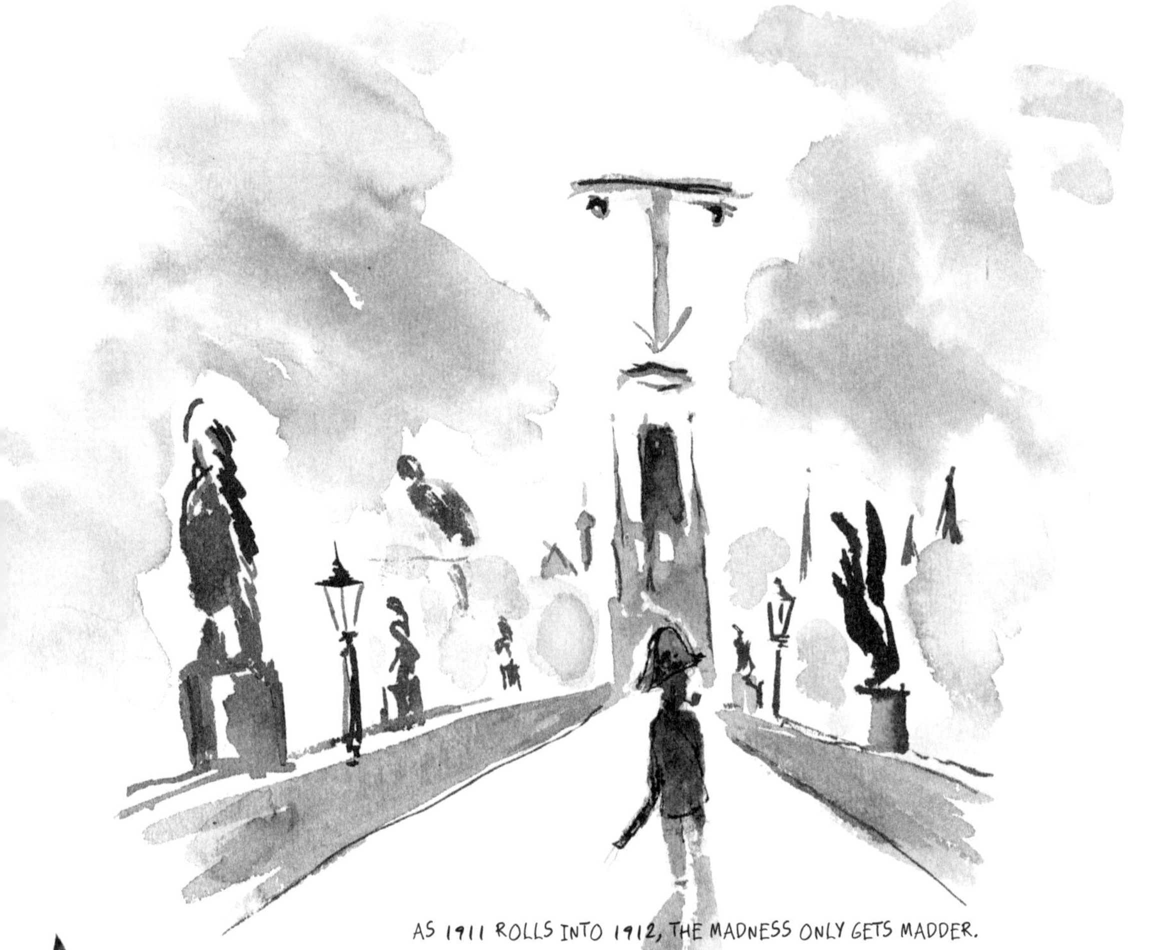

AS 1911 ROLLS INTO 1912, THE MADNESS ONLY GETS MADDER.

EINSTEIN ATTENDS THE SOLVAY CONFERENCE IN BRUSSELS, THE FIRST EVER GATHERING OF THE WORLD'S TOP THEORETICAL PHYSICISTS,* WHICH HE DISMISSES AS A "WITCHES' SABBATH."

THOMAS EDISON VISITS PRAGUE AND, I'M PROUD TO SAY, THE ONLY THING HE WANTS TO SEE IN TOWN IS THE MECHANICAL WORKINGS OF—**YOURS TRULY**, THANK YOU VERY MUCH.[26]
AS FOR SOLVING GRAVITY? WELL, NOT TO PUT TOO FINE A SPIN ON IT, EINSTEIN IS STILL FKTD. DRESSING UP IN A UNIFORM HADN'T WORKED OUT THE WAY HE PLANNED, AND HE'S FURIOUSLY SCRAPING TOGETHER FUNDS FOR THE 1914 ECLIPSE, EVEN DIPPING INTO HIS OWN BANK ACCOUNT.

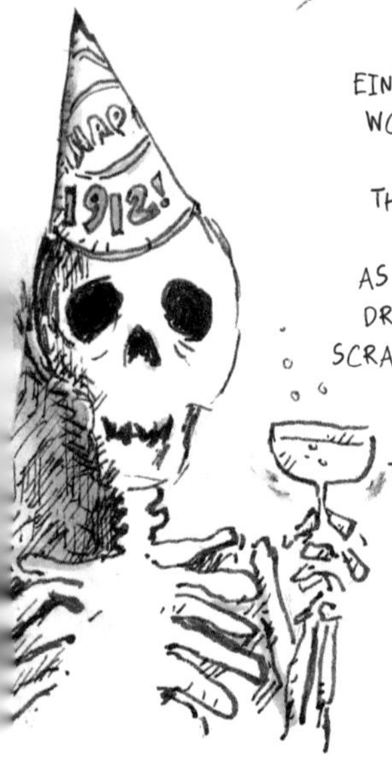
1912!

THE SAD TRUTH IS THAT THE HARDER EINSTEIN WORKS ON SOLVING GRAVITY, THE HARDER HE SEEMS TO BE BASHING HIS HEAD AGAINST THE WALL. SO ON NEW YEAR'S DAY 1912, POSSIBLY NURSING A HANGOVER FROM THE NIGHT BEFORE, AND DEFINITELY NURSING ONE FROM MAX ABRAHAM'S DECEMBER PAPER ON THE THEORY OF GRAVITY,[27] EINSTEIN EXTENDS AN INVITATION TO SLEEP ON HIS COUCH TO A RAMBLING, JOB-HUNTING AUSTRIAN PHYSICIST STRANDED IN RUSSIA BY THE NAME OF PAUL EHRENFEST.

AND JUST LIKE THAT, THE MADNESS ACCELERATES FROM THE COMIC TO THE COSMIC.

* No Max Abraham, however.

CHAPTER X

ALBERT & PAUL'S LOST WEEKEND
February 23rd, 1912

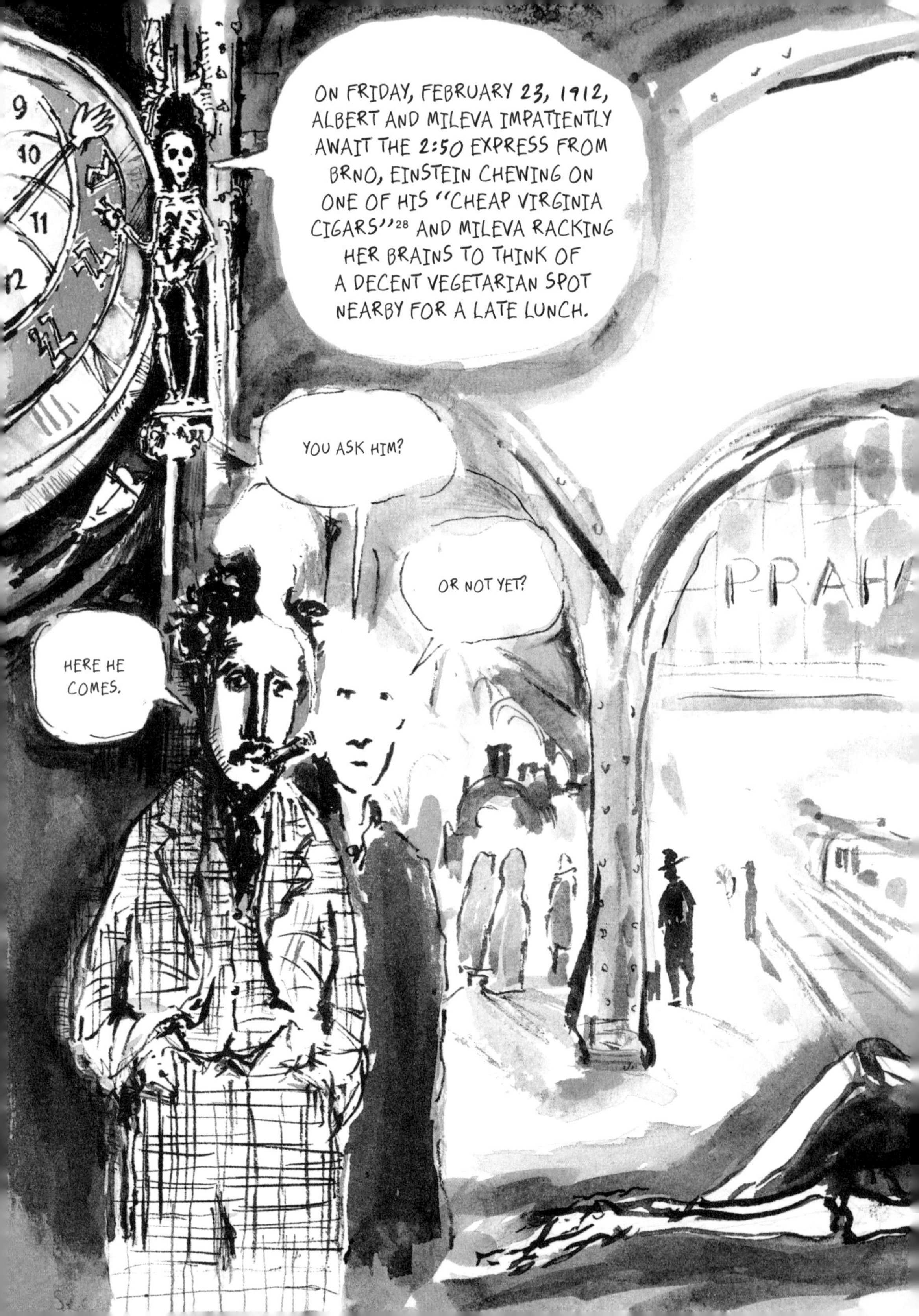
ON FRIDAY, FEBRUARY 23, 1912, ALBERT AND MILEVA IMPATIENTLY AWAIT THE 2:50 EXPRESS FROM BRNO, EINSTEIN CHEWING ON ONE OF HIS "CHEAP VIRGINIA CIGARS"[28] AND MILEVA RACKING HER BRAINS TO THINK OF A DECENT VEGETARIAN SPOT NEARBY FOR A LATE LUNCH.
YOU ASK HIM?
OR NOT YET?
HERE HE COMES.
PRAH

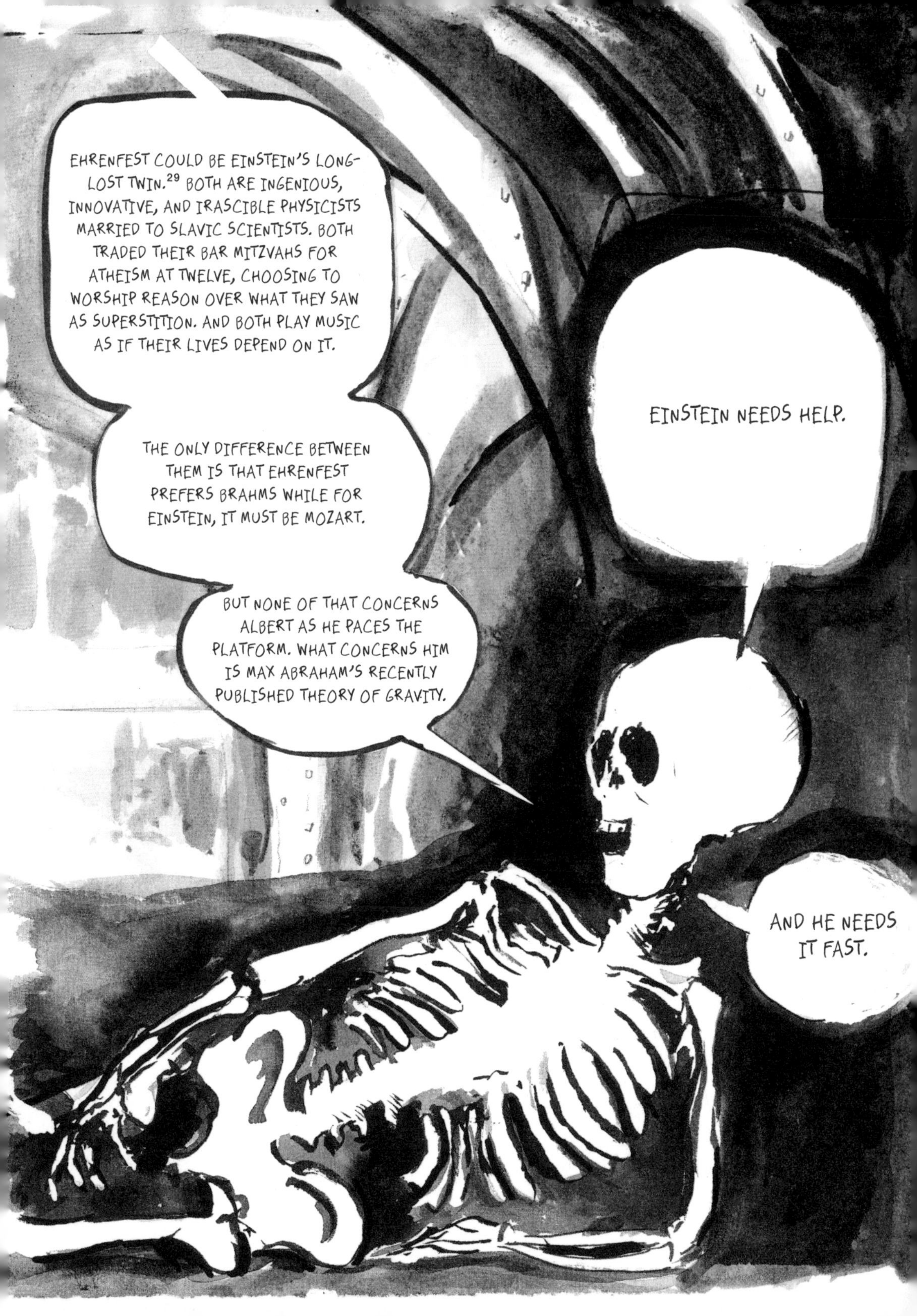
EHRENFEST COULD BE EINSTEIN'S LONG-LOST TWIN.[29] BOTH ARE INGENIOUS, INNOVATIVE, AND IRASCIBLE PHYSICISTS MARRIED TO SLAVIC SCIENTISTS. BOTH TRADED THEIR BAR MITZVAHS FOR ATHEISM AT TWELVE, CHOOSING TO WORSHIP REASON OVER WHAT THEY SAW AS SUPERSTITION. AND BOTH PLAY MUSIC AS IF THEIR LIVES DEPEND ON IT.
THE ONLY DIFFERENCE BETWEEN THEM IS THAT EHRENFEST PREFERS BRAHMS WHILE FOR EINSTEIN, IT MUST BE MOZART.
BUT NONE OF THAT CONCERNS ALBERT AS HE PACES THE PLATFORM. WHAT CONCERNS HIM IS MAX ABRAHAM'S RECENTLY PUBLISHED THEORY OF GRAVITY.
EINSTEIN NEEDS HELP.
AND HE NEEDS IT FAST.

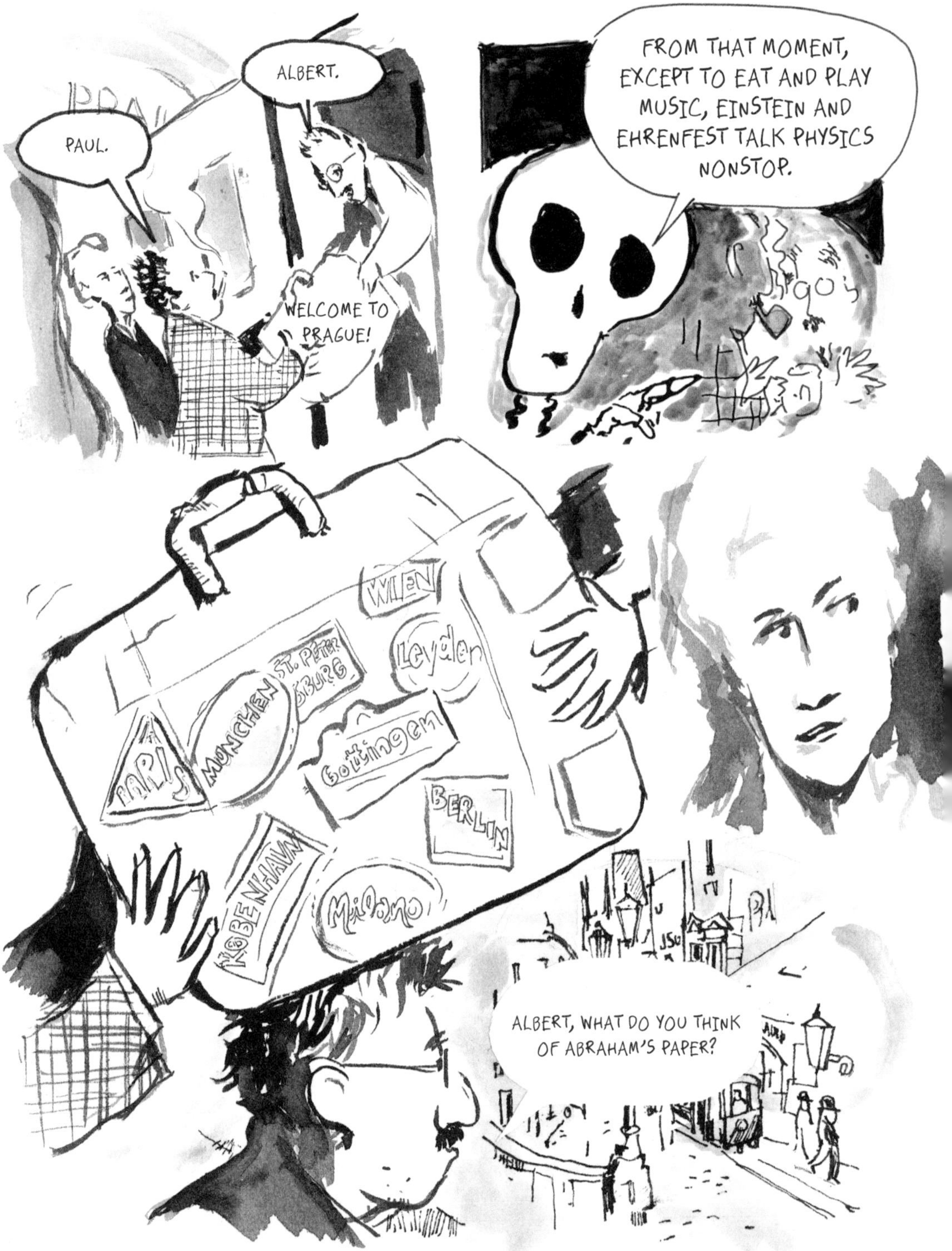
PAUL.
ALBERT.
WELCOME TO PRAGUE!
FROM THAT MOMENT, EXCEPT TO EAT AND PLAY MUSIC, EINSTEIN AND EHRENFEST TALK PHYSICS NONSTOP.
WIEN
Leyden
ST. PETERSBURG
MÜNCHEN
PARIS
Göttingen
BERLIN
KØBENHAVN
Milano
ALBERT, WHAT DO YOU THINK OF ABRAHAM'S PAPER?

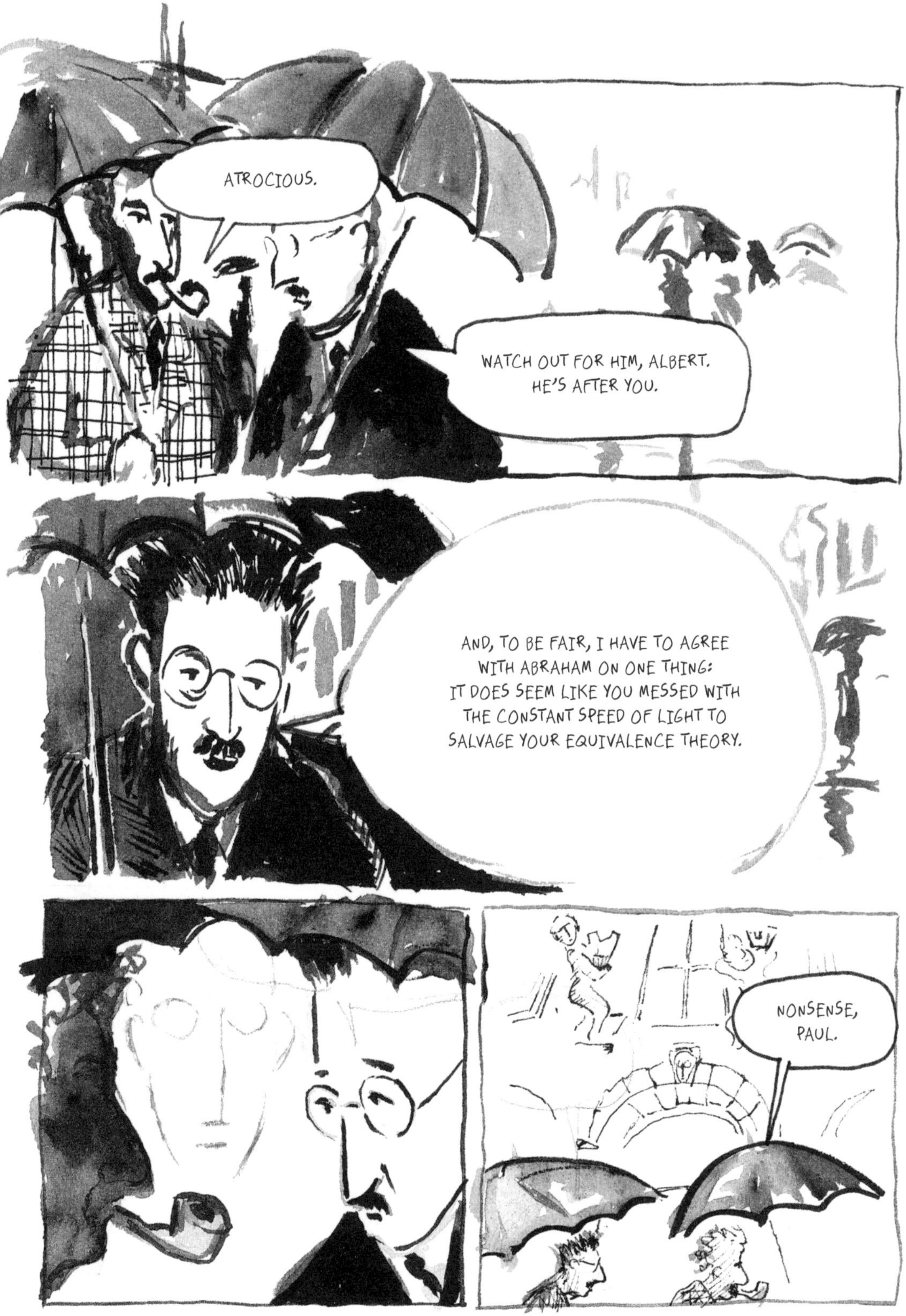
ATROCIOUS.
WATCH OUT FOR HIM, ALBERT. HE'S AFTER YOU.
AND, TO BE FAIR, I HAVE TO AGREE WITH ABRAHAM ON ONE THING: IT DOES SEEM LIKE YOU MESSED WITH THE CONSTANT SPEED OF LIGHT TO SALVAGE YOUR EQUIVALENCE THEORY.
NONSENSE, PAUL.

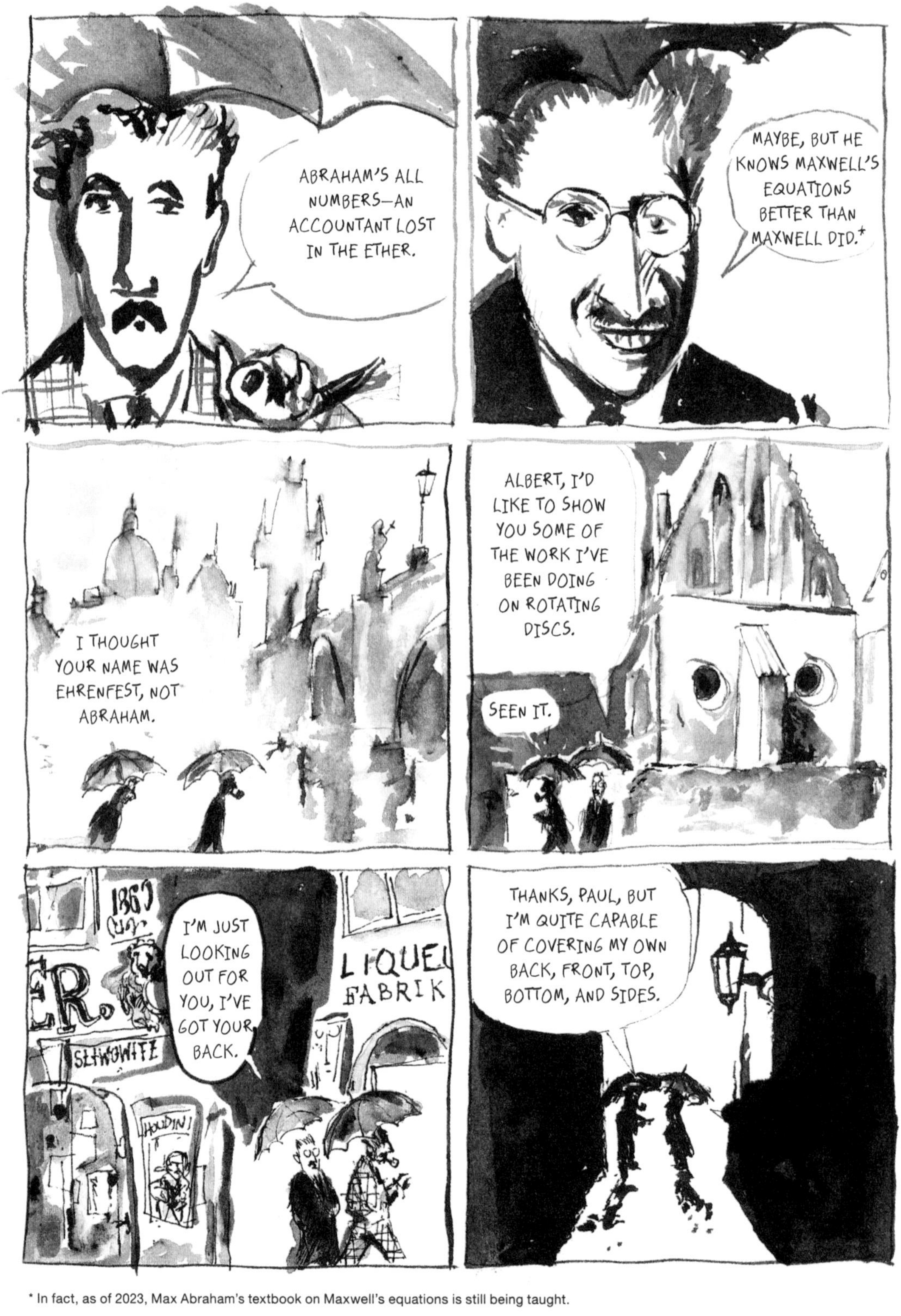
ABRAHAM'S ALL NUMBERS—AN ACCOUNTANT LOST IN THE ETHER.
MAYBE, BUT HE KNOWS MAXWELL'S EQUATIONS BETTER THAN MAXWELL DID.*
I THOUGHT YOUR NAME WAS EHRENFEST, NOT ABRAHAM.
ALBERT, I'D LIKE TO SHOW YOU SOME OF THE WORK I'VE BEEN DOING ON ROTATING DISCS.
SEEN IT.
I'M JUST LOOKING OUT FOR YOU, I'VE GOT YOUR BACK.
LIQUE
FABRIK
HOUDINI
THANKS, PAUL, BUT I'M QUITE CAPABLE OF COVERING MY OWN BACK, FRONT, TOP, BOTTOM, AND SIDES.

* In fact, as of 2023, Max Abraham's textbook on Maxwell's equations is still being taught.

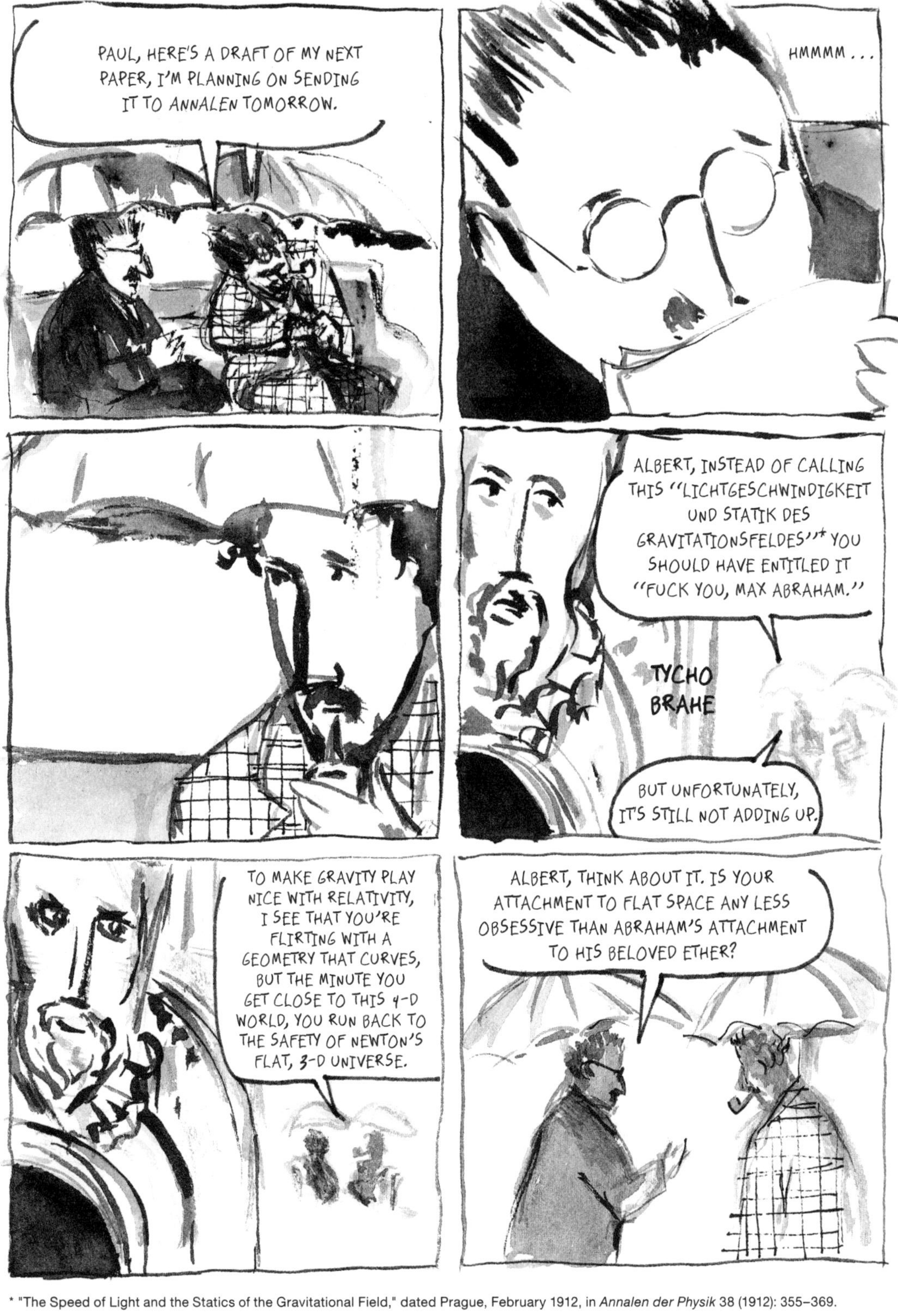
PAUL, HERE'S A DRAFT OF MY NEXT PAPER, I'M PLANNING ON SENDING IT TO ANNALEN TOMORROW.
HMMMM . . .
ALBERT, INSTEAD OF CALLING THIS "LICHTGESCHWINDIGKEIT UND STATIK DES GRAVITATIONSFELDES"* YOU SHOULD HAVE ENTITLED IT "FUCK YOU, MAX ABRAHAM."
TYCHO BRAHE
BUT UNFORTUNATELY, IT'S STILL NOT ADDING UP.
TO MAKE GRAVITY PLAY NICE WITH RELATIVITY, I SEE THAT YOU'RE FLIRTING WITH A GEOMETRY THAT CURVES, BUT THE MINUTE YOU GET CLOSE TO THIS 4-D WORLD, YOU RUN BACK TO THE SAFETY OF NEWTON'S FLAT, 3-D UNIVERSE.
ALBERT, THINK ABOUT IT. IS YOUR ATTACHMENT TO FLAT SPACE ANY LESS OBSESSIVE THAN ABRAHAM'S ATTACHMENT TO HIS BELOVED ETHER?

* "The Speed of Light and the Statics of the Gravitational Field," dated Prague, February 1912, in *Annalen der Physik* 38 (1912): 355–369.

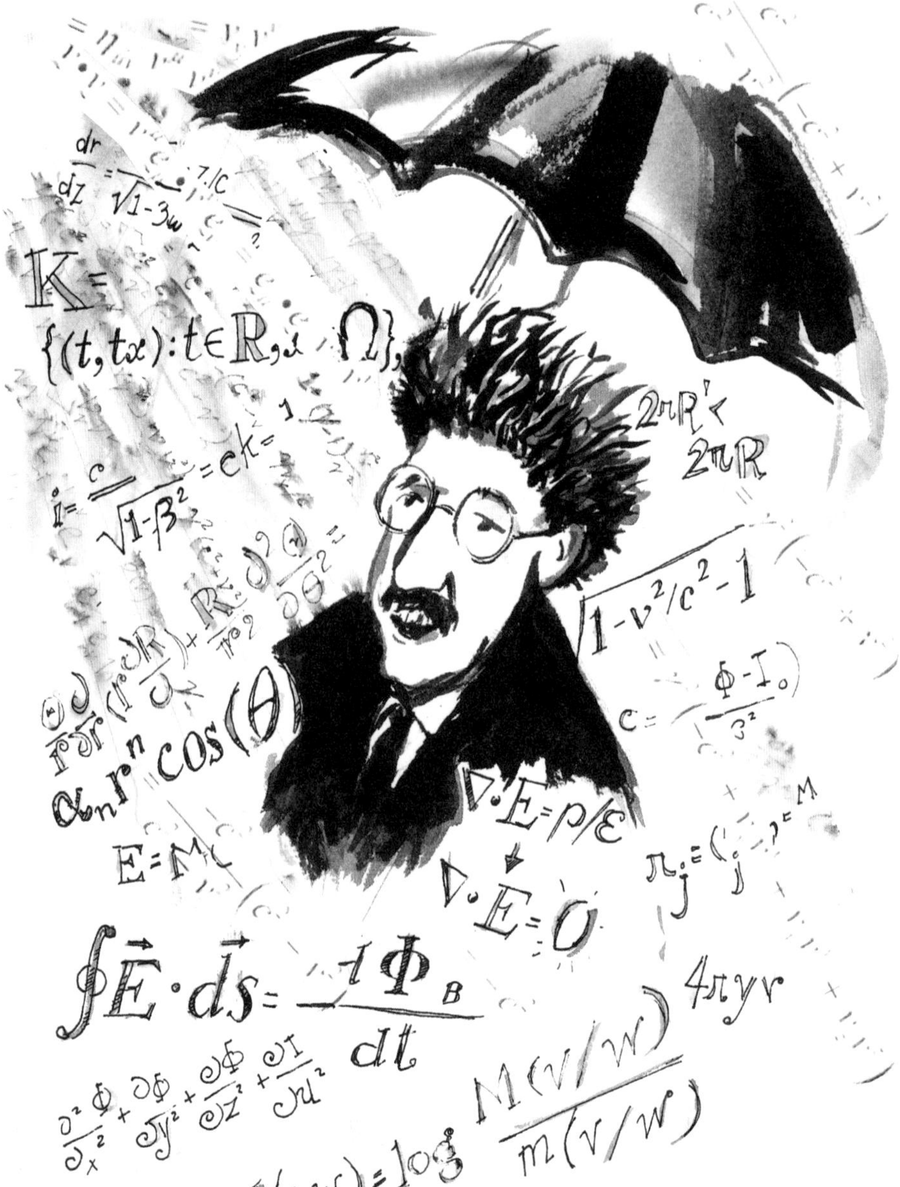
2πR
E=M

YOU'RE SO CLOSE. YOU'RE WALTZING AROUND IT. YOU DIP YOUR FINGERS IN THE TRUTH AND THEN YOU PULL THEM RIGHT BACK OUT.

ALBERT, TO WIN YOU HAVE TO BEAT ABRAHAM AT HIS OWN GAME. HAVE YOU EVER CONSIDERED THAT THE SHORTEST DISTANCE BETWEEN TWO POINTS ISN'T A STRAIGHT LINE? I'M JUST SAYING.

YEAH, PAUL, THAT'S ALL FINE ON A BLACKBOARD, BUT HOW DOES THAT HOCUS-POCUS MATH EXPLAIN ANYTHING THAT MEANS ANYTHING IN THE HERE AND NOW?

PAUL, COME ON, HOW THE HELL CAN SOMETHING EXPAND AND CONTRACT AT THE SAME TIME?

YOU SAY FLAT SPACE IS A STRAITJACKET. PERHAPS. BUT ONE THAT KEEPS THE MADNESS OF ENDLESS UNMEASURABLE DIMENSIONS IN CHECK, THAT ALLOWS US TO SEE THINGS HAVE A PLACE, THAT SAVES THE UNIVERSE FOR MEANING.

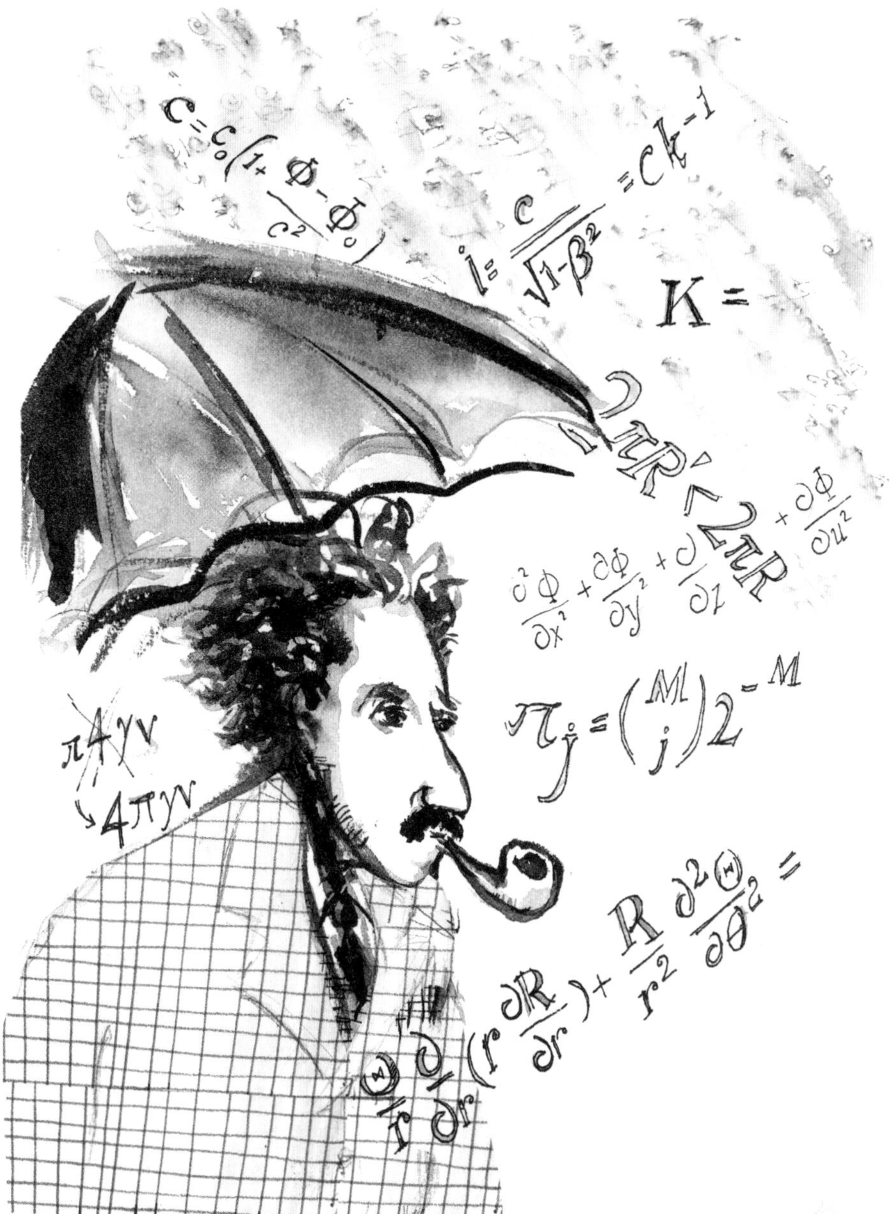

c = c₀(1 + (Φ − Φ₀)/c²)
i = c/√(1−β²) = ck⁻¹
K =
2πR′ < 2πR
π4γν
4πγν
π_j = (M j) 2^−M

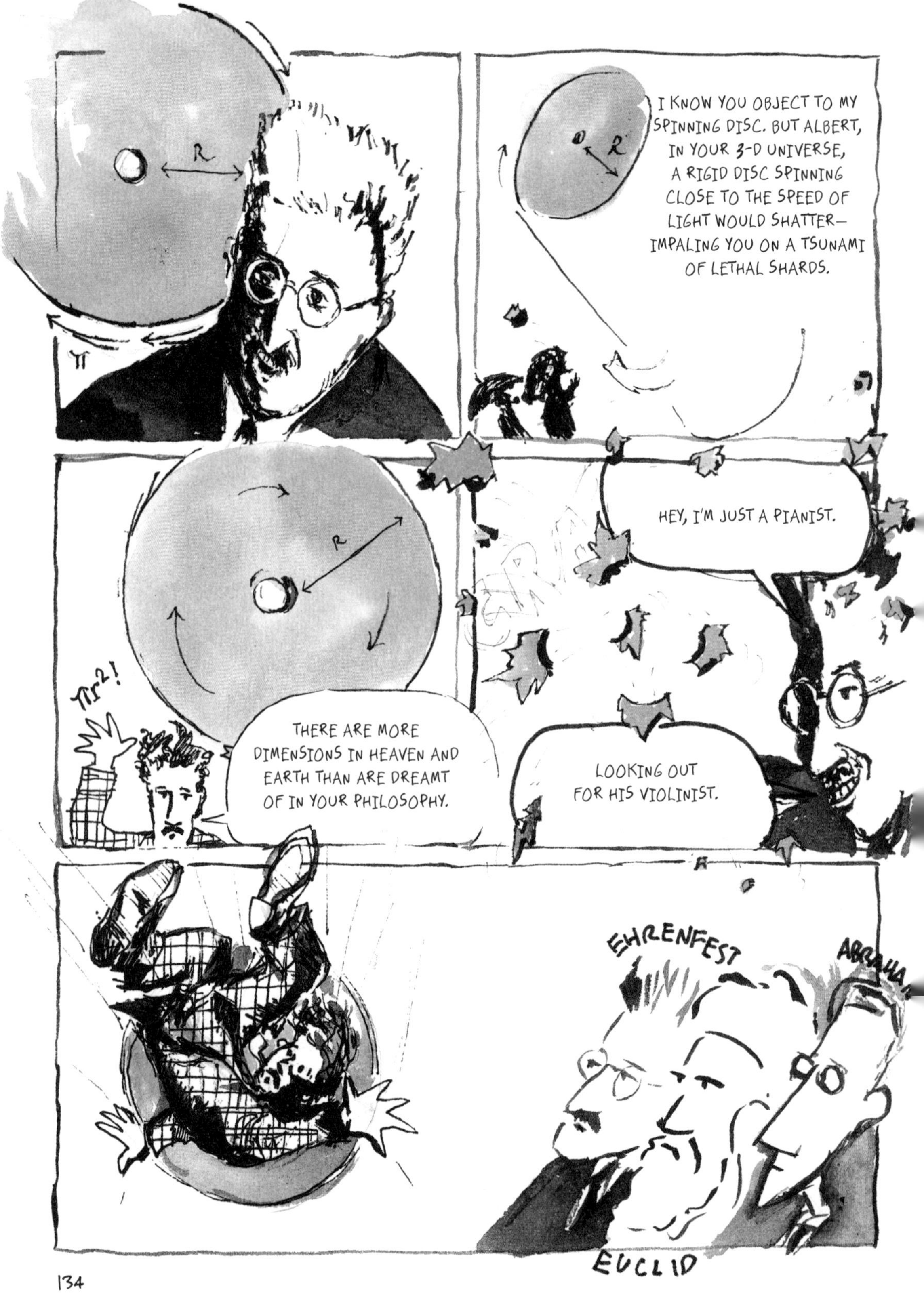
R
π
I KNOW YOU OBJECT TO MY SPINNING DISC. BUT ALBERT, IN YOUR 3-D UNIVERSE, A RIGID DISC SPINNING CLOSE TO THE SPEED OF LIGHT WOULD SHATTER— IMPALING YOU ON A TSUNAMI OF LETHAL SHARDS.
R
R
πr^2!
THERE ARE MORE DIMENSIONS IN HEAVEN AND EARTH THAN ARE DREAMT OF IN YOUR PHILOSOPHY.
HEY, I'M JUST A PIANIST.
LOOKING OUT FOR HIS VIOLINIST.
EHRENFEST
EUCLID

STILL I CONFESS, THE CLOSER I GET TO SOLVING GRAVITY, THE MORE I'M VISITED BY ALL SORTS OF NERVOUS CONFLICTS.

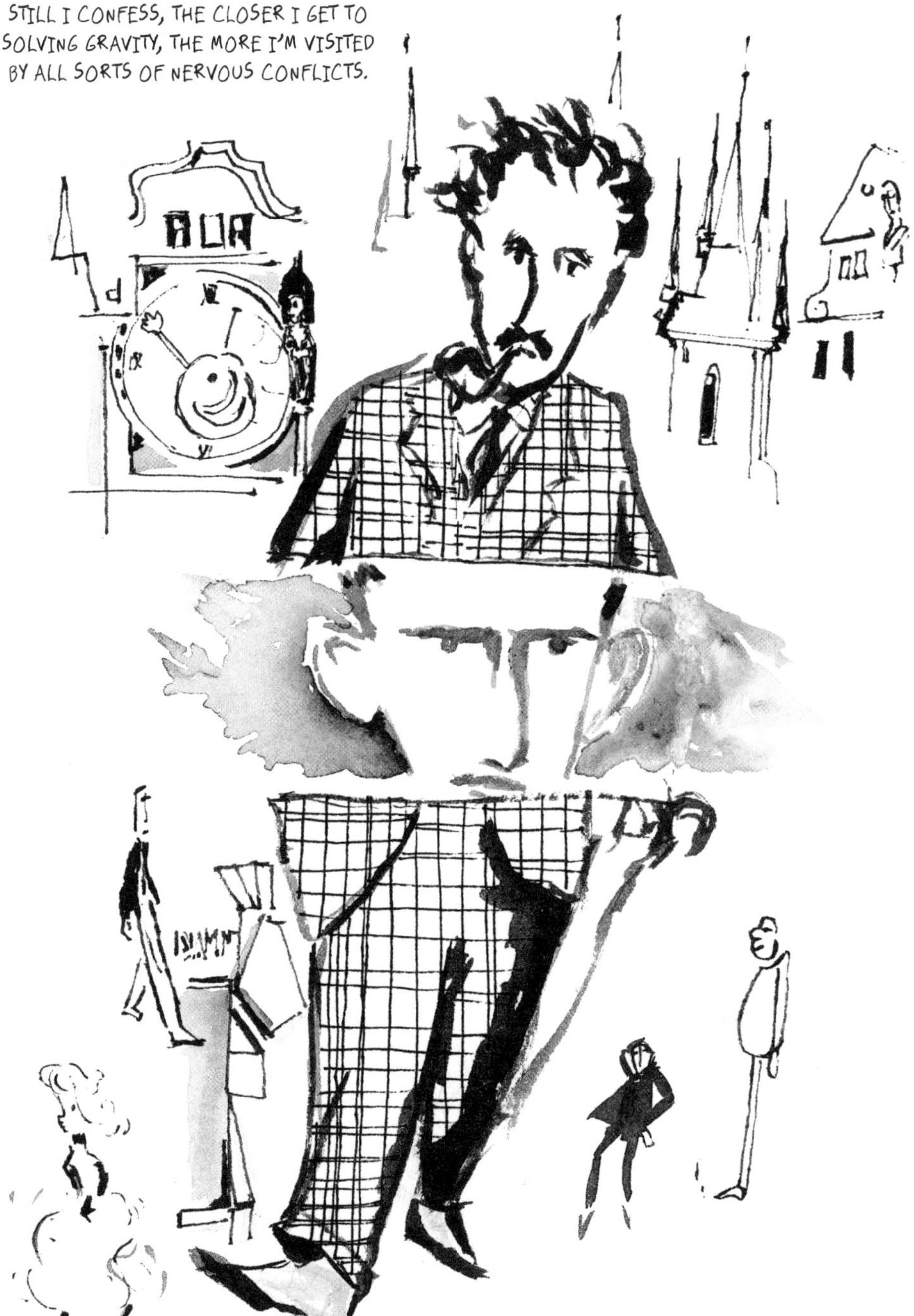

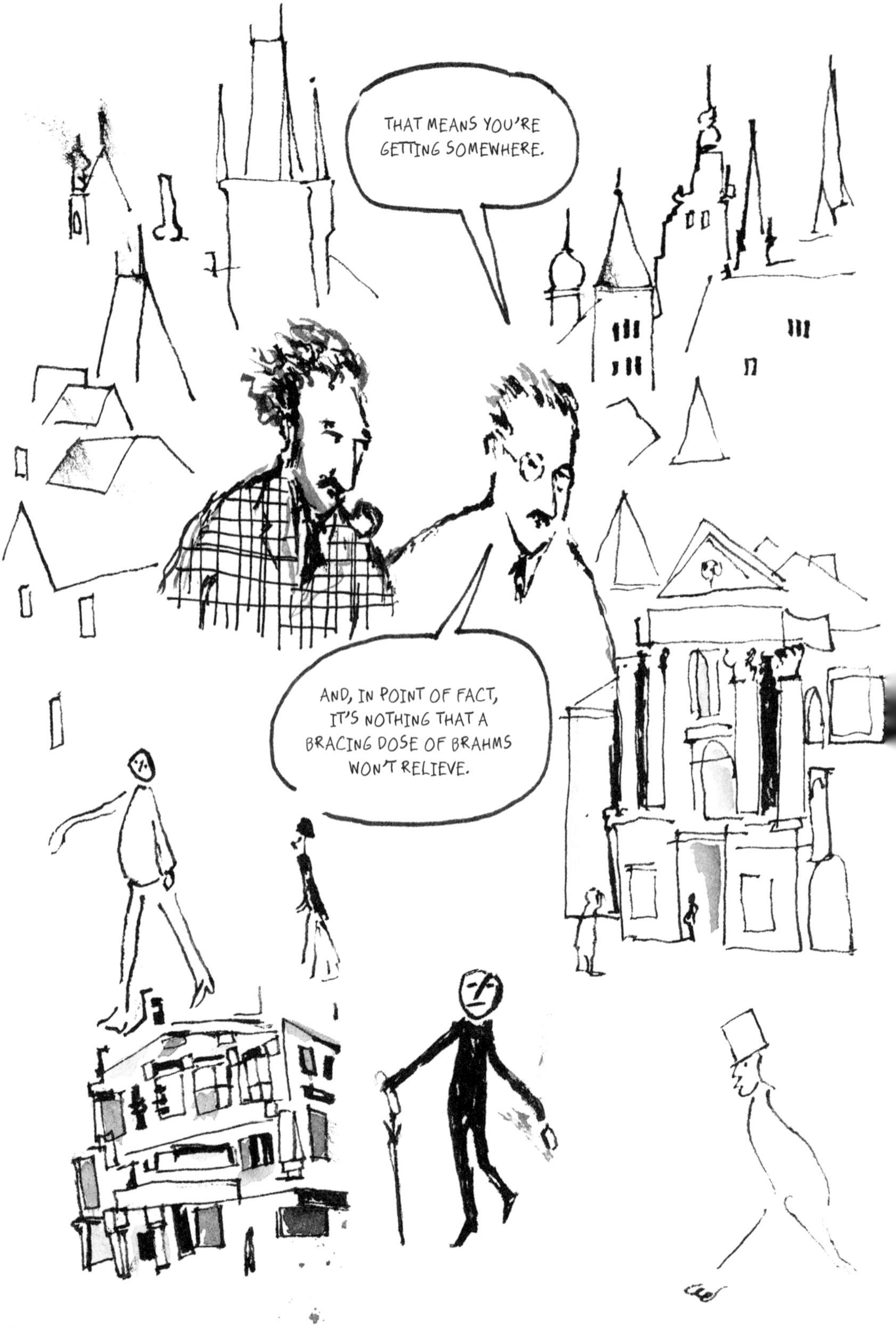
THAT MEANS YOU'RE GETTING SOMEWHERE.
AND, IN POINT OF FACT, IT'S NOTHING THAT A BRACING DOSE OF BRAHMS WON'T RELIEVE.

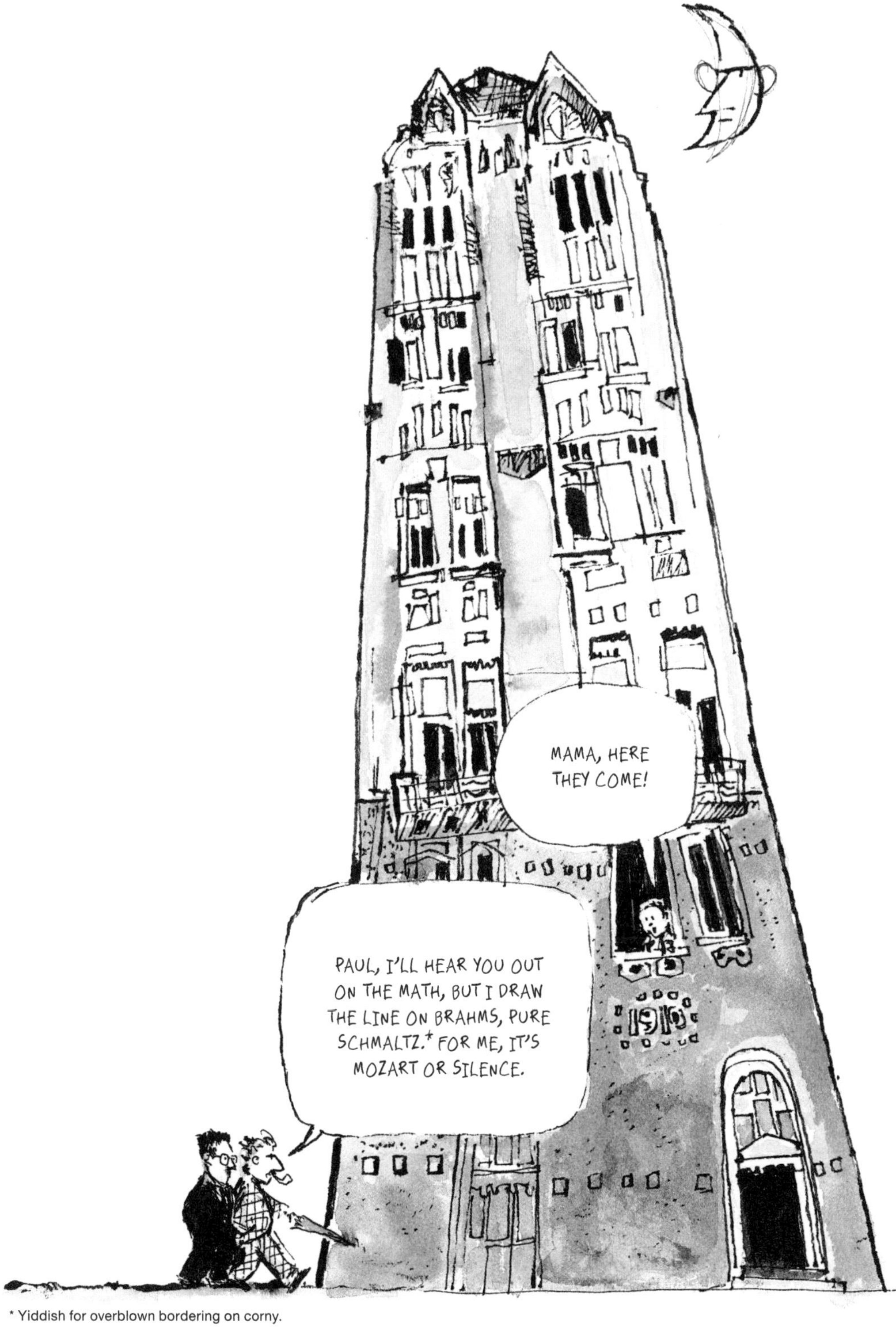
MAMA, HERE THEY COME!
PAUL, I'LL HEAR YOU OUT ON THE MATH, BUT I DRAW THE LINE ON BRAHMS, PURE SCHMALTZ.* FOR ME, IT'S MOZART OR SILENCE.
1916

* Yiddish for overblown bordering on corny.

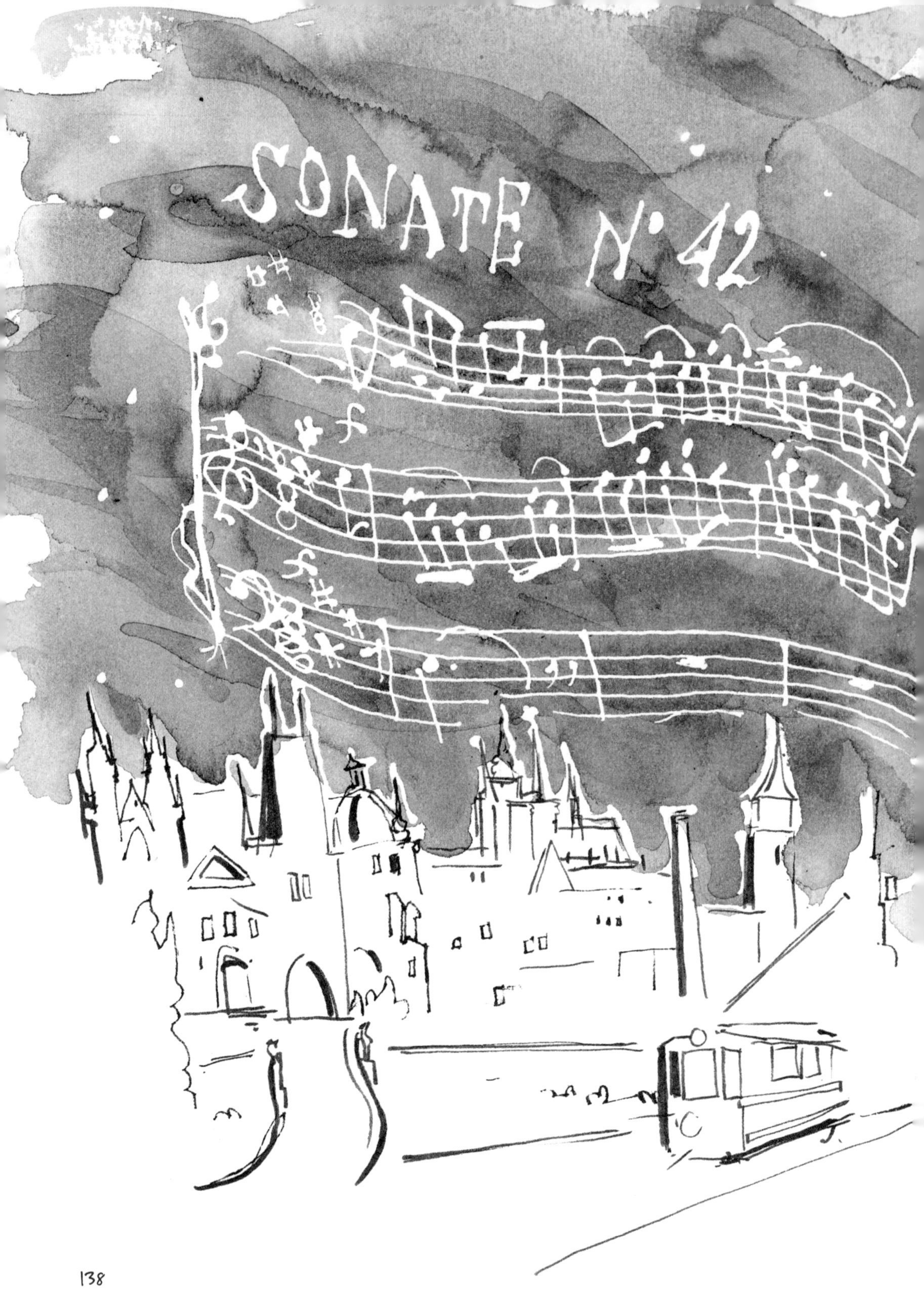
SONATE N° 42

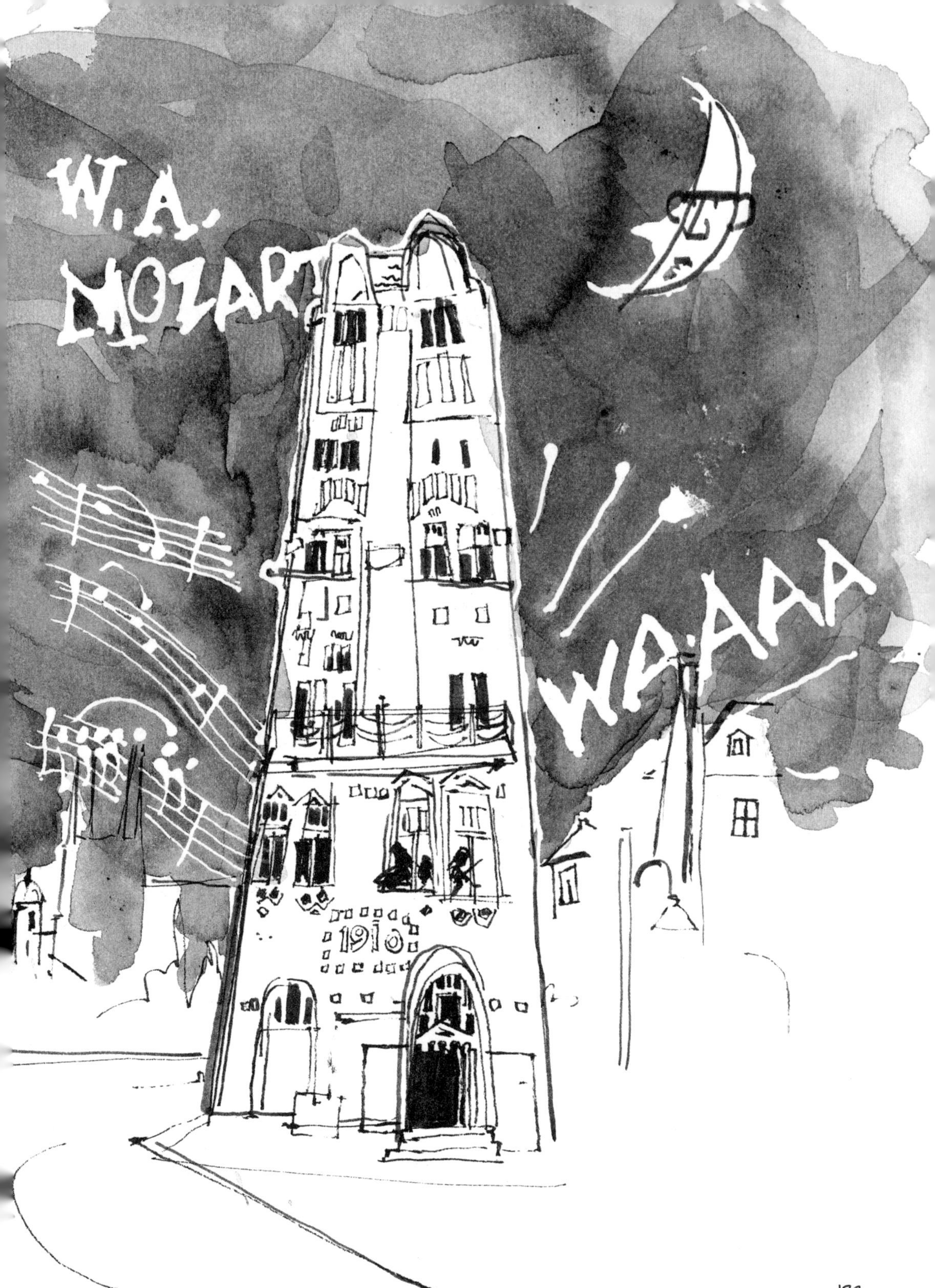
W.A. MOZART
WAAAA
1910

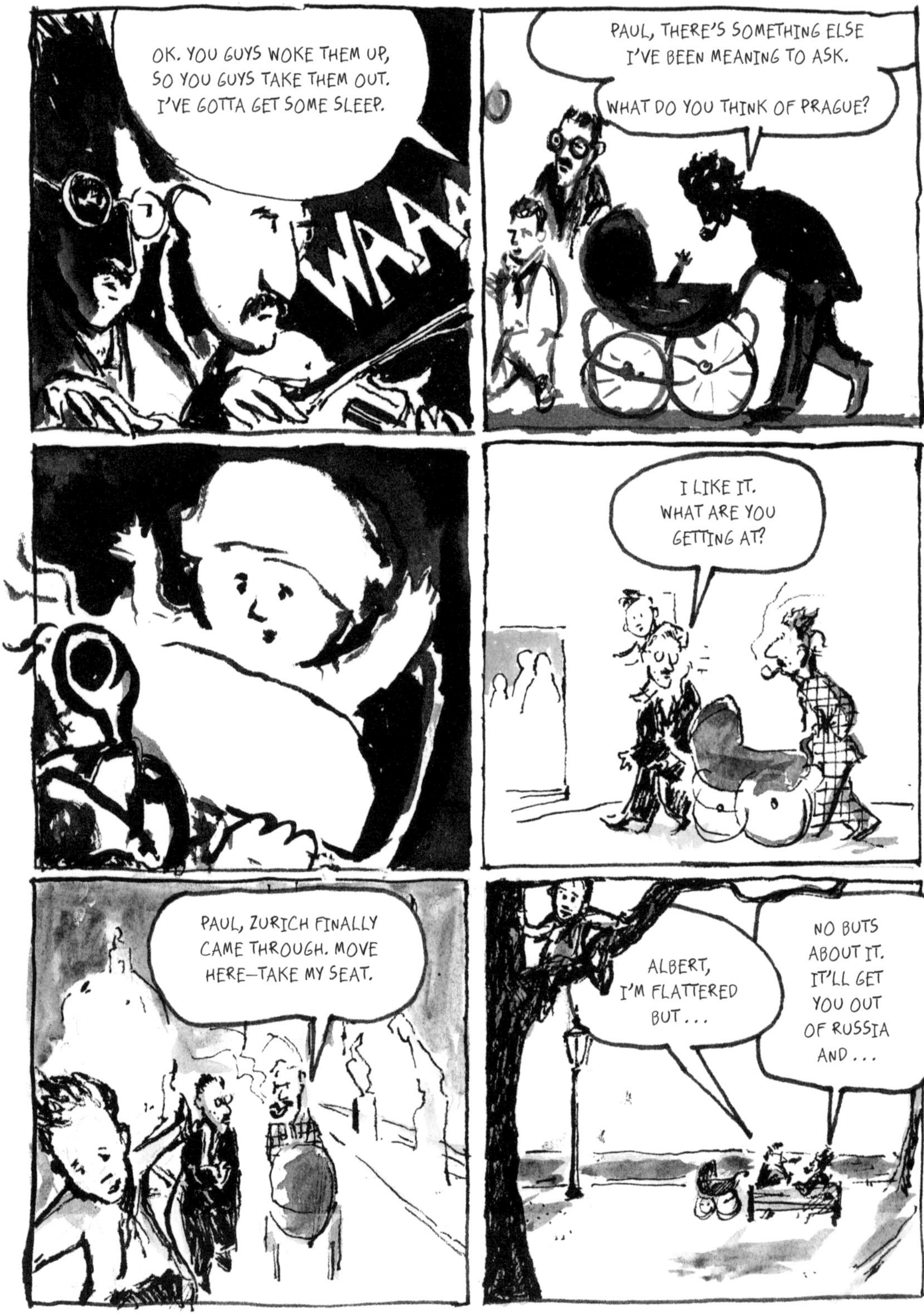
OK. YOU GUYS WOKE THEM UP, SO YOU GUYS TAKE THEM OUT. I'VE GOTTA GET SOME SLEEP.
WAA
PAUL, THERE'S SOMETHING ELSE I'VE BEEN MEANING TO ASK.
WHAT DO YOU THINK OF PRAGUE?
I LIKE IT. WHAT ARE YOU GETTING AT?
PAUL, ZURICH FINALLY CAME THROUGH. MOVE HERE—TAKE MY SEAT.
ALBERT, I'M FLATTERED BUT . . .
NO BUTS ABOUT IT. IT'LL GET YOU OUT OF RUSSIA AND . . .

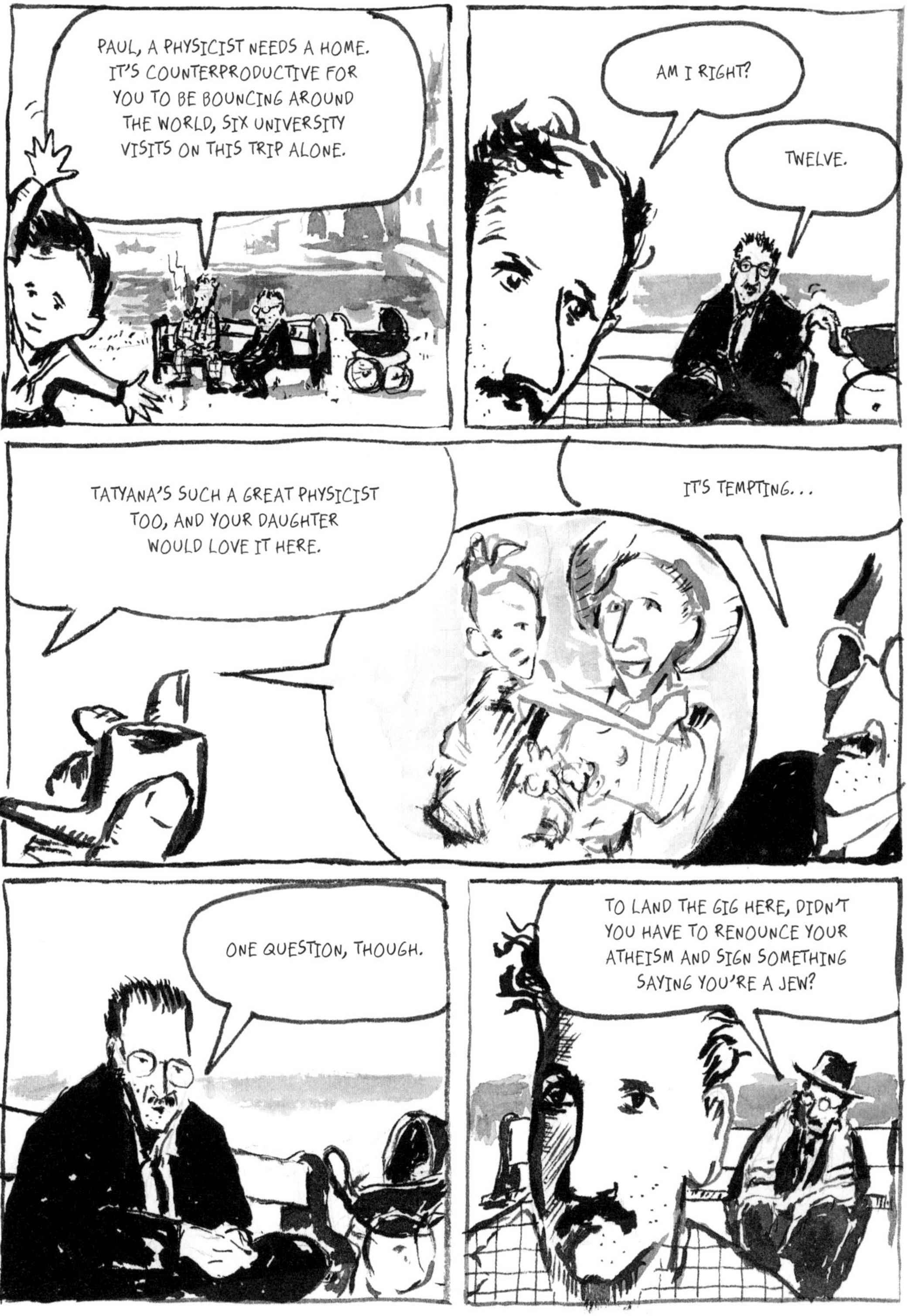
PAUL, A PHYSICIST NEEDS A HOME. IT'S COUNTERPRODUCTIVE FOR YOU TO BE BOUNCING AROUND THE WORLD, SIX UNIVERSITY VISITS ON THIS TRIP ALONE.
AM I RIGHT?
TWELVE.
TATYANA'S SUCH A GREAT PHYSICIST TOO, AND YOUR DAUGHTER WOULD LOVE IT HERE.
IT'S TEMPTING...
ONE QUESTION, THOUGH.
TO LAND THE GIG HERE, DIDN'T YOU HAVE TO RENOUNCE YOUR ATHEISM AND SIGN SOMETHING SAYING YOU'RE A JEW?

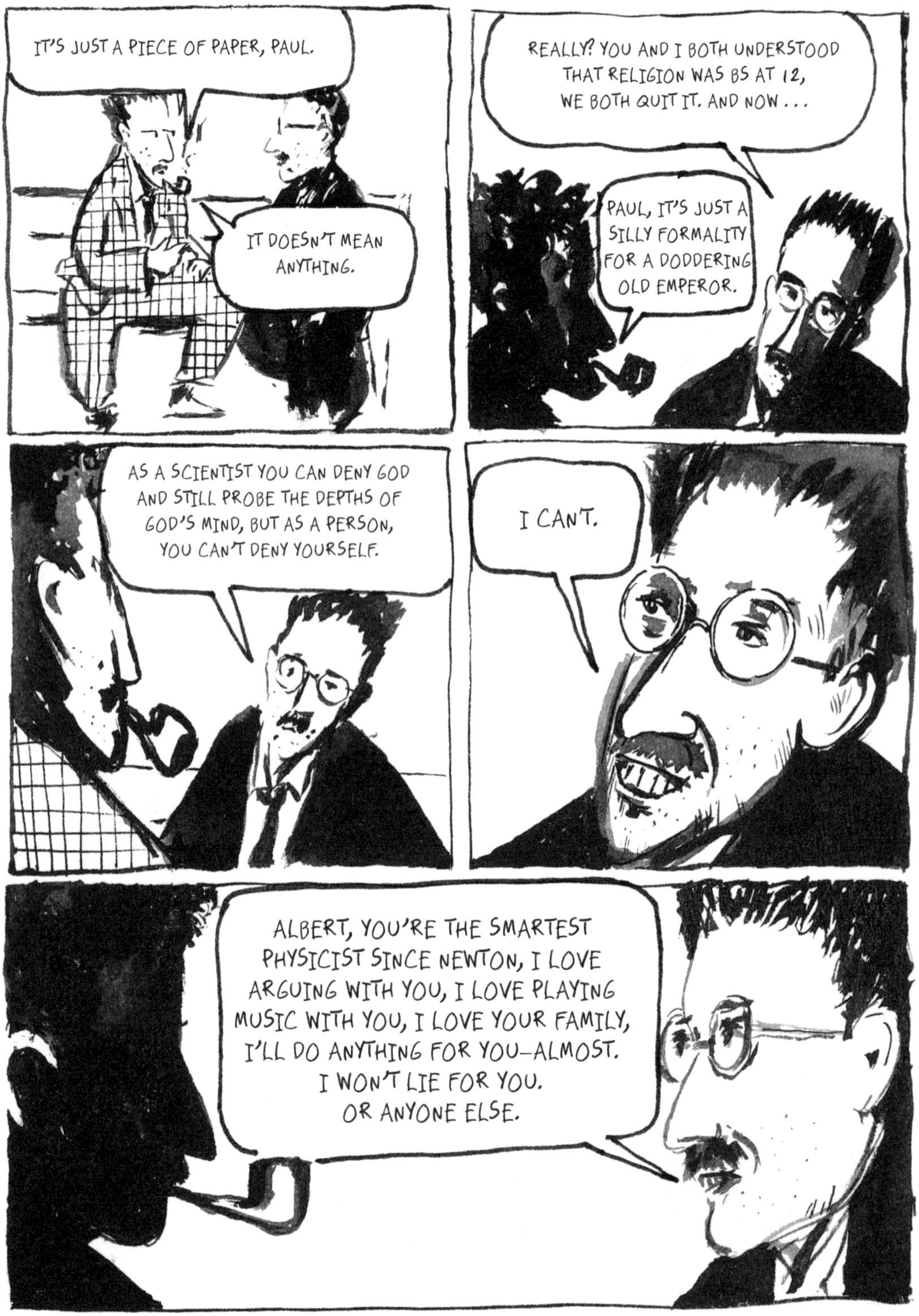
IT'S JUST A PIECE OF PAPER, PAUL.
IT DOESN'T MEAN ANYTHING.
REALLY? YOU AND I BOTH UNDERSTOOD THAT RELIGION WAS BS AT 12, WE BOTH QUIT IT. AND NOW . . .
PAUL, IT'S JUST A SILLY FORMALITY FOR A DODDERING OLD EMPEROR.
AS A SCIENTIST YOU CAN DENY GOD AND STILL PROBE THE DEPTHS OF GOD'S MIND, BUT AS A PERSON, YOU CAN'T DENY YOURSELF.
I CAN'T.
ALBERT, YOU'RE THE SMARTEST PHYSICIST SINCE NEWTON, I LOVE ARGUING WITH YOU, I LOVE PLAYING MUSIC WITH YOU, I LOVE YOUR FAMILY, I'LL DO ANYTHING FOR YOU–ALMOST. I WON'T LIE FOR YOU. OR ANYONE ELSE.

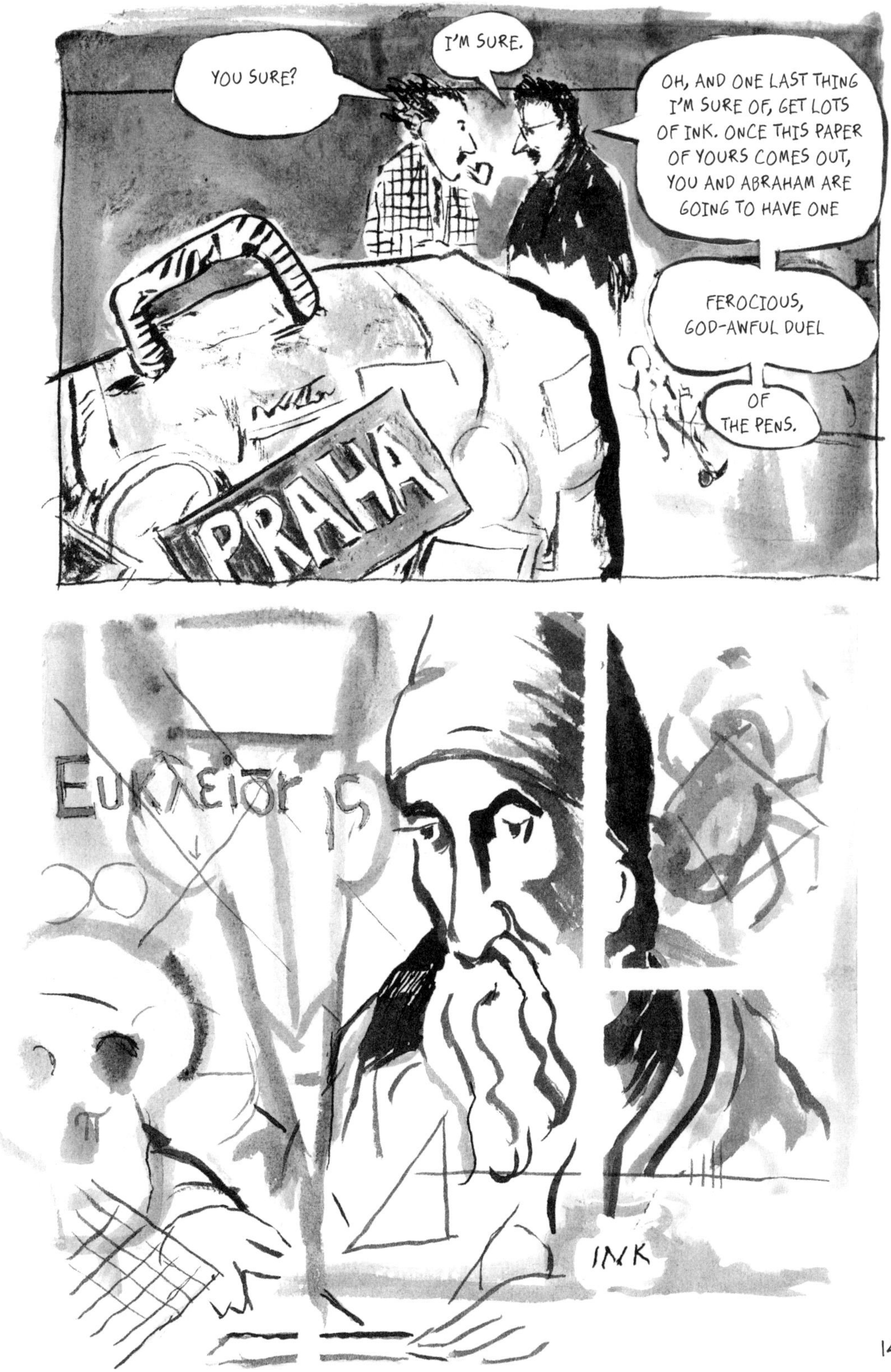
YOU SURE?
I'M SURE.
OH, AND ONE LAST THING I'M SURE OF, GET LOTS OF INK. ONCE THIS PAPER OF YOURS COMES OUT, YOU AND ABRAHAM ARE GOING TO HAVE ONE
FEROCIOUS, GOD-AWFUL DUEL
OF THE PENS.
PRAHA
INK

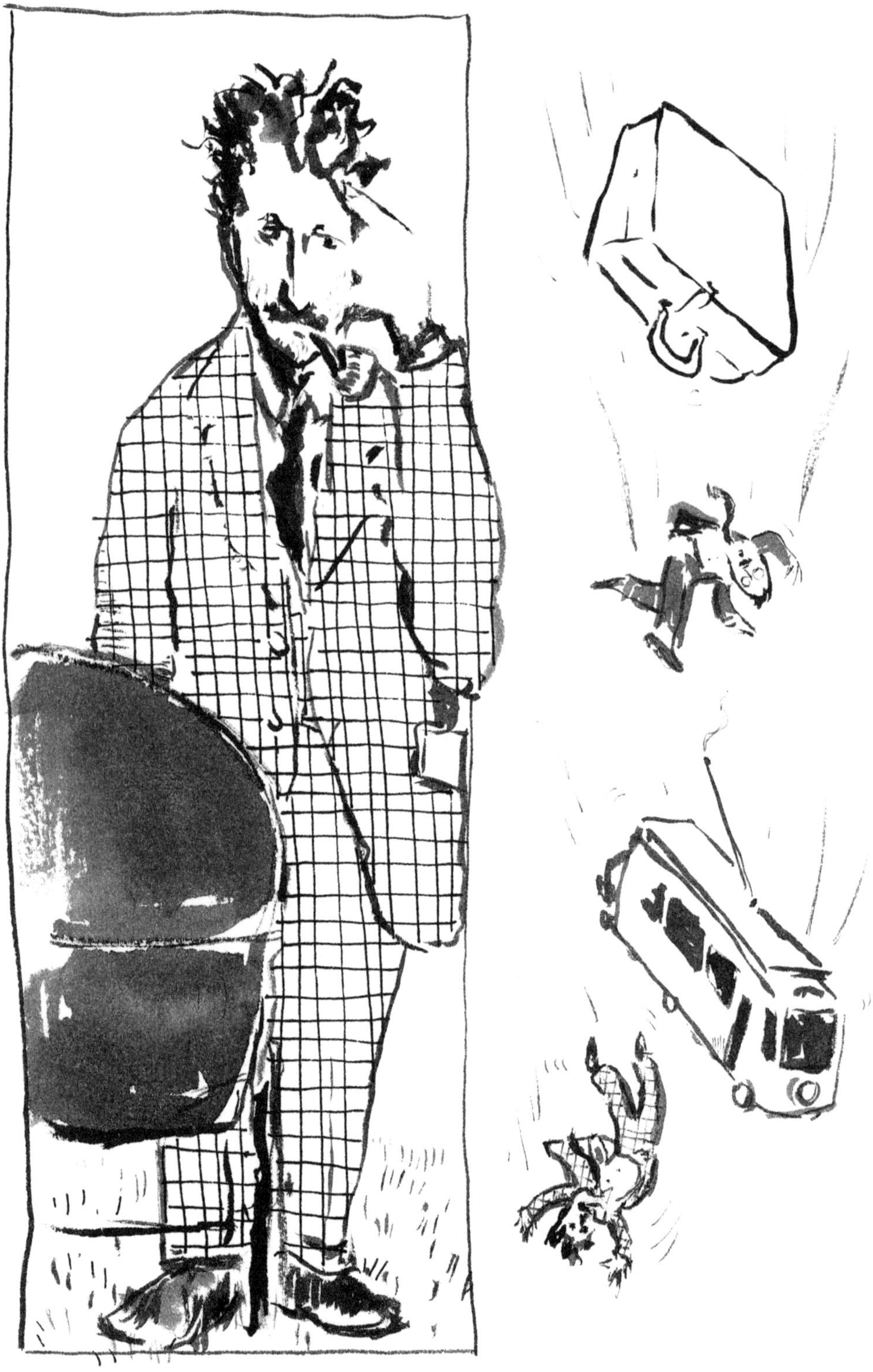

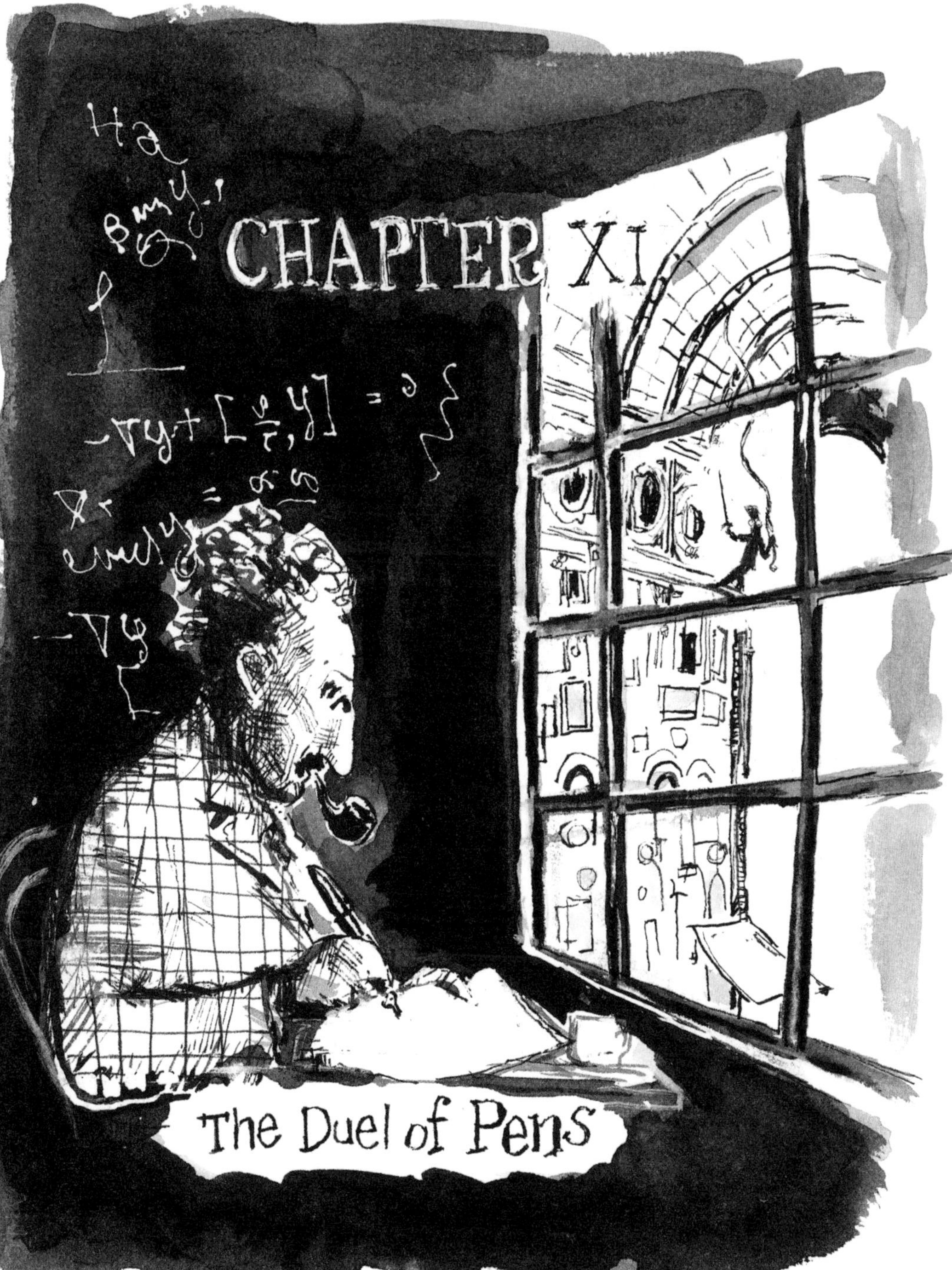
CHAPTER XI
The Duel of Pens

BOY, DID EHRENFEST EVER HAVE THAT BUSINESS
ABOUT THE DUEL OF THE PENS RIGHT!
AND SINCE ITS PRIZE IS NOTHING LESS THAN THE
SHAPE OF THE COSMOS, THE ARCHITECTURE OF
TIME, THE UNIFIED THEORY OF TOTALITY, AND,
YES, THAT MUCH-ABUSED CLICHÉ, THE MEANING
OF LIFE. ALLOW ME TO BUILD MY CASE:

THURS FEB. 29

AS SOON AS EHRENFAST'S TRAIN DEPARTS,
EINSTEIN POSTS A LETTER FROM THE TRAIN
STATION TO EDITOR WIEN, ASKING HIM TO
HOLD OFF RUNNING THE ARTICLE HE'D JUST
SENT, SAYING "I'M NOT 100% HAPPY WITH
IT." ABRAHAM HAS GIVEN HIM PAUSE.

SUN. MARCH 10

A WEEK AND A HALF OF INTENSE
WORK LATER, EINSTEIN SENDS A
TRIUMPHANT LETTER TO EHRENFEST.

"DEAR PAUL, I HAVE ANSWERED
ABRAHAM'S PROBLEMS, I HAVE
FOUND A PIECE OF THE TRUTH!"

NEVERTHELESS, THE VERY NEXT DAY,
EINSTEIN APPEARS TO PANIC, SENDING
TWO LETTERS TO WIEN, ONE IN THE
MORNING, AND THEN, RETRACTING IT
WITH ANOTHER IN THE AFTERNOON.

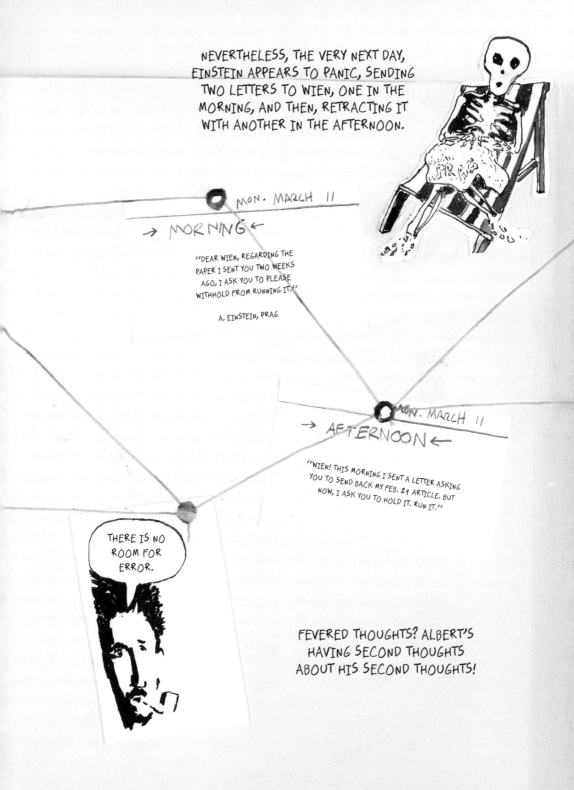

MON. MARCH 11

→ MORNING ←

"DEAR WIEN, REGARDING THE
PAPER I SENT YOU TWO WEEKS
AGO, I ASK YOU TO PLEASE
WITHHOLD FROM RUNNING IT."

A. EINSTEIN, PRAG

MON. MARCH 11

→ AFTERNOON ←

"WIEN! THIS MORNING I SENT A LETTER ASKING
YOU TO SEND BACK MY FEB. 24 ARTICLE. BUT
NOW, I ASK YOU TO HOLD IT. RUN IT."

THERE IS NO
ROOM FOR
ERROR.

FEVERED THOUGHTS? ALBERT'S
HAVING SECOND THOUGHTS
ABOUT HIS SECOND THOUGHTS!

I'M ONTO YOU, ALBERT,
I'M ONTO YOU.

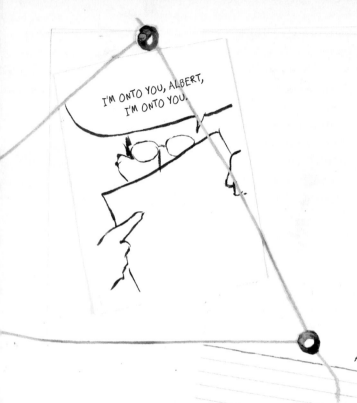

APRIL 5, 1912

AT LAST, STILL FEELING A NOTE OF TRIUMPH, BUT STILL MORE THAN A MONTH AWAY FROM PUBLICATION DAY FOR HIS "FUCK YOU, ABRAHAM" PAPER, EINSTEIN PENS THE SIX WORDS THAT WILL EVENTUALLY CHANGE IT ALL.

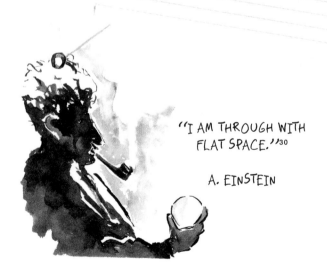

"I AM THROUGH WITH FLAT SPACE."[30]

A. EINSTEIN

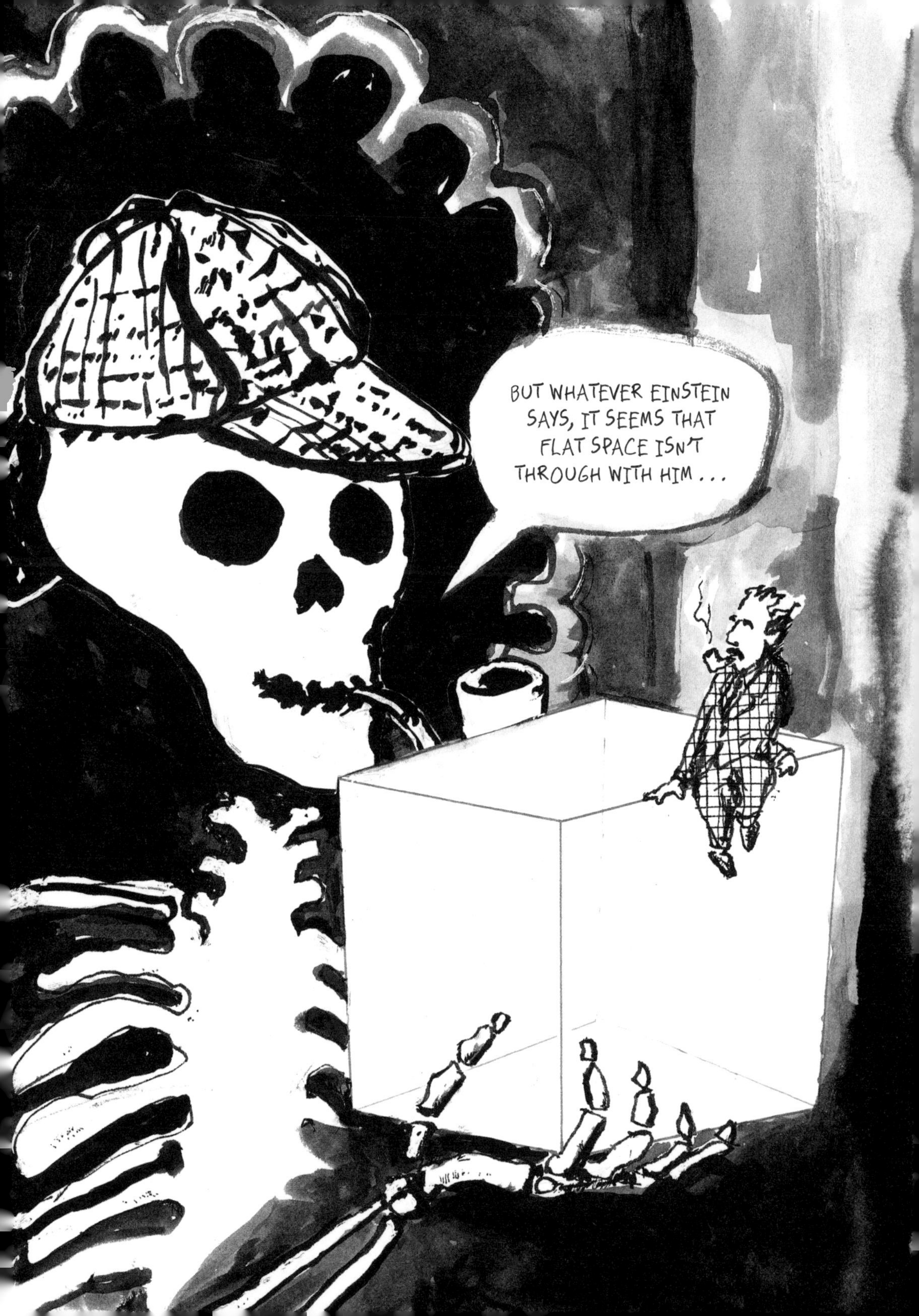
BUT WHATEVER EINSTEIN SAYS, IT SEEMS THAT FLAT SPACE ISN'T THROUGH WITH HIM . . .

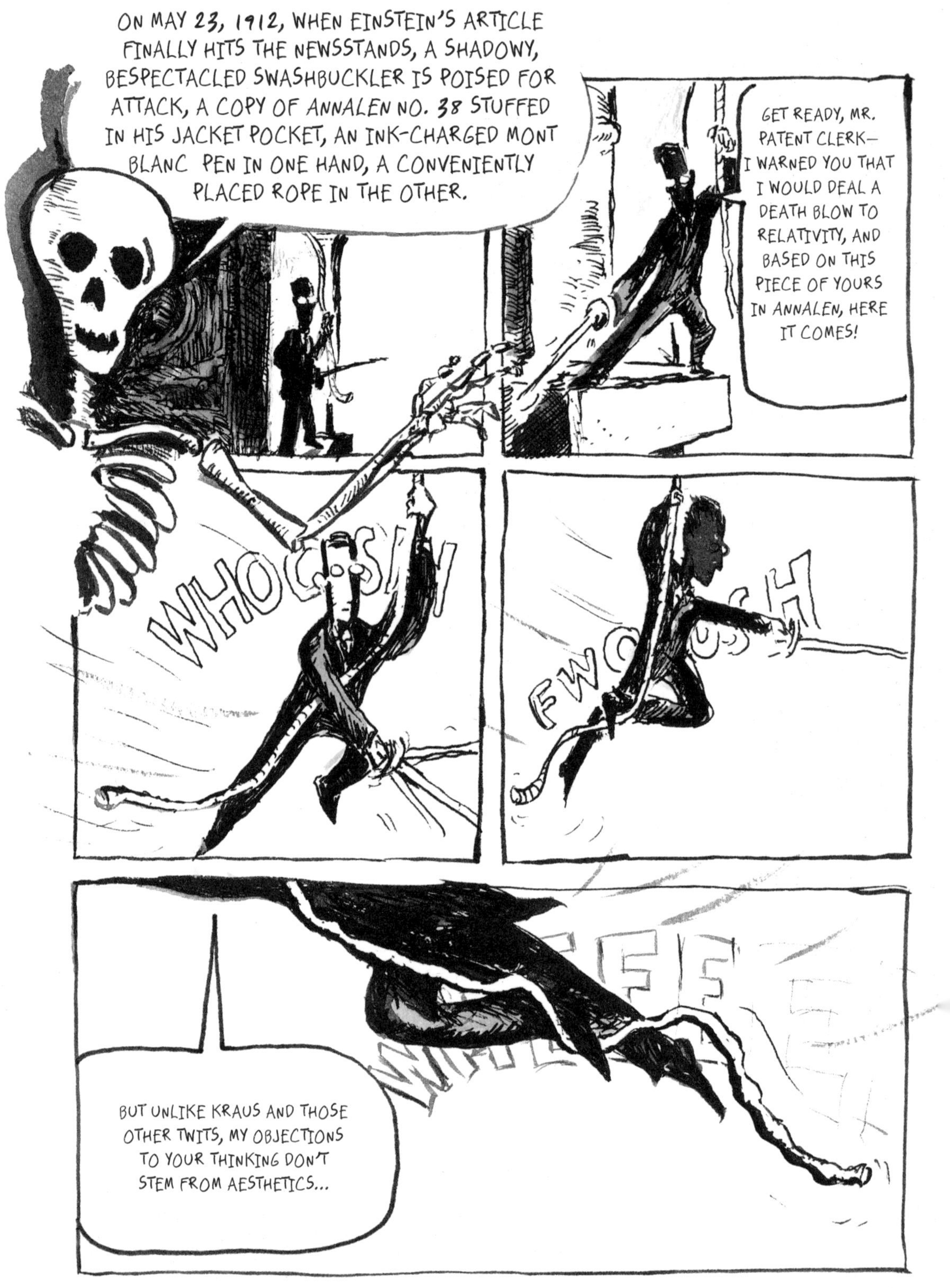
ON MAY 23, 1912, WHEN EINSTEIN'S ARTICLE FINALLY HITS THE NEWSSTANDS, A SHADOWY, BESPECTACLED SWASHBUCKLER IS POISED FOR ATTACK, A COPY OF ANNALEN NO. 38 STUFFED IN HIS JACKET POCKET, AN INK-CHARGED MONT BLANC PEN IN ONE HAND, A CONVENIENTLY PLACED ROPE IN THE OTHER.
GET READY, MR. PATENT CLERK— I WARNED YOU THAT I WOULD DEAL A DEATH BLOW TO RELATIVITY, AND BASED ON THIS PIECE OF YOURS IN ANNALEN, HERE IT COMES!
BUT UNLIKE KRAUS AND THOSE OTHER TWITS, MY OBJECTIONS TO YOUR THINKING DON'T STEM FROM AESTHETICS...

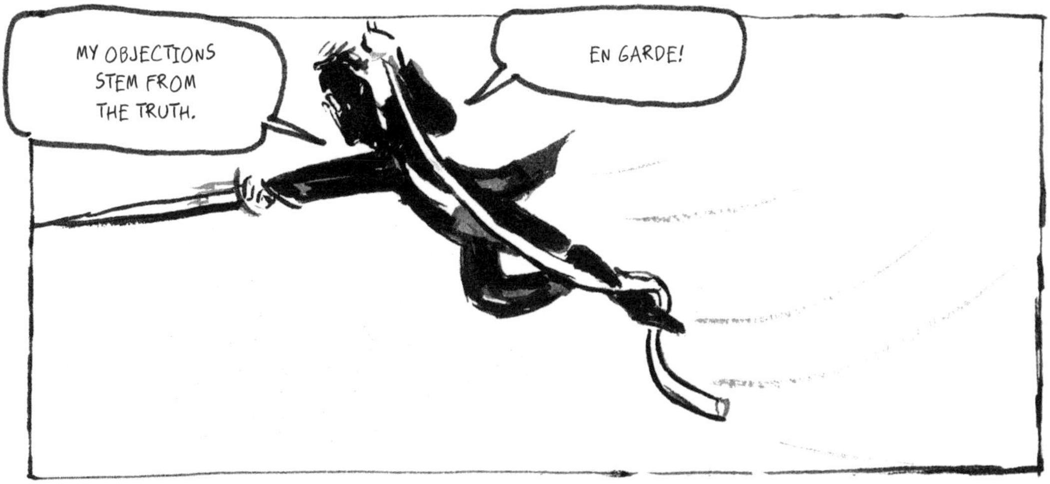
MY OBJECTIONS STEM FROM THE TRUTH.
EN GARDE!

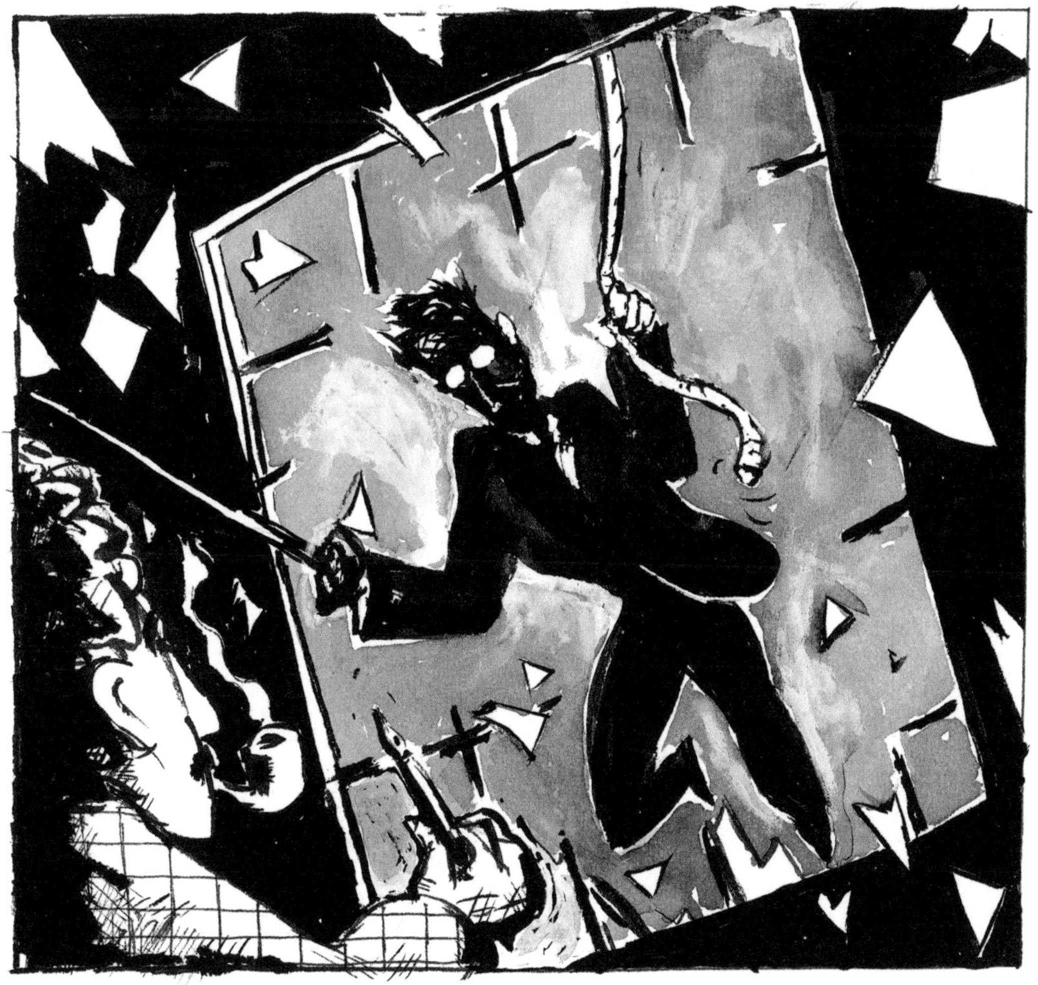

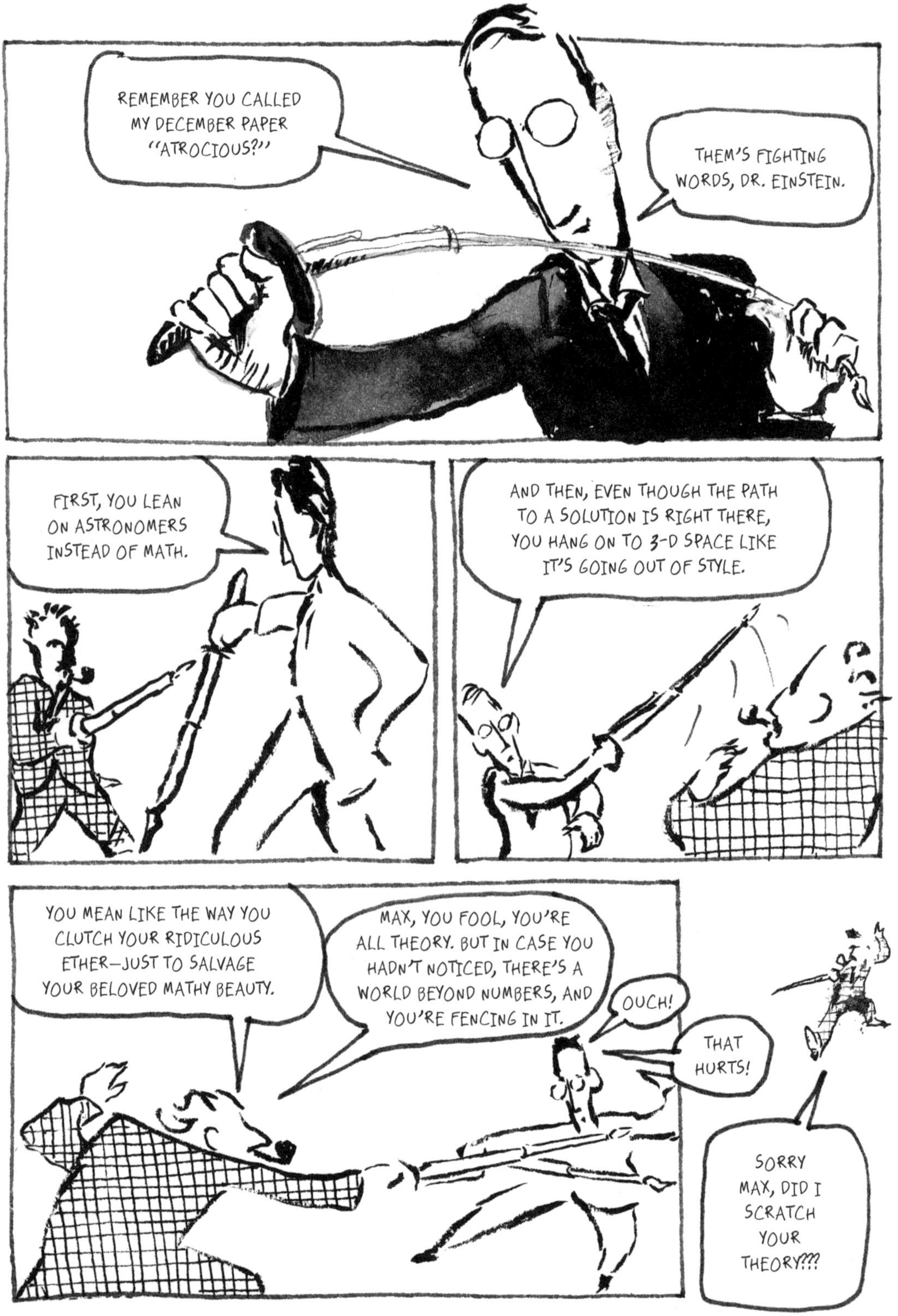
REMEMBER YOU CALLED MY DECEMBER PAPER "ATROCIOUS?"
THEM'S FIGHTING WORDS, DR. EINSTEIN.
FIRST, YOU LEAN ON ASTRONOMERS INSTEAD OF MATH.
AND THEN, EVEN THOUGH THE PATH TO A SOLUTION IS RIGHT THERE, YOU HANG ON TO 3-D SPACE LIKE IT'S GOING OUT OF STYLE.
YOU MEAN LIKE THE WAY YOU CLUTCH YOUR RIDICULOUS ETHER—JUST TO SALVAGE YOUR BELOVED MATHY BEAUTY.
MAX, YOU FOOL, YOU'RE ALL THEORY. BUT IN CASE YOU HADN'T NOTICED, THERE'S A WORLD BEYOND NUMBERS, AND YOU'RE FENCING IN IT.
OUCH!
THAT HURTS!
SORRY MAX, DID I SCRATCH YOUR THEORY???

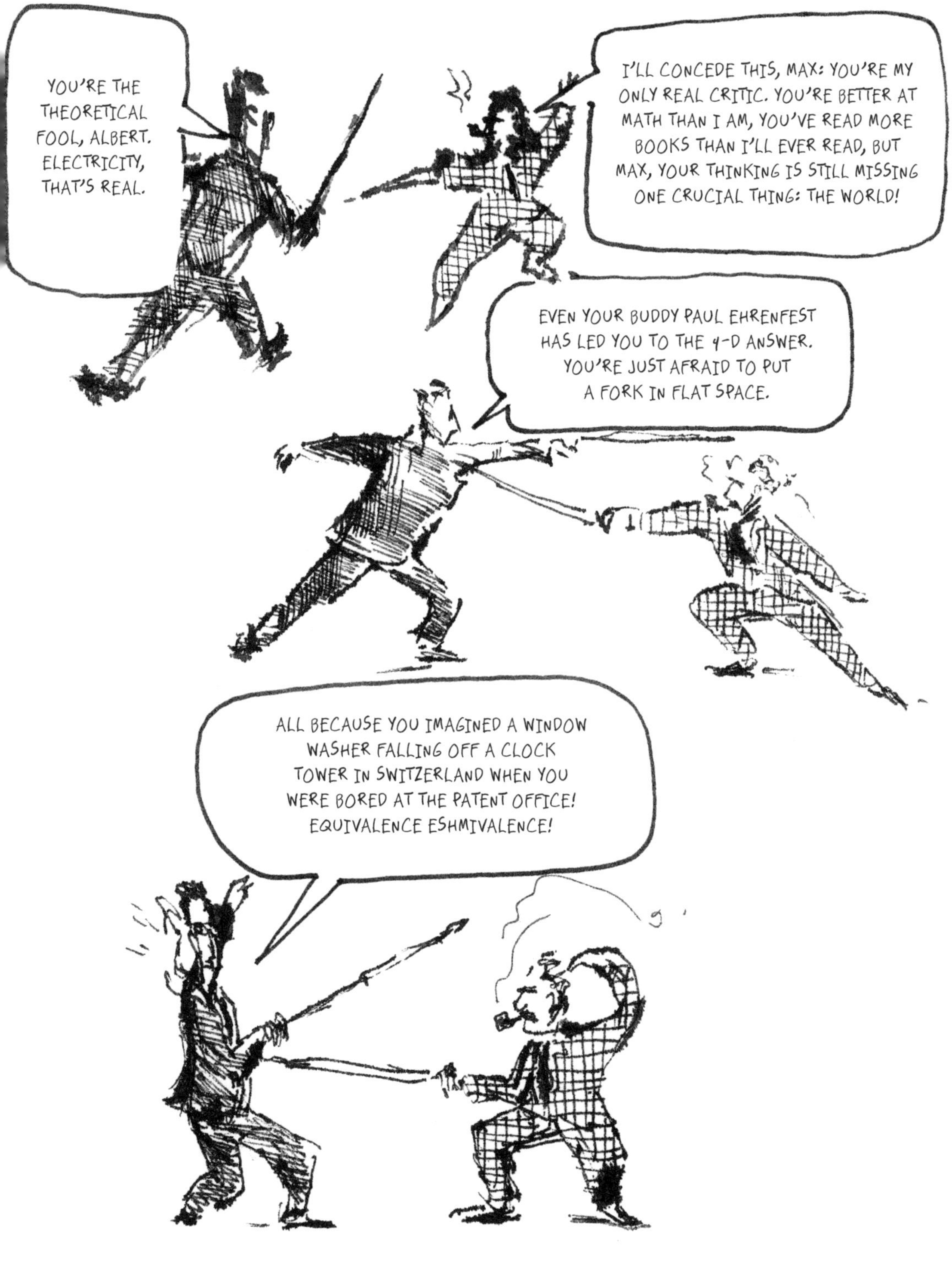
YOU'RE THE THEORETICAL FOOL, ALBERT. ELECTRICITY, THAT'S REAL.
I'LL CONCEDE THIS, MAX: YOU'RE MY ONLY REAL CRITIC. YOU'RE BETTER AT MATH THAN I AM, YOU'VE READ MORE BOOKS THAN I'LL EVER READ, BUT MAX, YOUR THINKING IS STILL MISSING ONE CRUCIAL THING: THE WORLD!
EVEN YOUR BUDDY PAUL EHRENFEST HAS LED YOU TO THE 4-D ANSWER. YOU'RE JUST AFRAID TO PUT A FORK IN FLAT SPACE.
ALL BECAUSE YOU IMAGINED A WINDOW WASHER FALLING OFF A CLOCK TOWER IN SWITZERLAND WHEN YOU WERE BORED AT THE PATENT OFFICE! EQUIVALENCE ESHMIVALENCE!

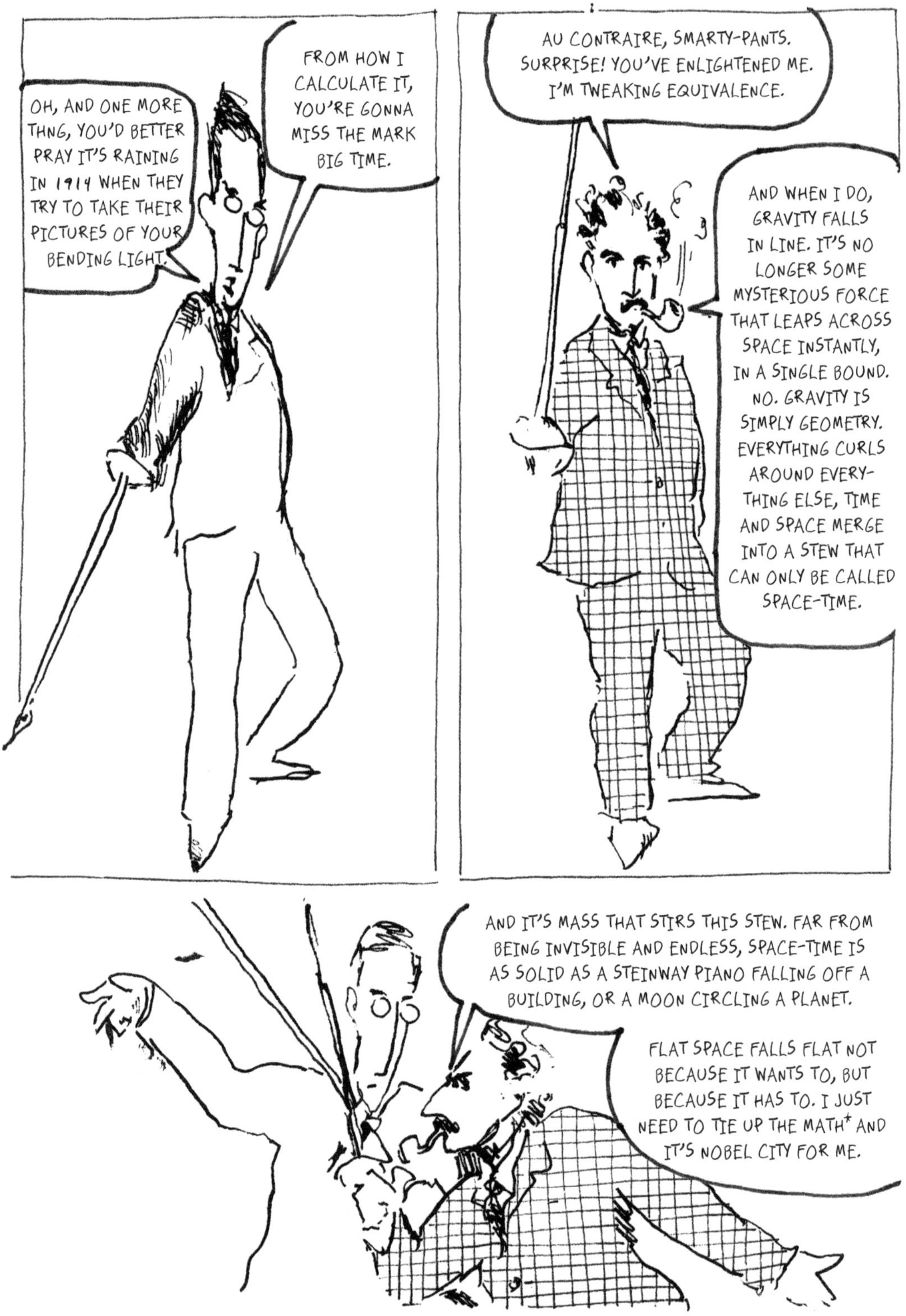
OH, AND ONE MORE THNG, YOU'D BETTER PRAY IT'S RAINING IN 1914 WHEN THEY TRY TO TAKE THEIR PICTURES OF YOUR BENDING LIGHT.
FROM HOW I CALCULATE IT, YOU'RE GONNA MISS THE MARK BIG TIME.
AU CONTRAIRE, SMARTY-PANTS. SURPRISE! YOU'VE ENLIGHTENED ME. I'M TWEAKING EQUIVALENCE.
AND WHEN I DO, GRAVITY FALLS IN LINE. IT'S NO LONGER SOME MYSTERIOUS FORCE THAT LEAPS ACROSS SPACE INSTANTLY, IN A SINGLE BOUND. NO. GRAVITY IS SIMPLY GEOMETRY. EVERYTHING CURLS AROUND EVERY-THING ELSE, TIME AND SPACE MERGE INTO A STEW THAT CAN ONLY BE CALLED SPACE-TIME.
AND IT'S MASS THAT STIRS THIS STEW. FAR FROM BEING INVISIBLE AND ENDLESS, SPACE-TIME IS AS SOLID AS A STEINWAY PIANO FALLING OFF A BUILDING, OR A MOON CIRCLING A PLANET.
FLAT SPACE FALLS FLAT NOT BECAUSE IT WANTS TO, BUT BECAUSE IT HAS TO. I JUST NEED TO TIE UP THE MATH* AND IT'S NOBEL CITY FOR ME.

* Easier said than done. Indeed, one of the key forces driving Einstein back to Zurich was to lean in to the non-Euclidean math chops of his old school pal Marcel Grossmann, a task that took nearly three arduous years.

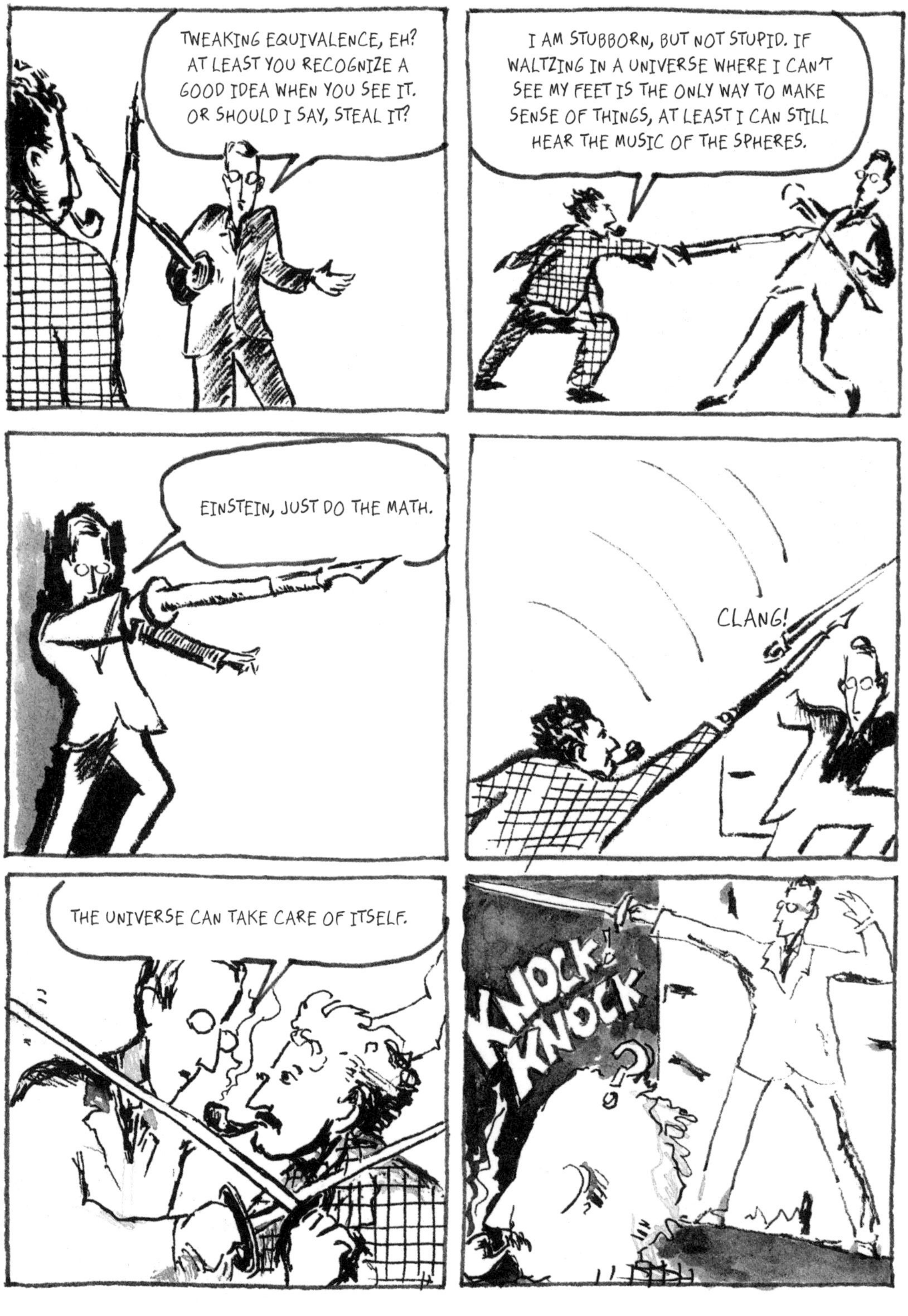
TWEAKING EQUIVALENCE, EH? AT LEAST YOU RECOGNIZE A GOOD IDEA WHEN YOU SEE IT. OR SHOULD I SAY, STEAL IT?
I AM STUBBORN, BUT NOT STUPID. IF WALTZING IN A UNIVERSE WHERE I CAN'T SEE MY FEET IS THE ONLY WAY TO MAKE SENSE OF THINGS, AT LEAST I CAN STILL HEAR THE MUSIC OF THE SPHERES.
EINSTEIN, JUST DO THE MATH.
CLANG!
THE UNIVERSE CAN TAKE CARE OF ITSELF.
KNOCK! KNOCK
?!

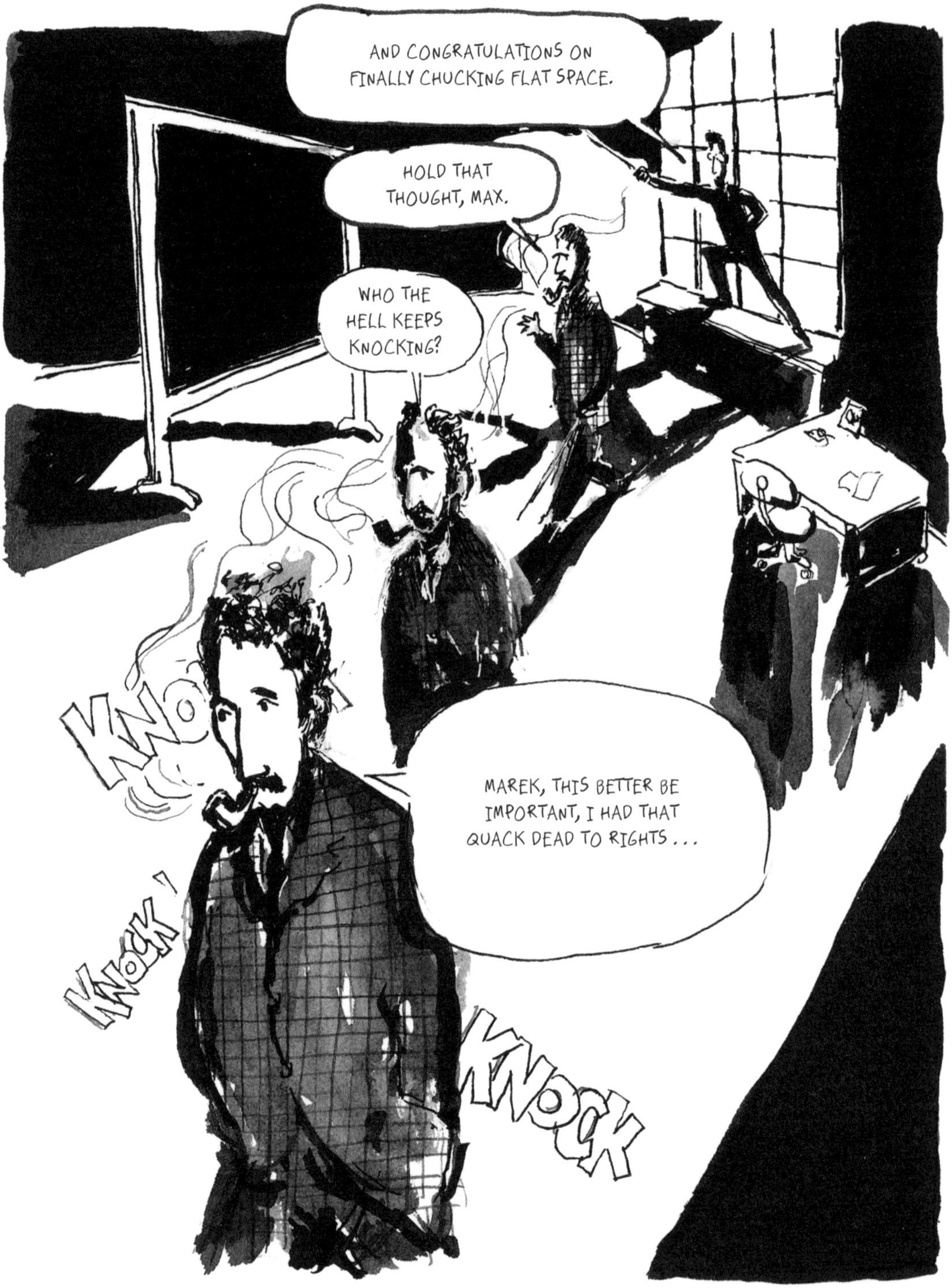
AND CONGRATULATIONS ON FINALLY CHUCKING FLAT SPACE.
HOLD THAT THOUGHT, MAX.
WHO THE HELL KEEPS KNOCKING?
KNOCK
MAREK, THIS BETTER BE IMPORTANT, I HAD THAT QUACK DEAD TO RIGHTS . . .
KNOCK
KNOCK

CHAPTER XII

Einstein

A Knock at the Door

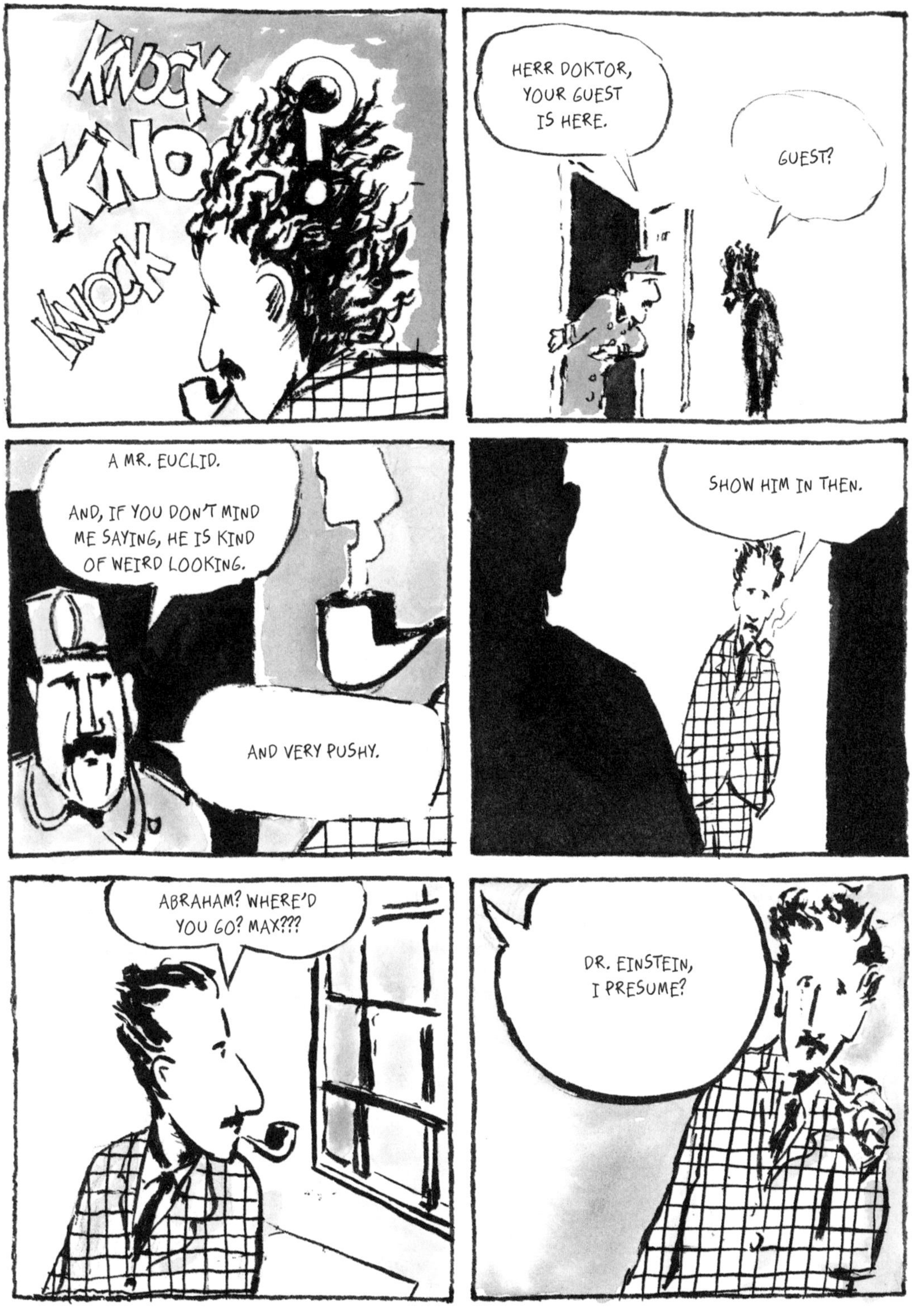
KNOCK
KNOCK
KNOCK
HERR DOKTOR, YOUR GUEST IS HERE.
GUEST?
A MR. EUCLID.
AND, IF YOU DON'T MIND ME SAYING, HE IS KIND OF WEIRD LOOKING.
AND VERY PUSHY.
SHOW HIM IN THEN.
ABRAHAM? WHERE'D YOU GO? MAX???
DR. EINSTEIN, I PRESUME?

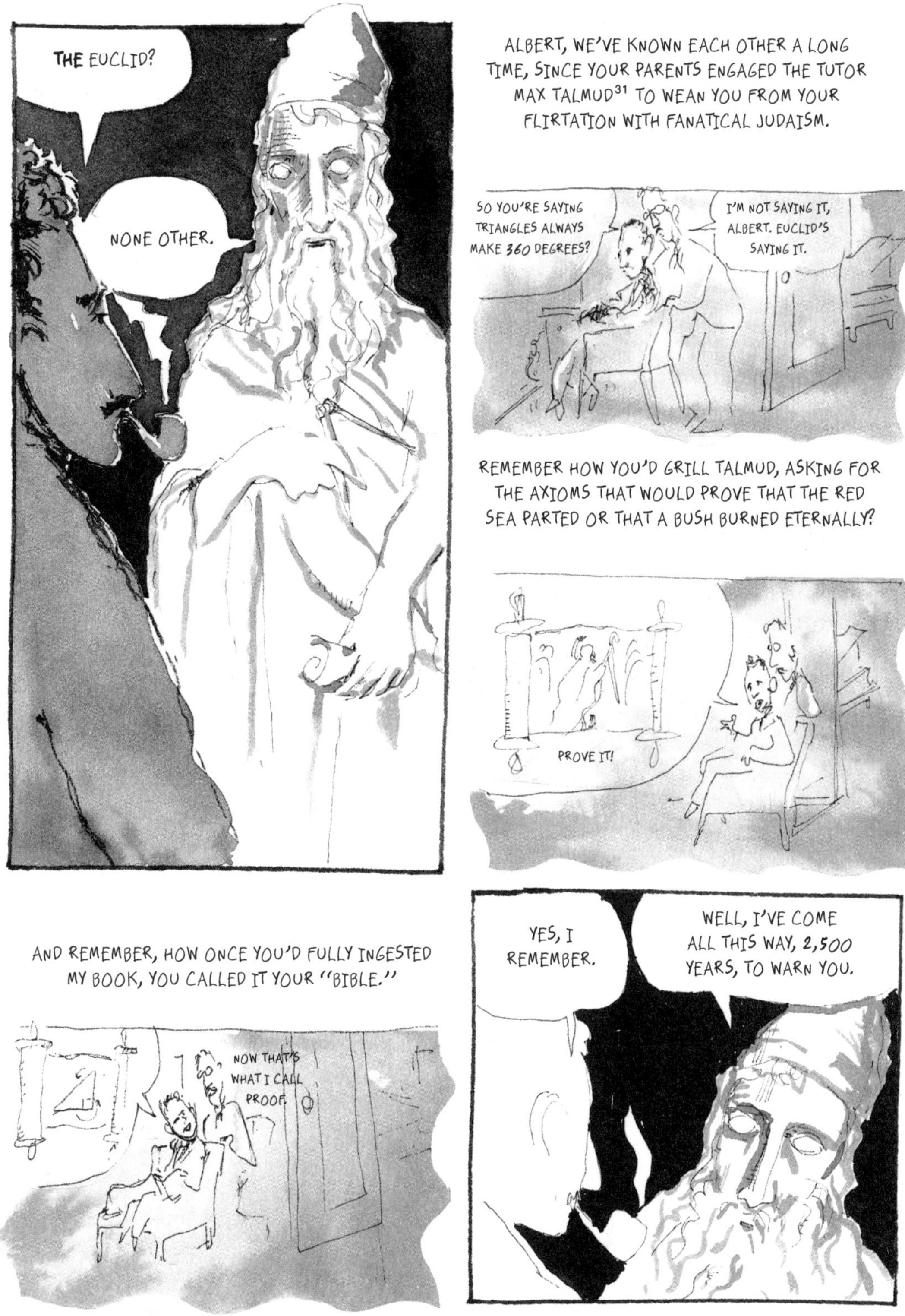
THE EUCLID?
NONE OTHER.
ALBERT, WE'VE KNOWN EACH OTHER A LONG TIME, SINCE YOUR PARENTS ENGAGED THE TUTOR MAX TALMUD[31] TO WEAN YOU FROM YOUR FLIRTATION WITH FANATICAL JUDAISM.
SO YOU'RE SAYING TRIANGLES ALWAYS MAKE 360 DEGREES?
I'M NOT SAYING IT, ALBERT. EUCLID'S SAYING IT.
REMEMBER HOW YOU'D GRILL TALMUD, ASKING FOR THE AXIOMS THAT WOULD PROVE THAT THE RED SEA PARTED OR THAT A BUSH BURNED ETERNALLY?
PROVE IT!
AND REMEMBER, HOW ONCE YOU'D FULLY INGESTED MY BOOK, YOU CALLED IT YOUR "BIBLE."
NOW THAT'S WHAT I CALL PROOF.
YES, I REMEMBER.
WELL, I'VE COME ALL THIS WAY, 2,500 YEARS, TO WARN YOU.

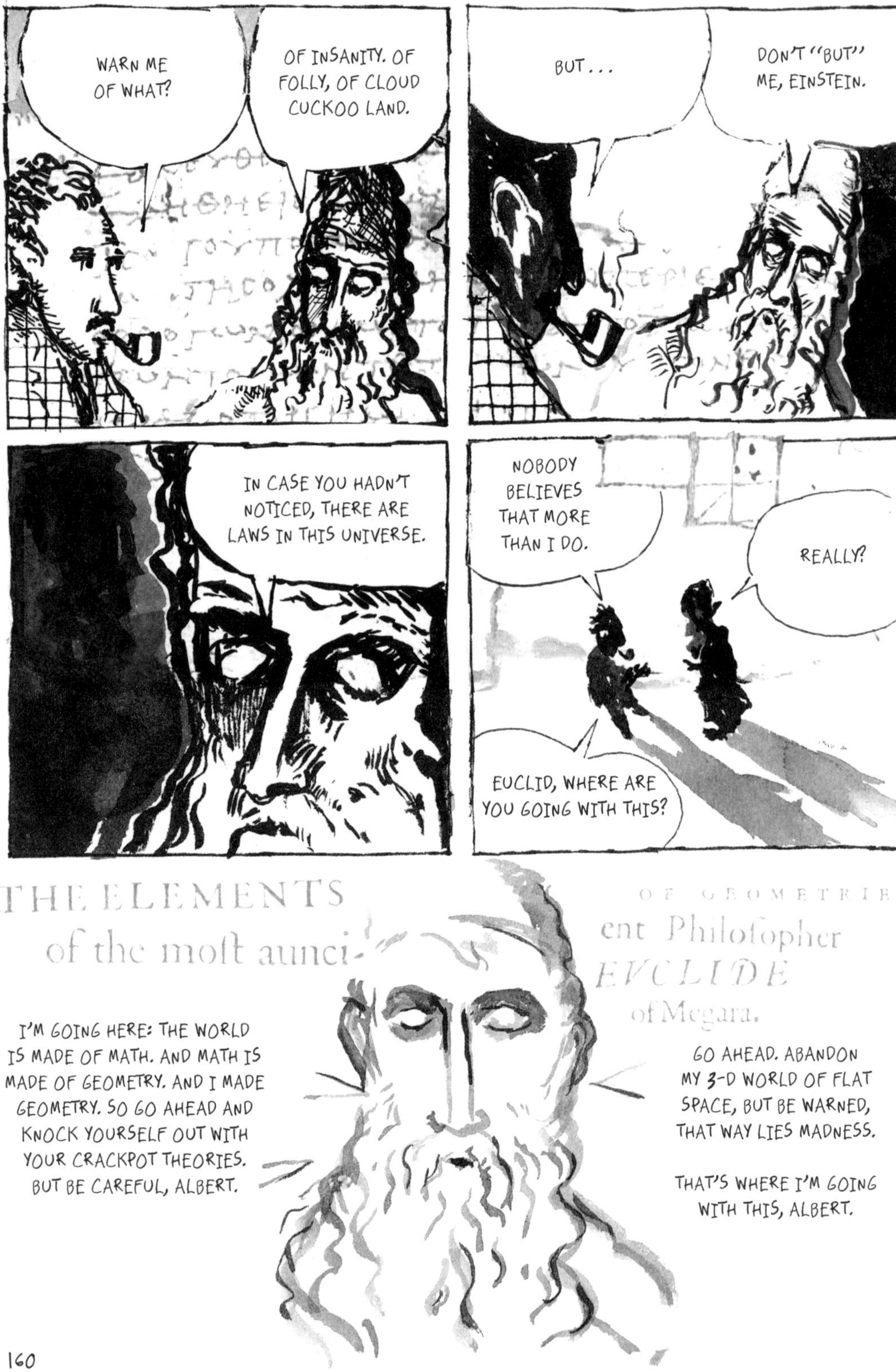
WARN ME OF WHAT?
OF INSANITY. OF FOLLY, OF CLOUD CUCKOO LAND.
BUT . . .
DON'T "BUT" ME, EINSTEIN.
IN CASE YOU HADN'T NOTICED, THERE ARE LAWS IN THIS UNIVERSE.
NOBODY BELIEVES THAT MORE THAN I DO.
REALLY?
EUCLID, WHERE ARE YOU GOING WITH THIS?
THE ELEMENTS OF GEOMETRIE of the moſt aunci-ent Philoſopher EVCLIDE of Megara.
I'M GOING HERE: THE WORLD IS MADE OF MATH. AND MATH IS MADE OF GEOMETRY. AND I MADE GEOMETRY. SO GO AHEAD AND KNOCK YOURSELF OUT WITH YOUR CRACKPOT THEORIES. BUT BE CAREFUL, ALBERT.
GO AHEAD. ABANDON MY 3-D WORLD OF FLAT SPACE, BUT BE WARNED, THAT WAY LIES MADNESS.
THAT'S WHERE I'M GOING WITH THIS, ALBERT.

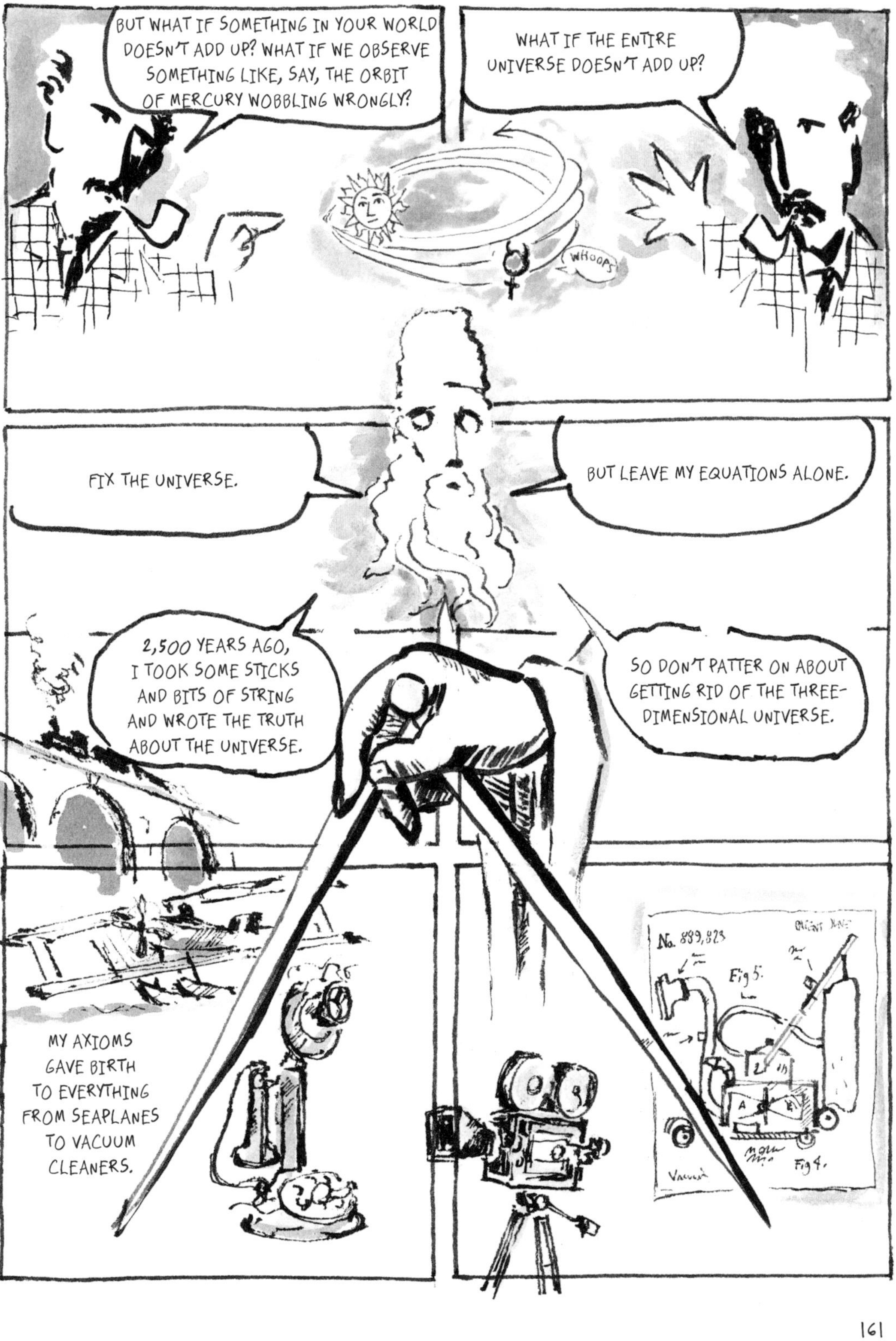

BUT WHAT IF SOMETHING IN YOUR WORLD DOESN'T ADD UP? WHAT IF WE OBSERVE SOMETHING LIKE, SAY, THE ORBIT OF MERCURY WOBBLING WRONGLY?
WHAT IF THE ENTIRE UNIVERSE DOESN'T ADD UP?
WHOOPS
FIX THE UNIVERSE.
BUT LEAVE MY EQUATIONS ALONE.
2,500 YEARS AGO, I TOOK SOME STICKS AND BITS OF STRING AND WROTE THE TRUTH ABOUT THE UNIVERSE.
SO DON'T PATTER ON ABOUT GETTING RID OF THE THREE-DIMENSIONAL UNIVERSE.
MY AXIOMS GAVE BIRTH TO EVERYTHING FROM SEAPLANES TO VACUUM CLEANERS.
No. 889,823
Fig 5.
Fig 4.

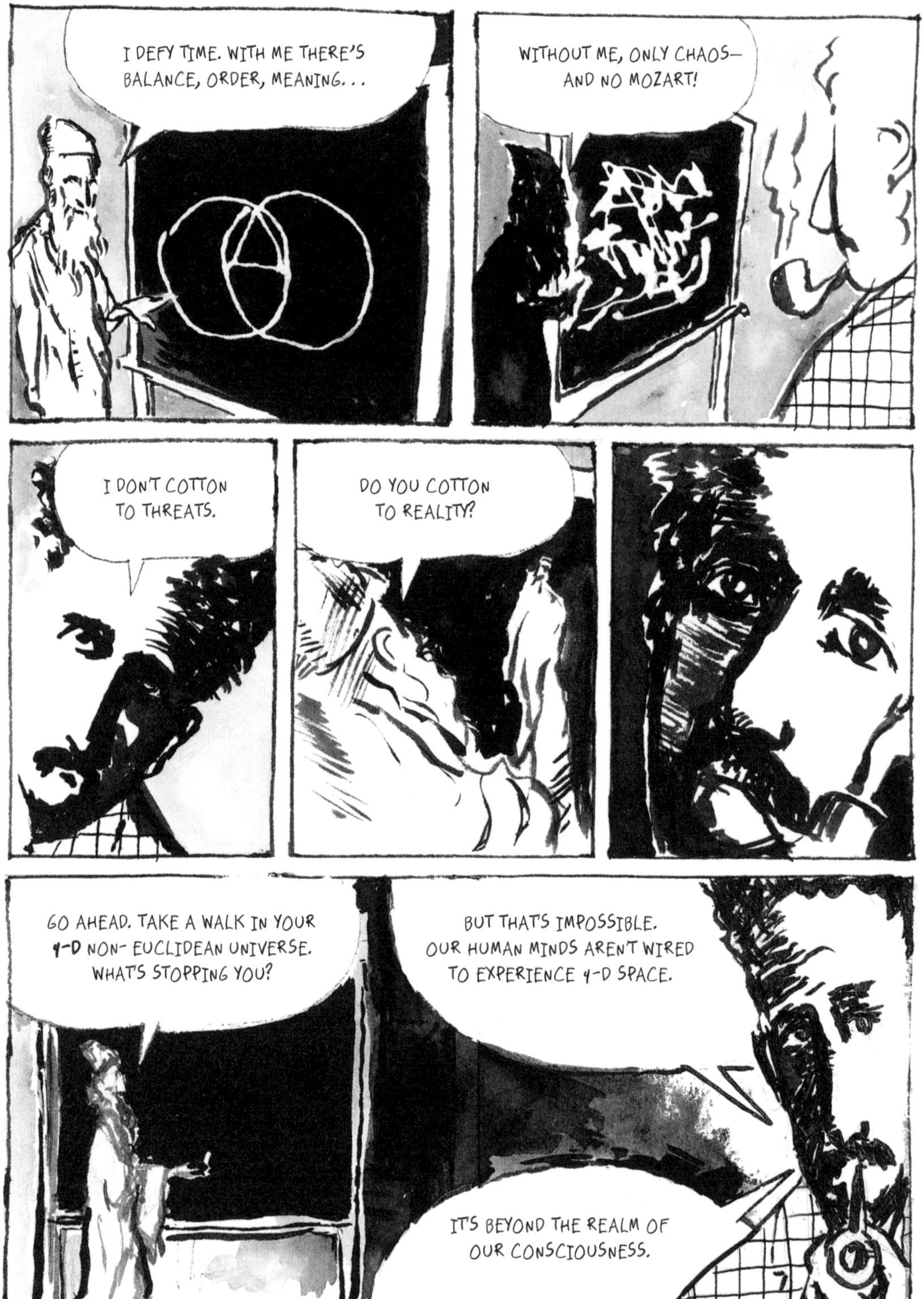
I DEFY TIME. WITH ME THERE'S BALANCE, ORDER, MEANING. . .
WITHOUT ME, ONLY CHAOS— AND NO MOZART!
I DON'T COTTON TO THREATS.
DO YOU COTTON TO REALITY?
GO AHEAD. TAKE A WALK IN YOUR 4-D NON-EUCLIDEAN UNIVERSE. WHAT'S STOPPING YOU?
BUT THAT'S IMPOSSIBLE. OUR HUMAN MINDS AREN'T WIRED TO EXPERIENCE 4-D SPACE.
IT'S BEYOND THE REALM OF OUR CONSCIOUSNESS.

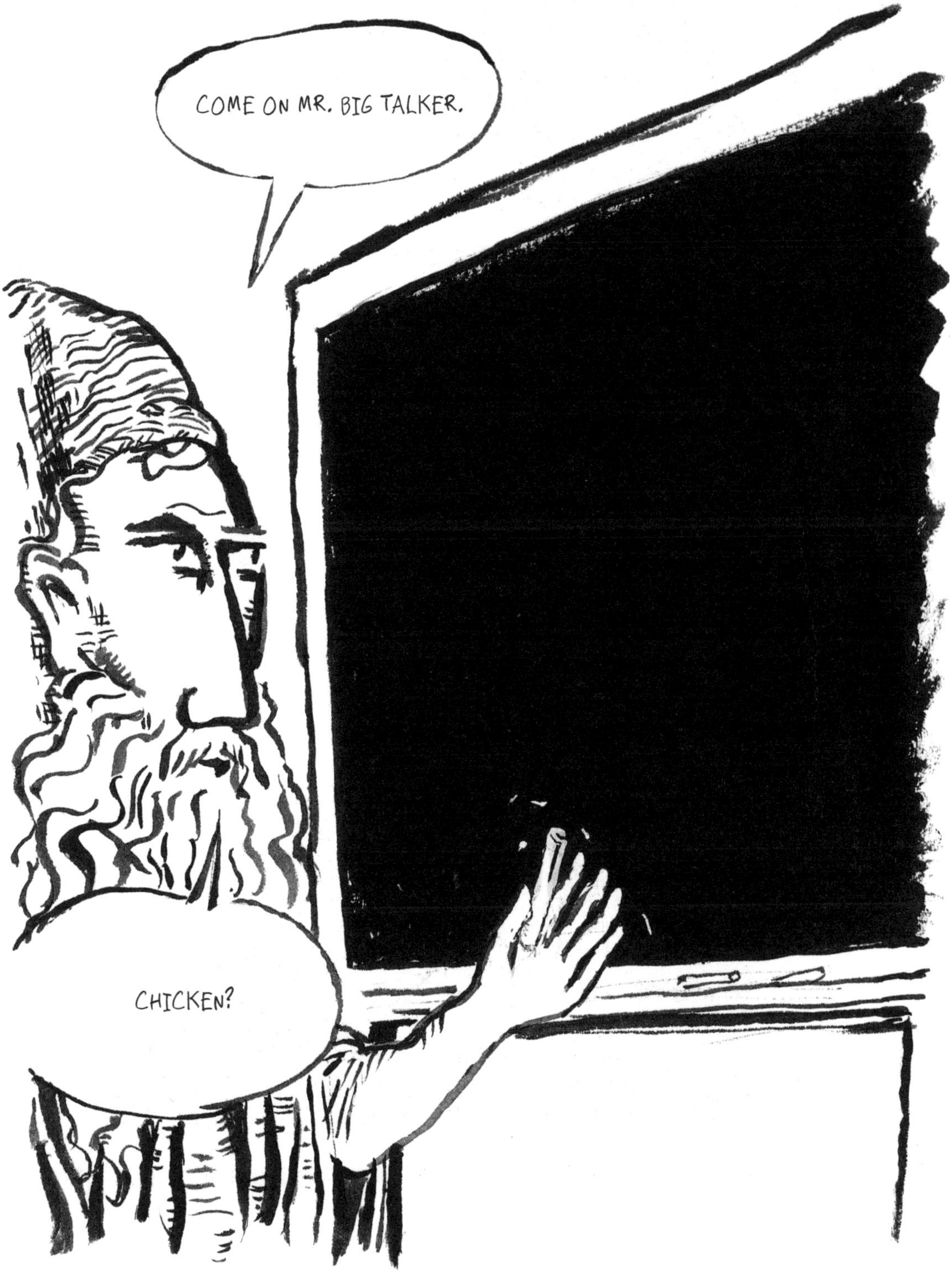
COME ON MR. BIG TALKER.
CHICKEN?

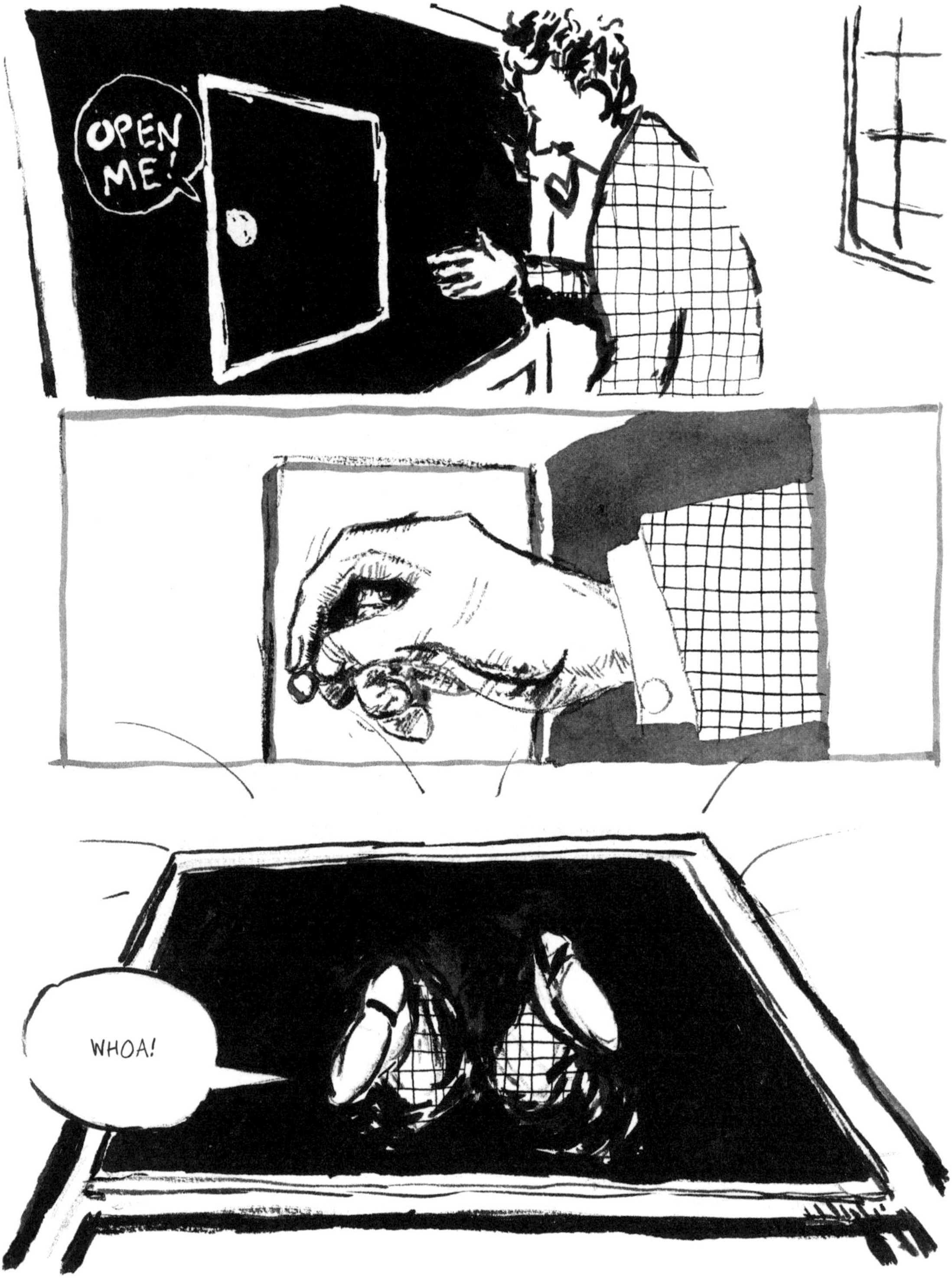
OPEN ME!
WHOA!

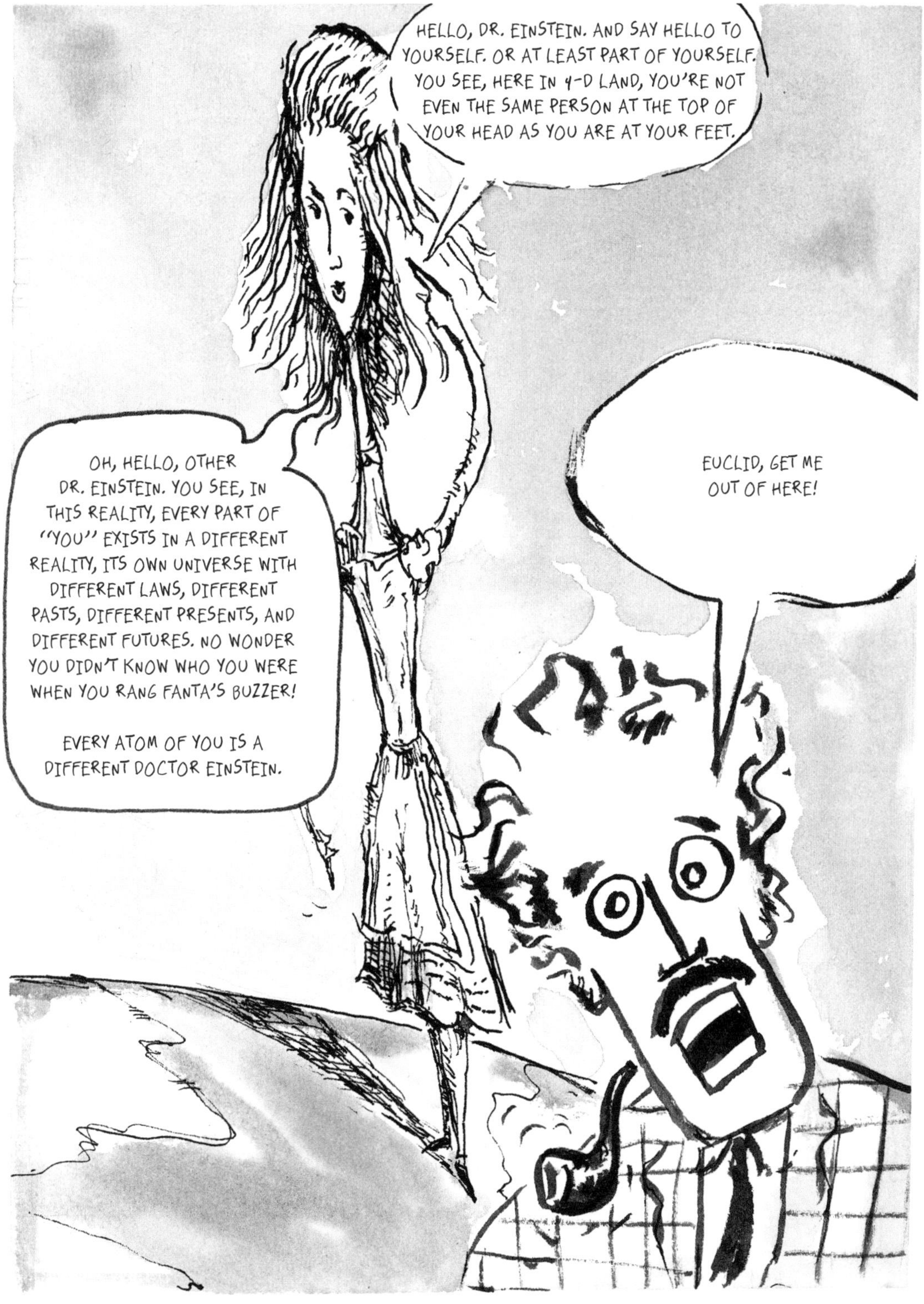

HELLO, DR. EINSTEIN. AND SAY HELLO TO YOURSELF. OR AT LEAST PART OF YOURSELF. YOU SEE, HERE IN 4-D LAND, YOU'RE NOT EVEN THE SAME PERSON AT THE TOP OF YOUR HEAD AS YOU ARE AT YOUR FEET.
OH, HELLO, OTHER DR. EINSTEIN. YOU SEE, IN THIS REALITY, EVERY PART OF "YOU" EXISTS IN A DIFFERENT REALITY, ITS OWN UNIVERSE WITH DIFFERENT LAWS, DIFFERENT PASTS, DIFFERENT PRESENTS, AND DIFFERENT FUTURES. NO WONDER YOU DIDN'T KNOW WHO YOU WERE WHEN YOU RANG FANTA'S BUZZER!
EVERY ATOM OF YOU IS A DIFFERENT DOCTOR EINSTEIN.
EUCLID, GET ME OUT OF HERE!

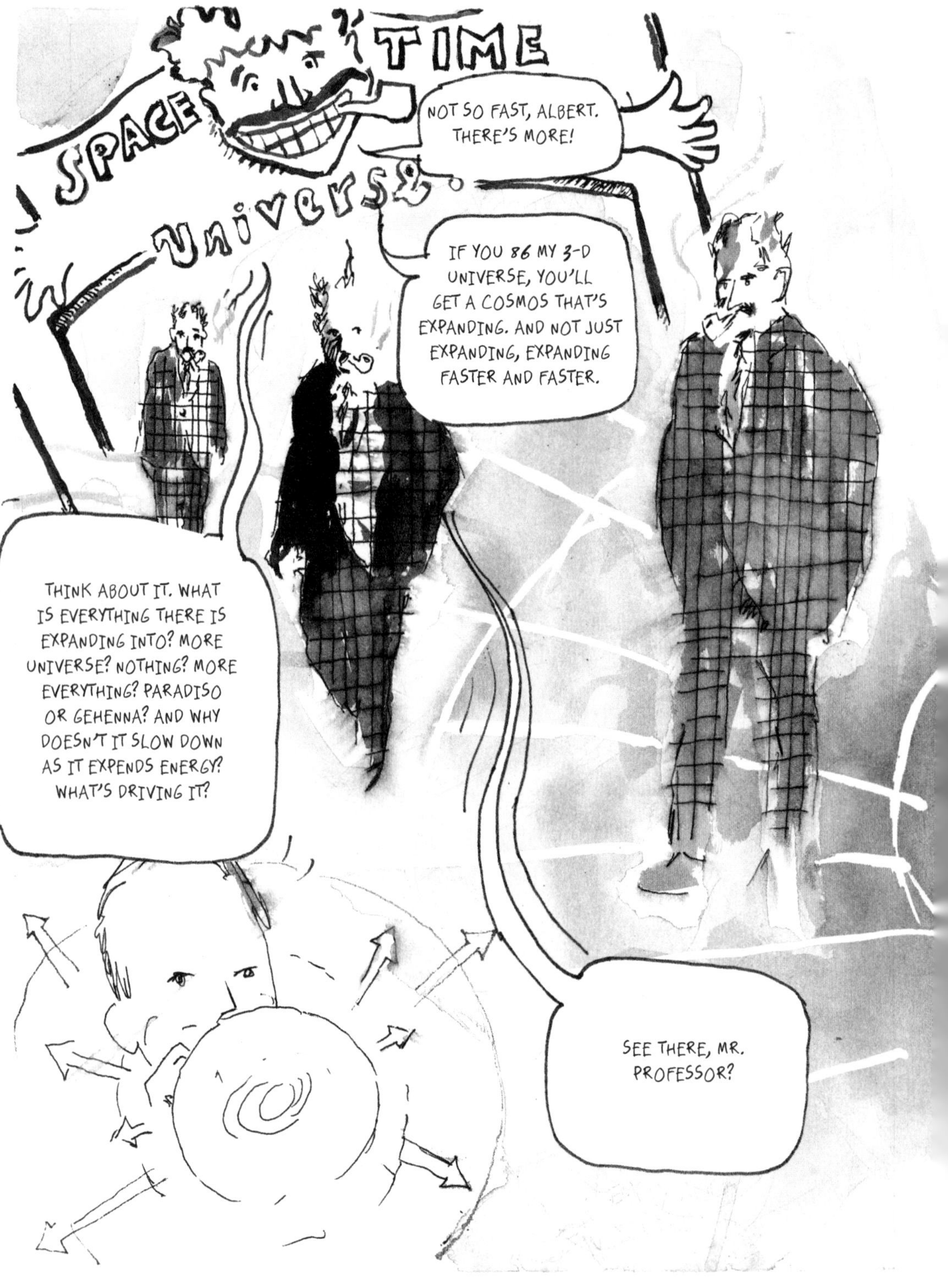
SPACE
TIME
UNIVERSE
NOT SO FAST, ALBERT. THERE'S MORE!
IF YOU 86 MY 3-D UNIVERSE, YOU'LL GET A COSMOS THAT'S EXPANDING. AND NOT JUST EXPANDING, EXPANDING FASTER AND FASTER.
THINK ABOUT IT. WHAT IS EVERYTHING THERE IS EXPANDING INTO? MORE UNIVERSE? NOTHING? MORE EVERYTHING? PARADISO OR GEHENNA? AND WHY DOESN'T IT SLOW DOWN AS IT EXPENDS ENERGY? WHAT'S DRIVING IT?
SEE THERE, MR. PROFESSOR?

SEE THERE?

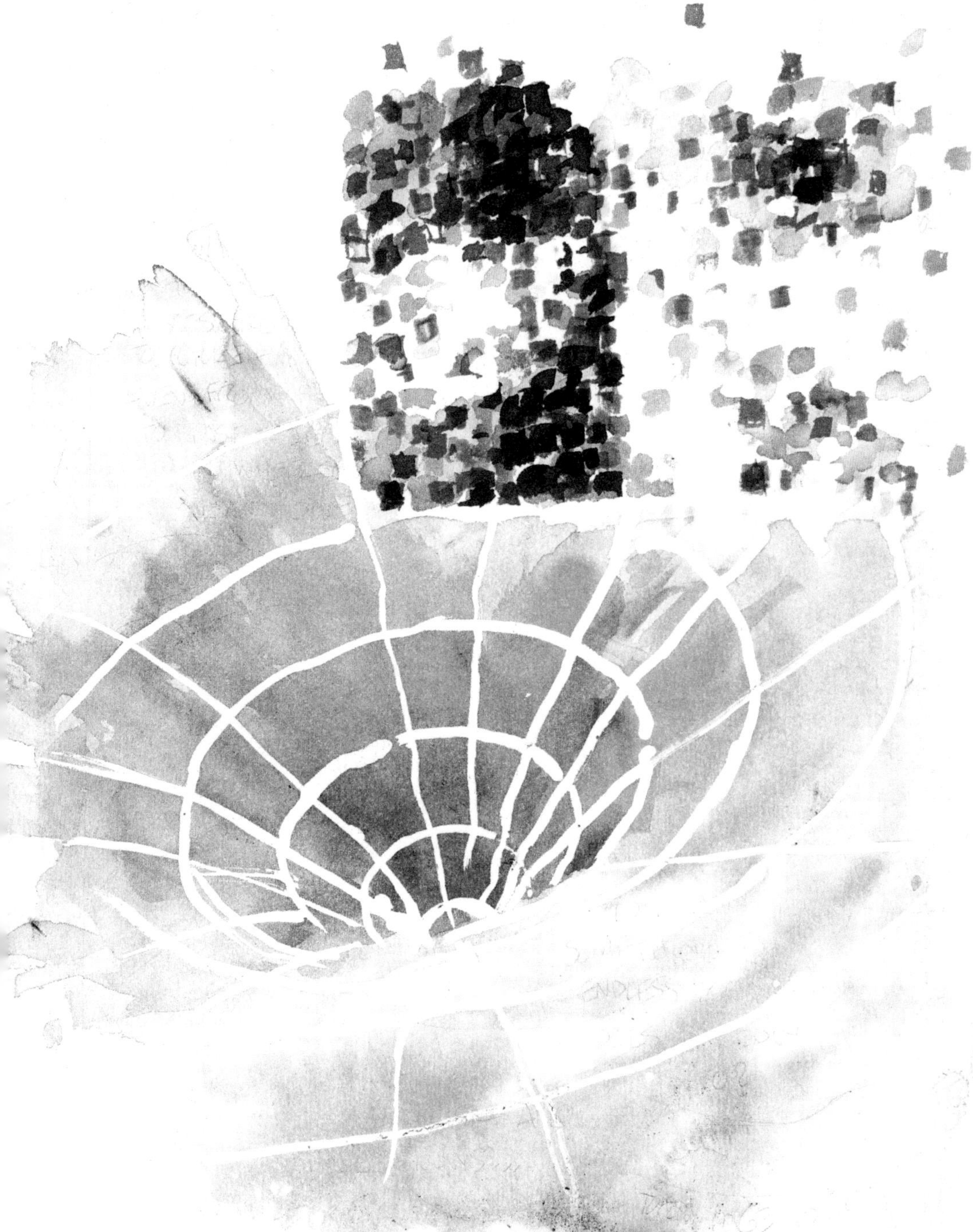

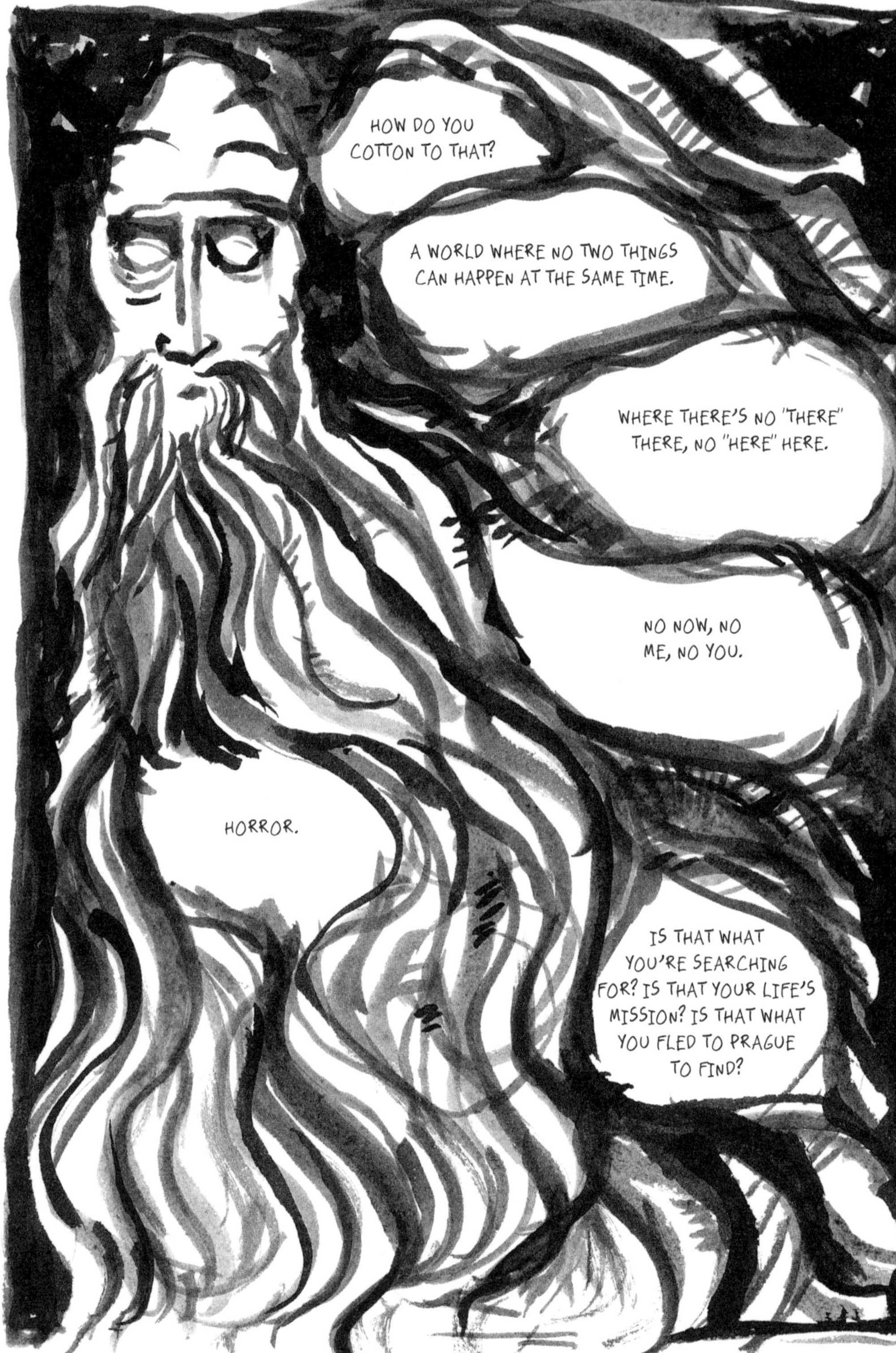
HOW DO YOU COTTON TO THAT?
A WORLD WHERE NO TWO THINGS CAN HAPPEN AT THE SAME TIME.
WHERE THERE'S NO "THERE" THERE, NO "HERE" HERE.
NO NOW, NO ME, NO YOU.
HORROR.
IS THAT WHAT YOU'RE SEARCHING FOR? IS THAT YOUR LIFE'S MISSION? IS THAT WHAT YOU FLED TO PRAGUE TO FIND?

CHAPTER XIII

wtf

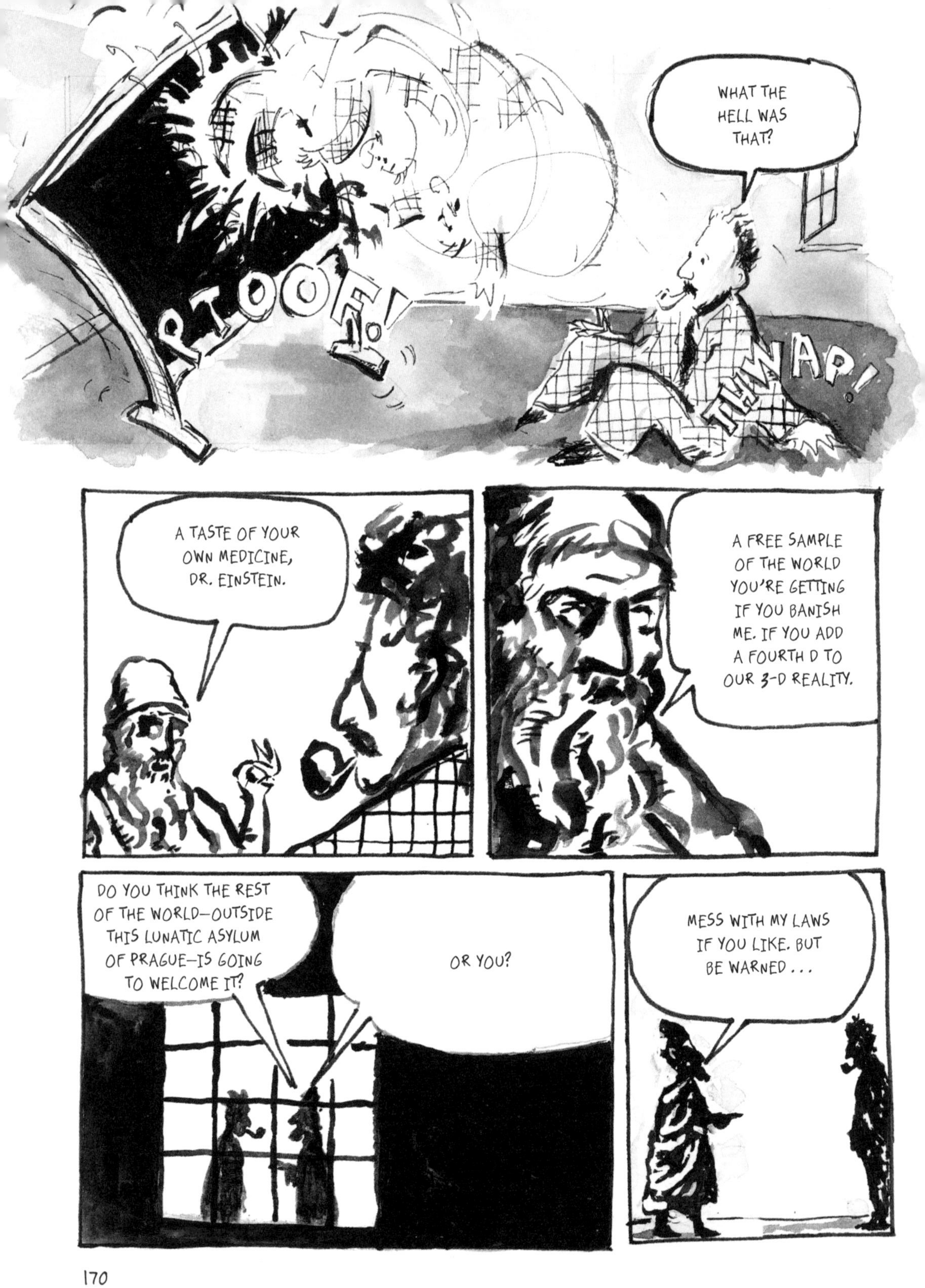
PTOOF!
WHAT THE HELL WAS THAT?
THWAP!
A TASTE OF YOUR OWN MEDICINE, DR. EINSTEIN.
A FREE SAMPLE OF THE WORLD YOU'RE GETTING IF YOU BANISH ME. IF YOU ADD A FOURTH D TO OUR 3-D REALITY.
DO YOU THINK THE REST OF THE WORLD—OUTSIDE THIS LUNATIC ASYLUM OF PRAGUE—IS GOING TO WELCOME IT?
OR YOU?
MESS WITH MY LAWS IF YOU LIKE. BUT BE WARNED . . .

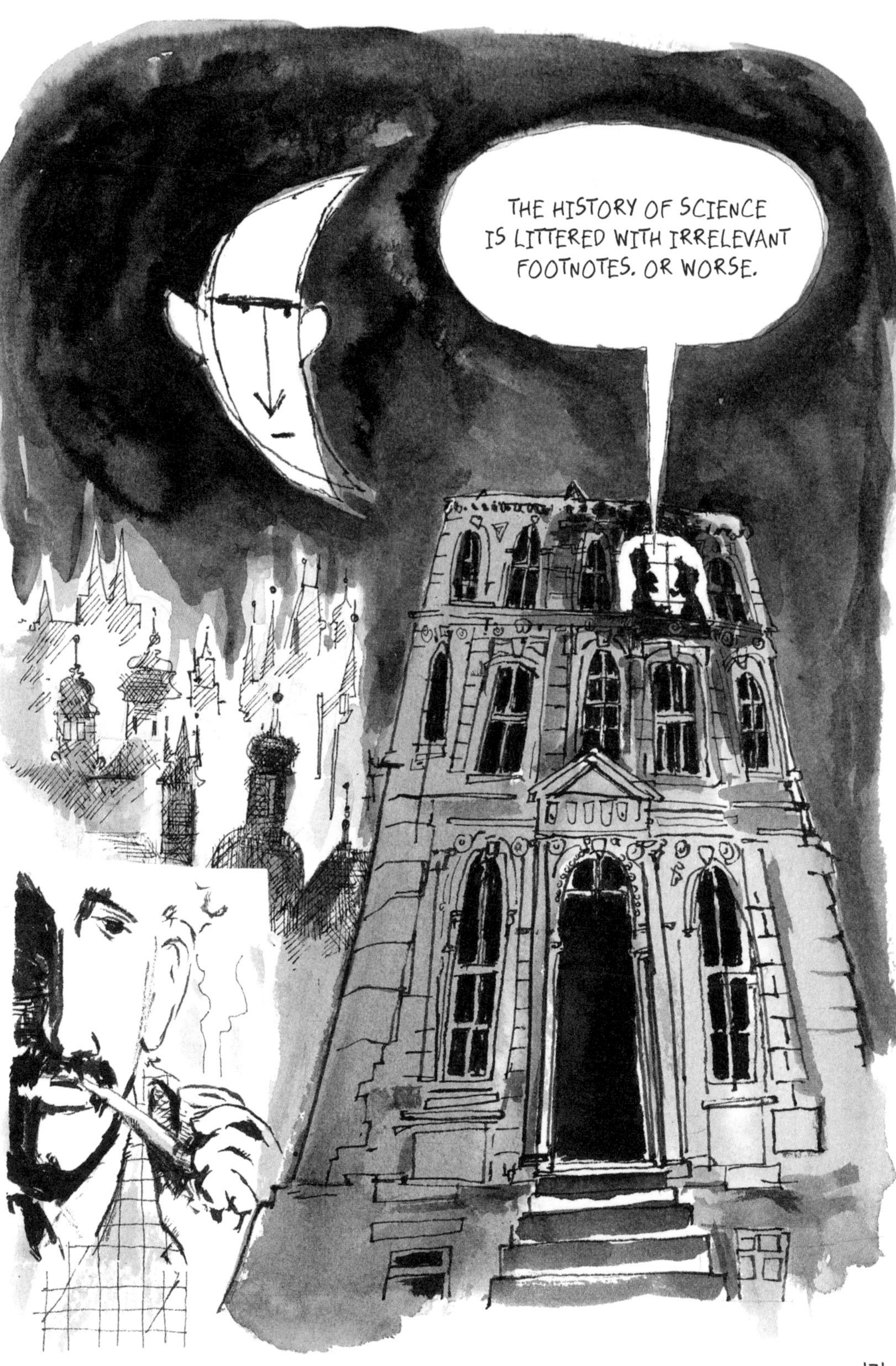
THE HISTORY OF SCIENCE IS LITTERED WITH IRRELEVANT FOOTNOTES. OR WORSE.

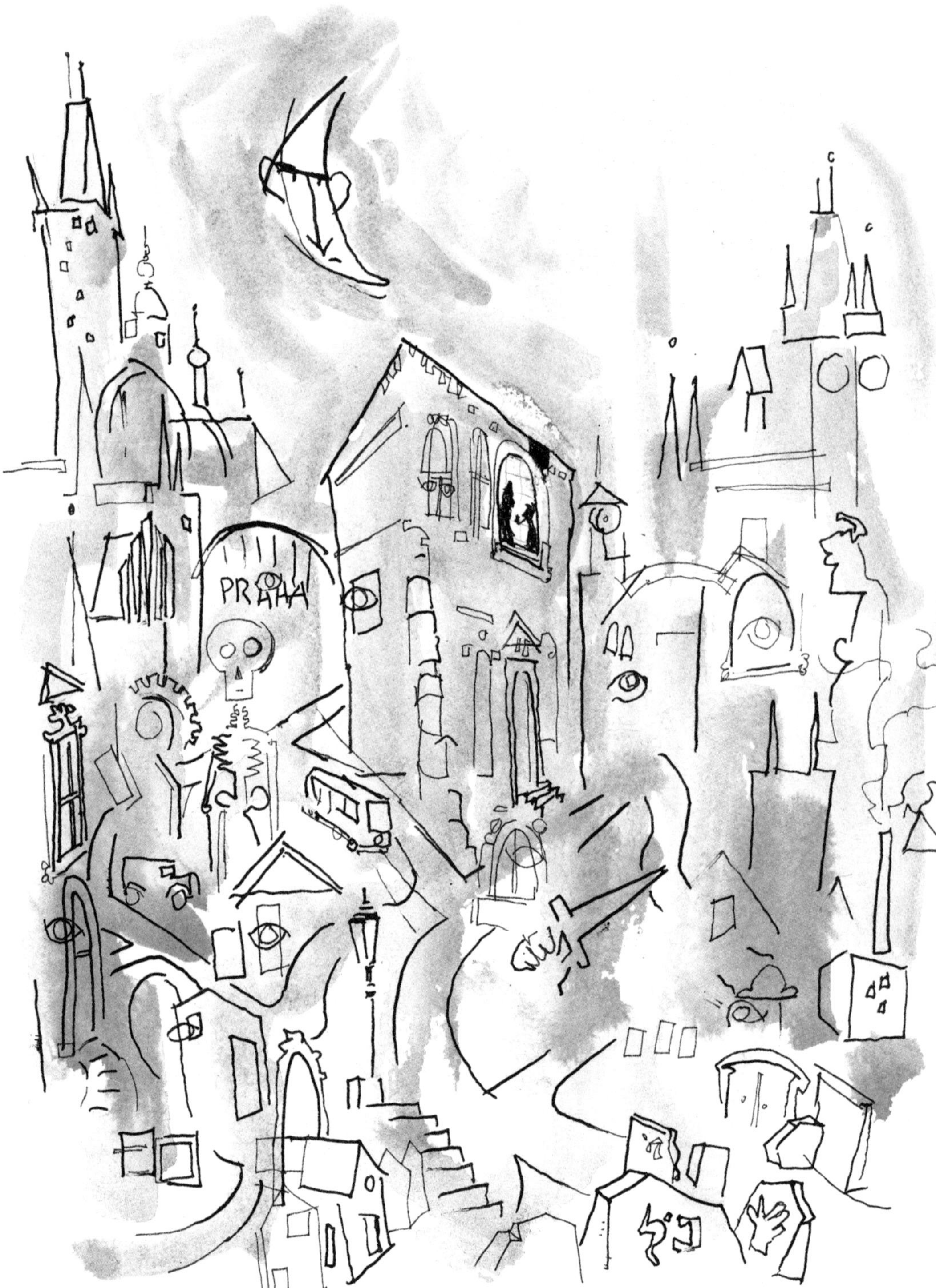
PRAHA

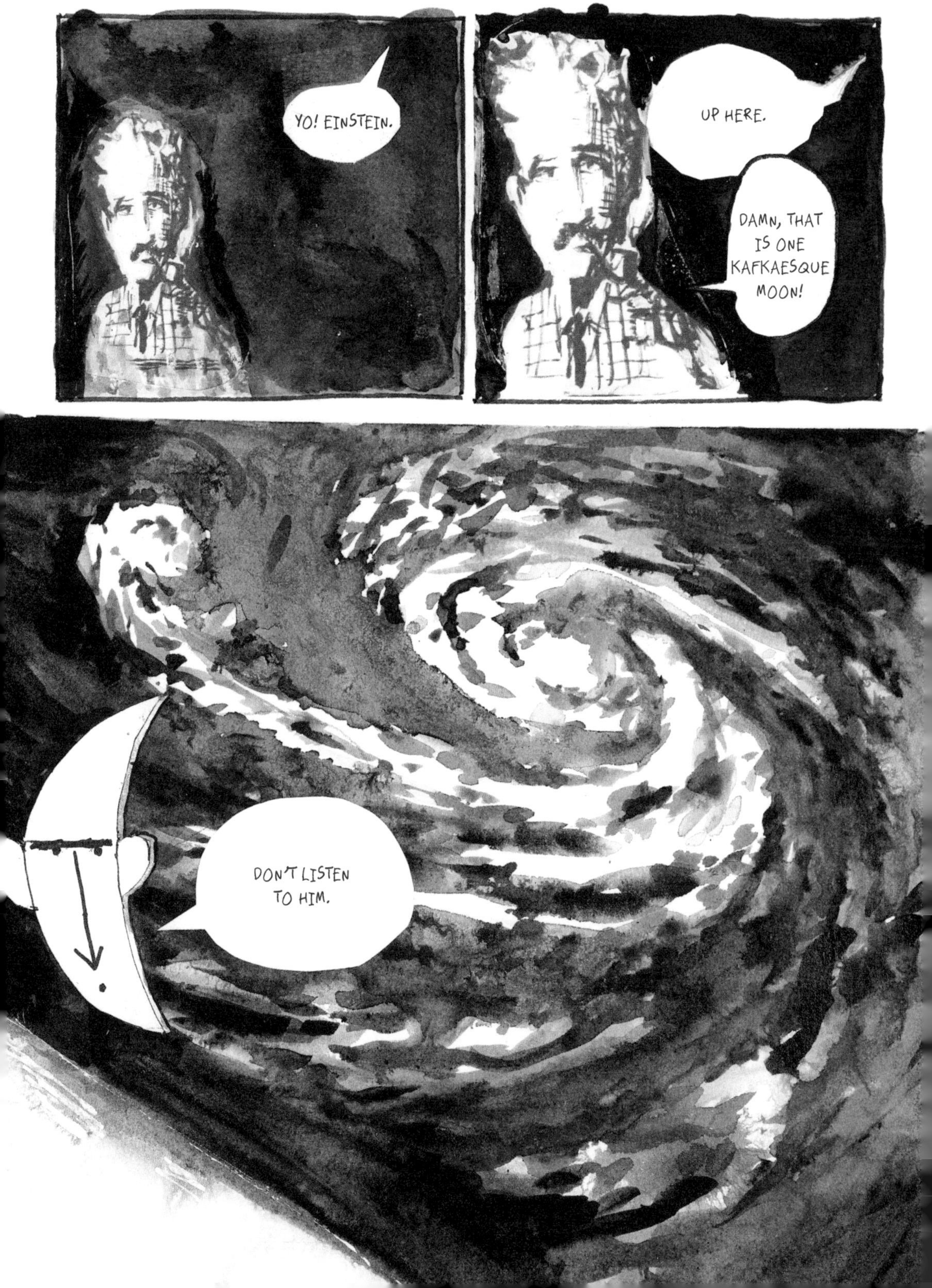
YO! EINSTEIN.
UP HERE.
DAMN, THAT IS ONE KAFKAESQUE MOON!
DON'T LISTEN TO HIM.

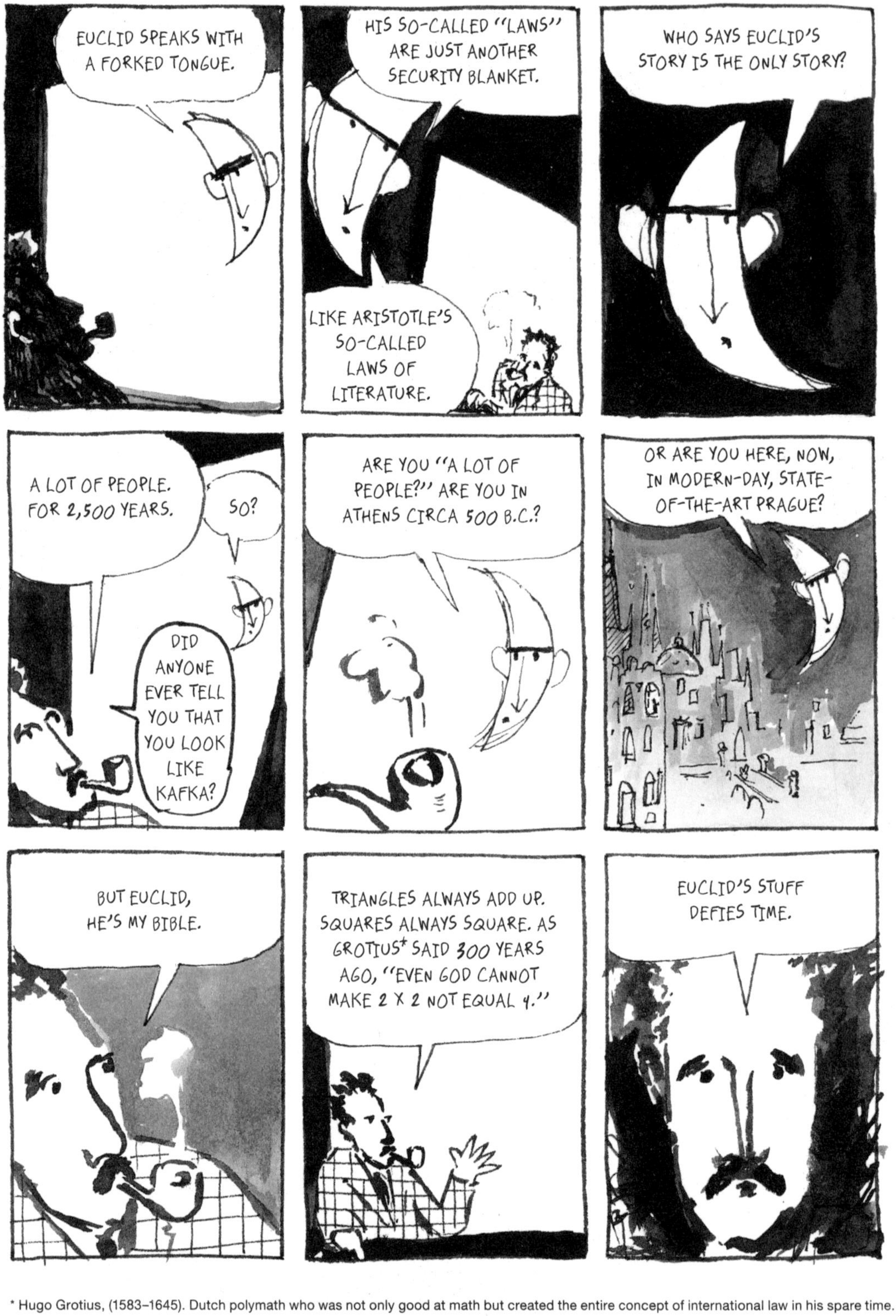
EUCLID SPEAKS WITH A FORKED TONGUE.
HIS SO-CALLED "LAWS" ARE JUST ANOTHER SECURITY BLANKET.
LIKE ARISTOTLE'S SO-CALLED LAWS OF LITERATURE.
WHO SAYS EUCLID'S STORY IS THE ONLY STORY?
A LOT OF PEOPLE. FOR 2,500 YEARS.
SO?
DID ANYONE EVER TELL YOU THAT YOU LOOK LIKE KAFKA?
ARE YOU "A LOT OF PEOPLE?" ARE YOU IN ATHENS CIRCA 500 B.C.?
OR ARE YOU HERE, NOW, IN MODERN-DAY, STATE-OF-THE-ART PRAGUE?
BUT EUCLID, HE'S MY BIBLE.
TRIANGLES ALWAYS ADD UP. SQUARES ALWAYS SQUARE. AS GROTIUS* SAID 300 YEARS AGO, "EVEN GOD CANNOT MAKE 2 X 2 NOT EQUAL 4."
EUCLID'S STUFF DEFIES TIME.

* Hugo Grotius, (1583–1645). Dutch polymath who was not only good at math but created the entire concept of international law in his spare time.

IT'S ETERNAL.

FUNNY.

THAT'S EXACTLY WHAT ARISTOTLE SAID ABOUT HIS STUFF JUST BEFORE I KILLED HIM.

YOU KILLED ARISTOTLE?

I KILLED ARISTOTLE.

HOW?

SIMPLE. I WROTE HIM OUT OF THE EQUATION.

REMEMBER THAT NEWS STORY LAST SUMMER ABOUT THE "LONELY INDIAN" WHO WANDERED OUT OF THE WILDERNESS IN CALIFORNIA?

IT RINGS A BELL.

I COULDN'T GET IT OUT OF MY HEAD. IT SPUN ROUND AND ROUND AND ROUND. SO I RETOLD IT. BUT MY WAY. FRANZ KAFKA'S WAY.

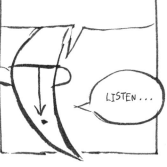

INSTEAD OF SETTING THE CASE IN ARISTOTLE'S UNIVERSE, I SET IT IN MINE.

LISTEN . . .

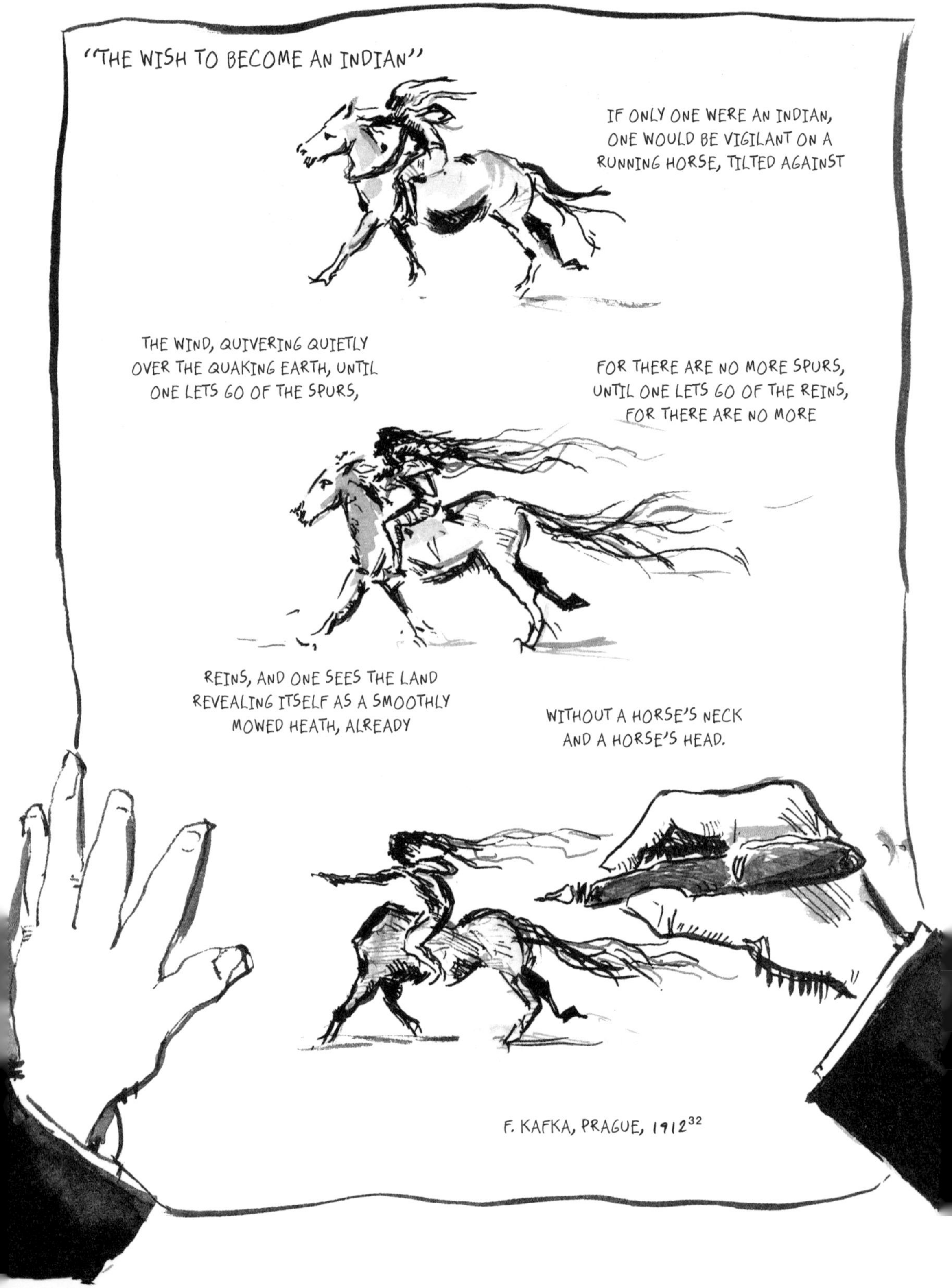
"THE WISH TO BECOME AN INDIAN"
IF ONLY ONE WERE AN INDIAN, ONE WOULD BE VIGILANT ON A RUNNING HORSE, TILTED AGAINST
THE WIND, QUIVERING QUIETLY OVER THE QUAKING EARTH, UNTIL ONE LETS GO OF THE SPURS,
FOR THERE ARE NO MORE SPURS, UNTIL ONE LETS GO OF THE REINS, FOR THERE ARE NO MORE
REINS, AND ONE SEES THE LAND REVEALING ITSELF AS A SMOOTHLY MOWED HEATH, ALREADY
WITHOUT A HORSE'S NECK AND A HORSE'S HEAD.
F. KAFKA, PRAGUE, 1912[32]

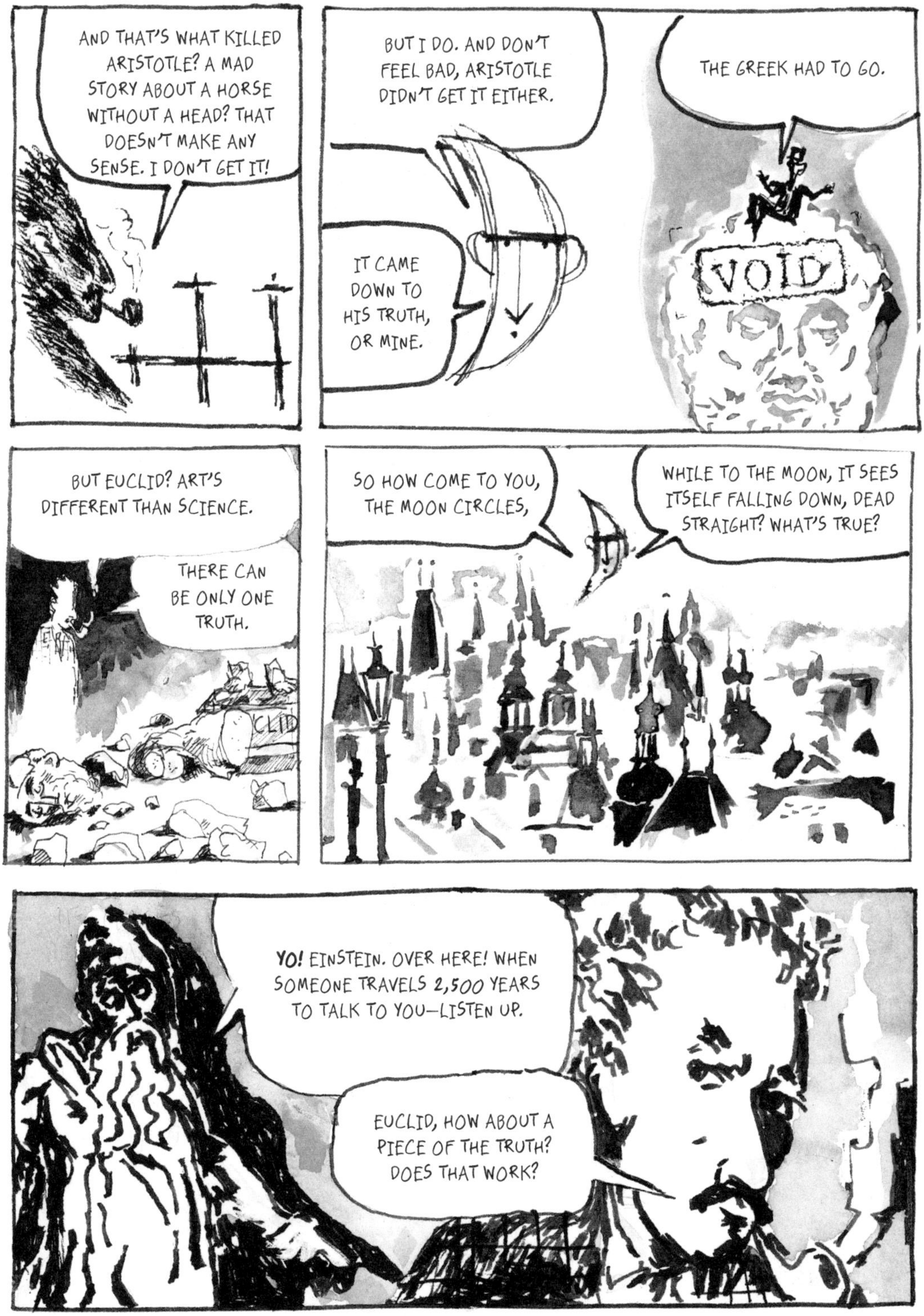

AND THAT'S WHAT KILLED ARISTOTLE? A MAD STORY ABOUT A HORSE WITHOUT A HEAD? THAT DOESN'T MAKE ANY SENSE. I DON'T GET IT!
BUT I DO. AND DON'T FEEL BAD, ARISTOTLE DIDN'T GET IT EITHER.
IT CAME DOWN TO HIS TRUTH, OR MINE.
THE GREEK HAD TO GO.
VOID
BUT EUCLID? ART'S DIFFERENT THAN SCIENCE.
THERE CAN BE ONLY ONE TRUTH.
SO HOW COME TO YOU, THE MOON CIRCLES,
WHILE TO THE MOON, IT SEES ITSELF FALLING DOWN, DEAD STRAIGHT? WHAT'S TRUE?
YO! EINSTEIN. OVER HERE! WHEN SOMEONE TRAVELS 2,500 YEARS TO TALK TO YOU—LISTEN UP.
EUCLID, HOW ABOUT A PIECE OF THE TRUTH? DOES THAT WORK?

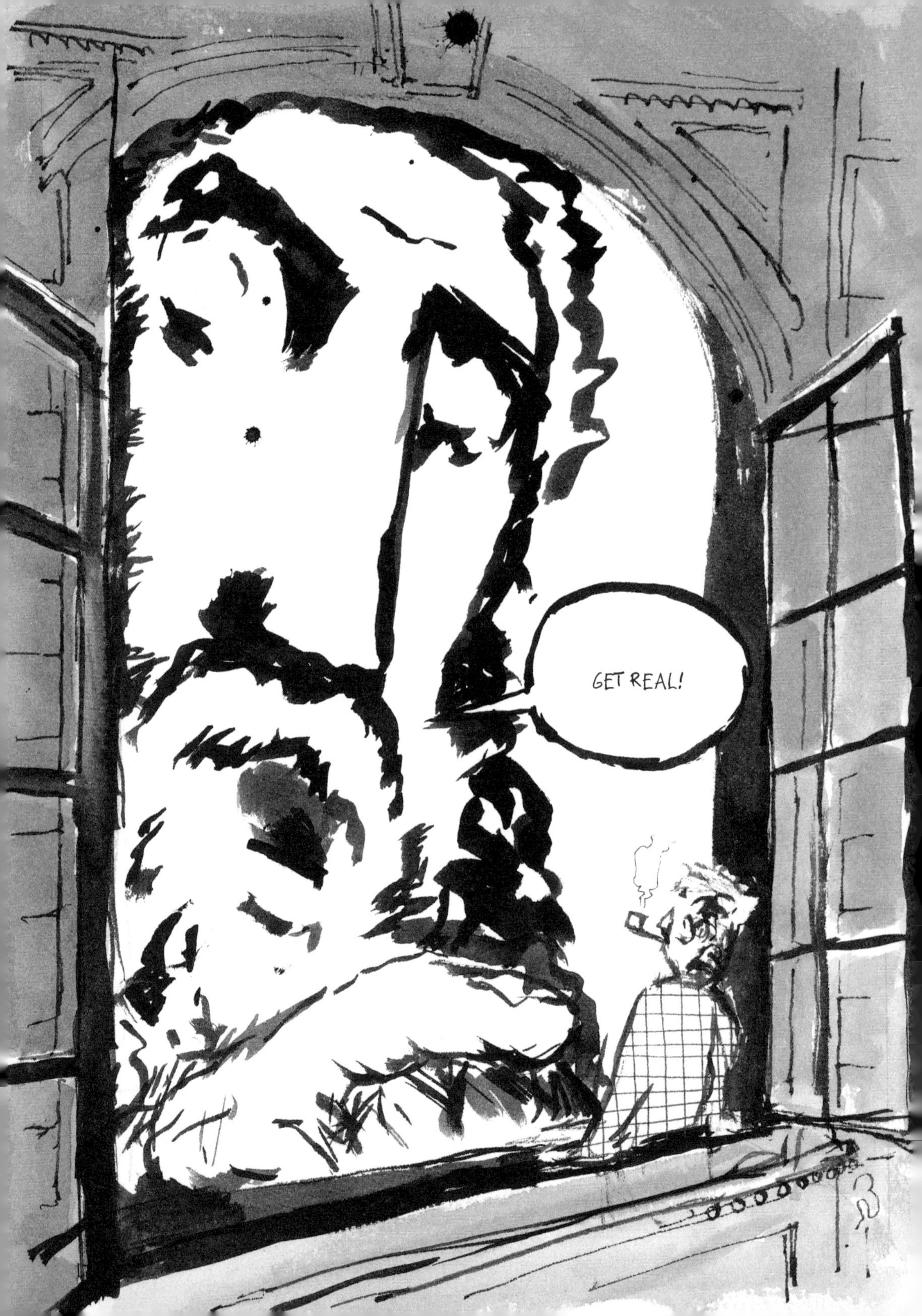
GET REAL!

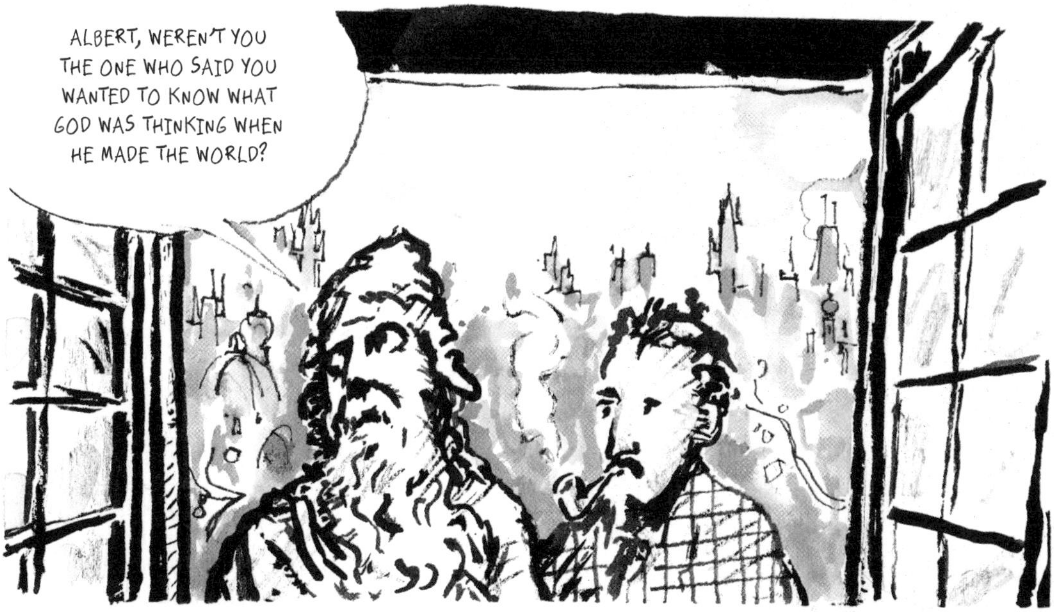
ALBERT, WEREN'T YOU THE ONE WHO SAID YOU WANTED TO KNOW WHAT GOD WAS THINKING WHEN HE MADE THE WORLD?

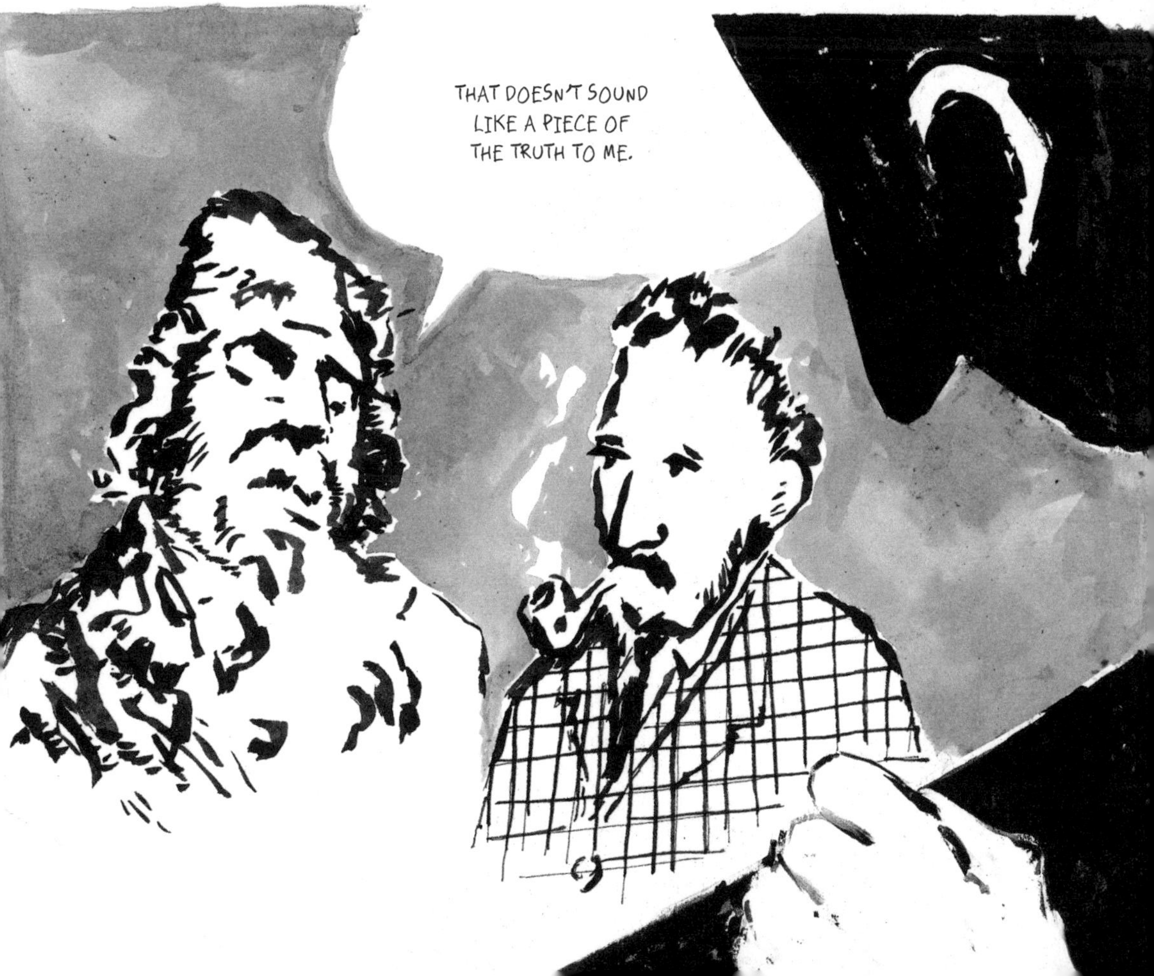
THAT DOESN'T SOUND LIKE A PIECE OF THE TRUTH TO ME.

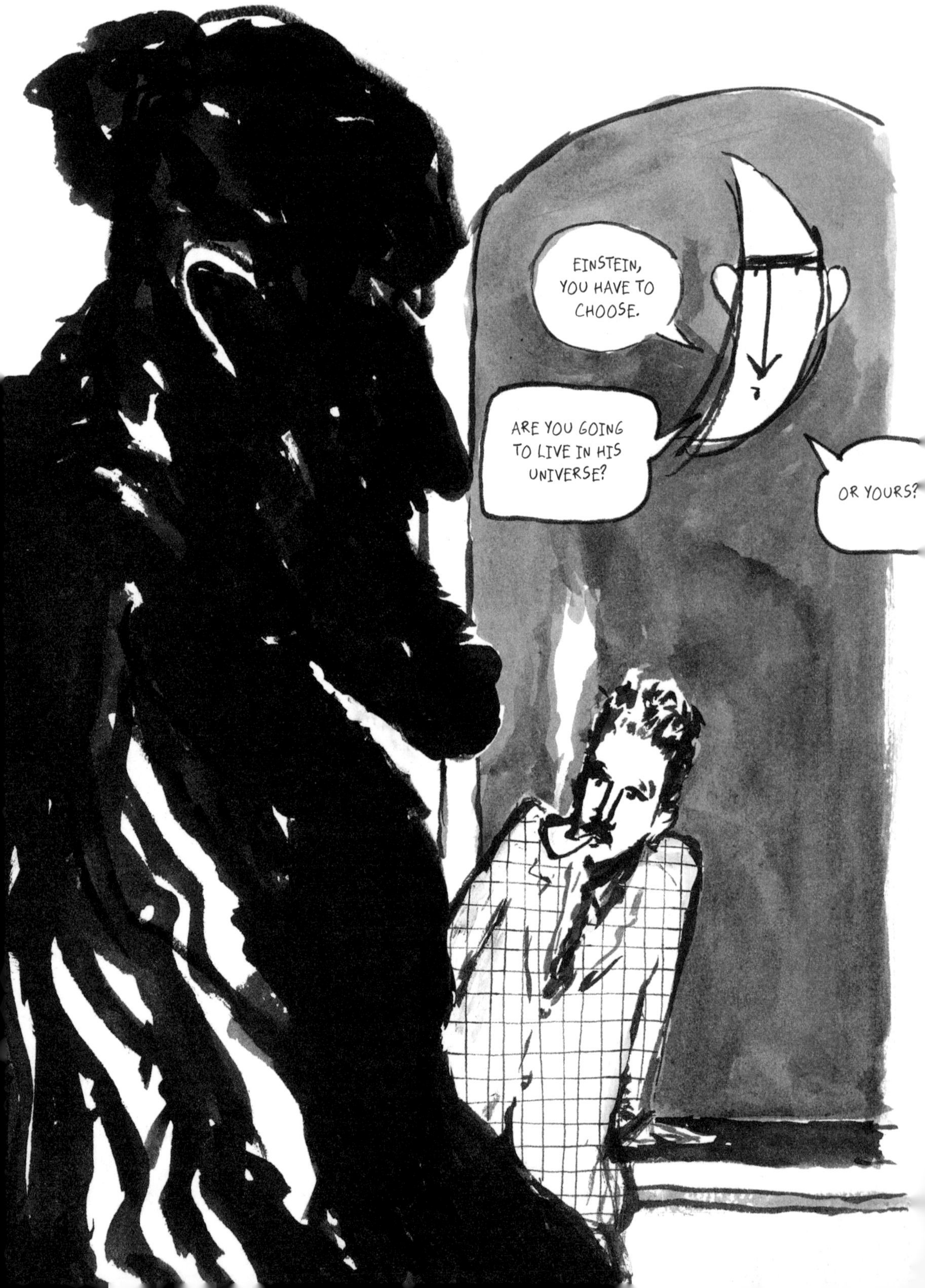
EINSTEIN, YOU HAVE TO CHOOSE.
ARE YOU GOING TO LIVE IN HIS UNIVERSE?
OR YOURS?

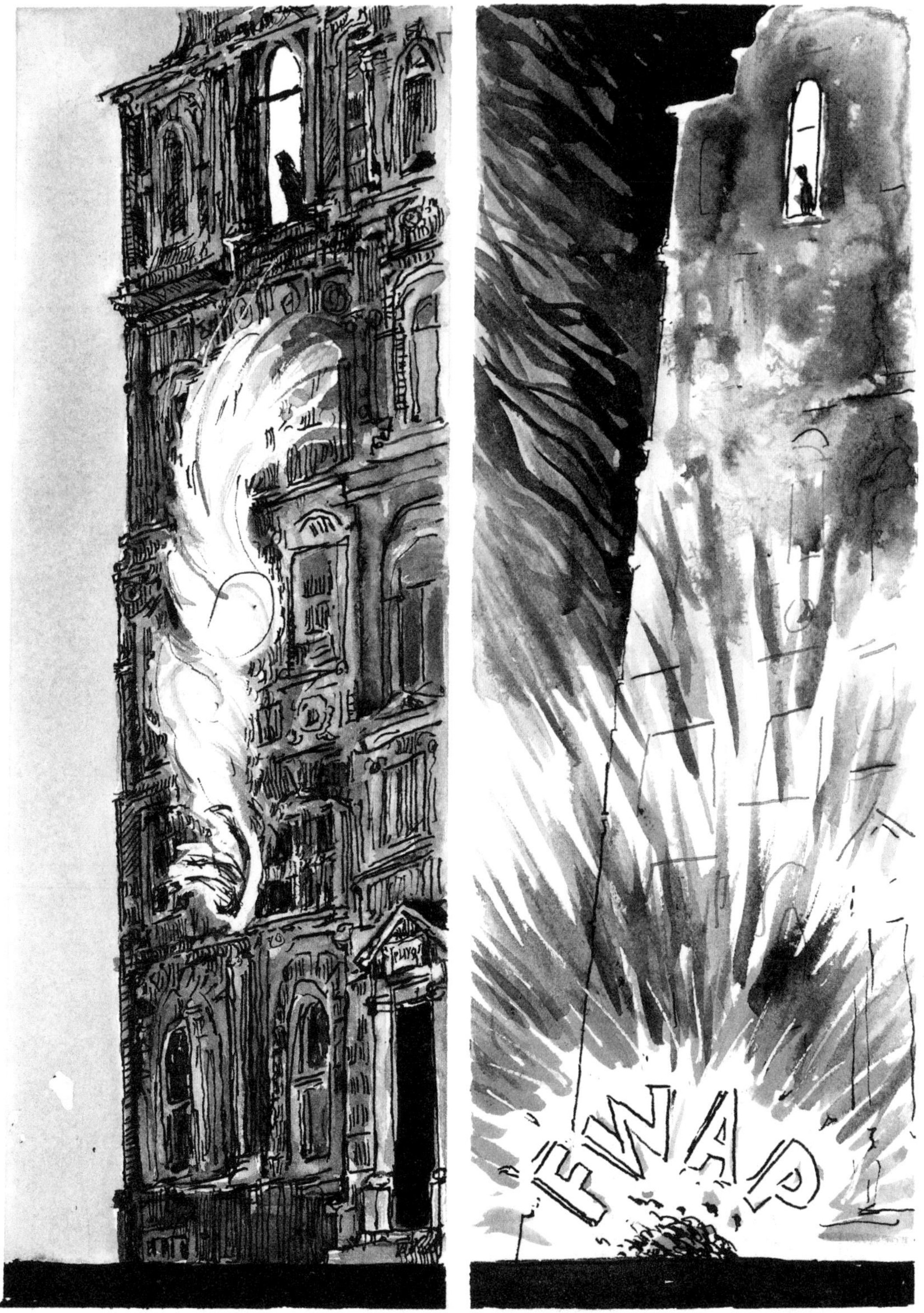
FWAP

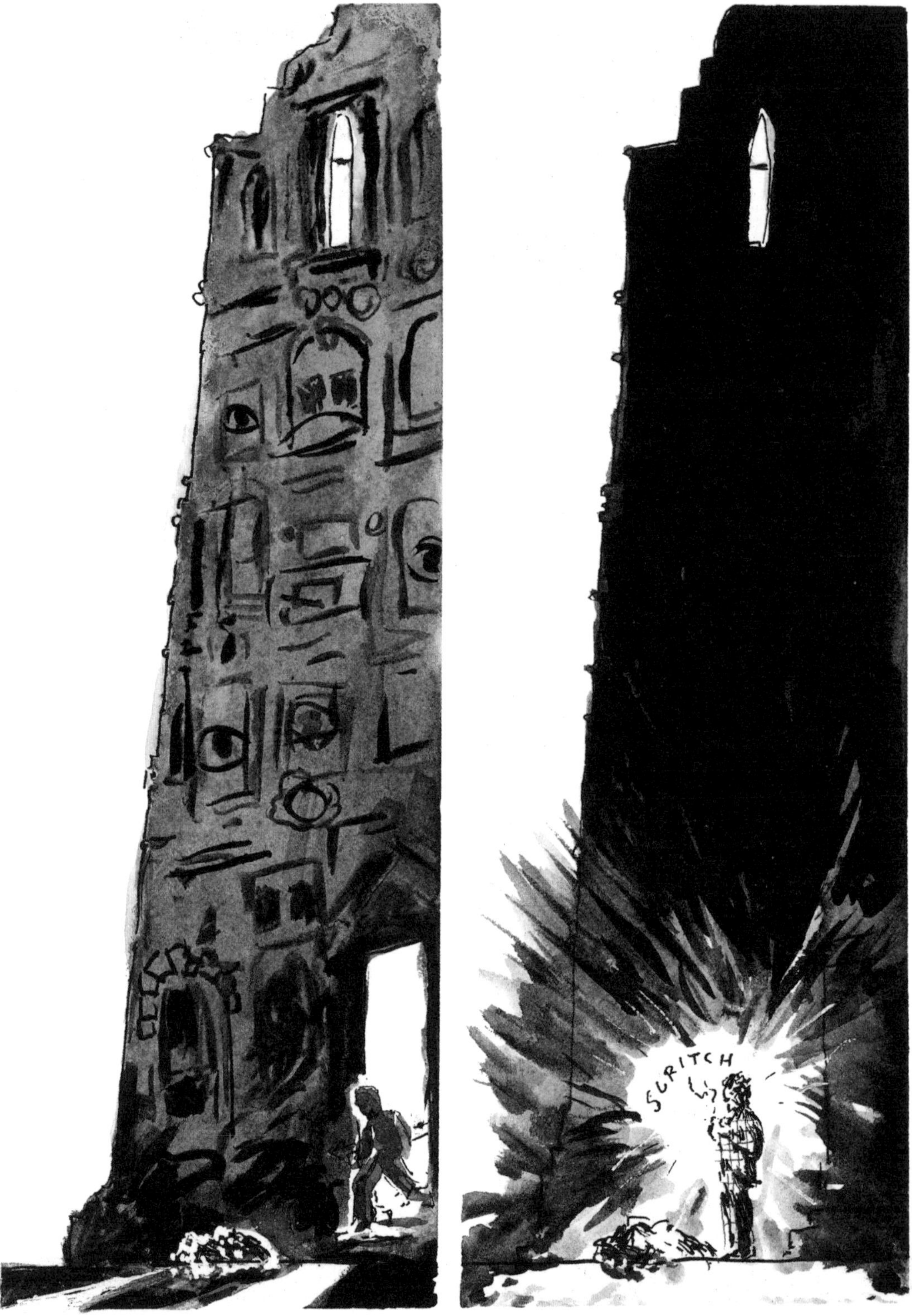
SCRITCH

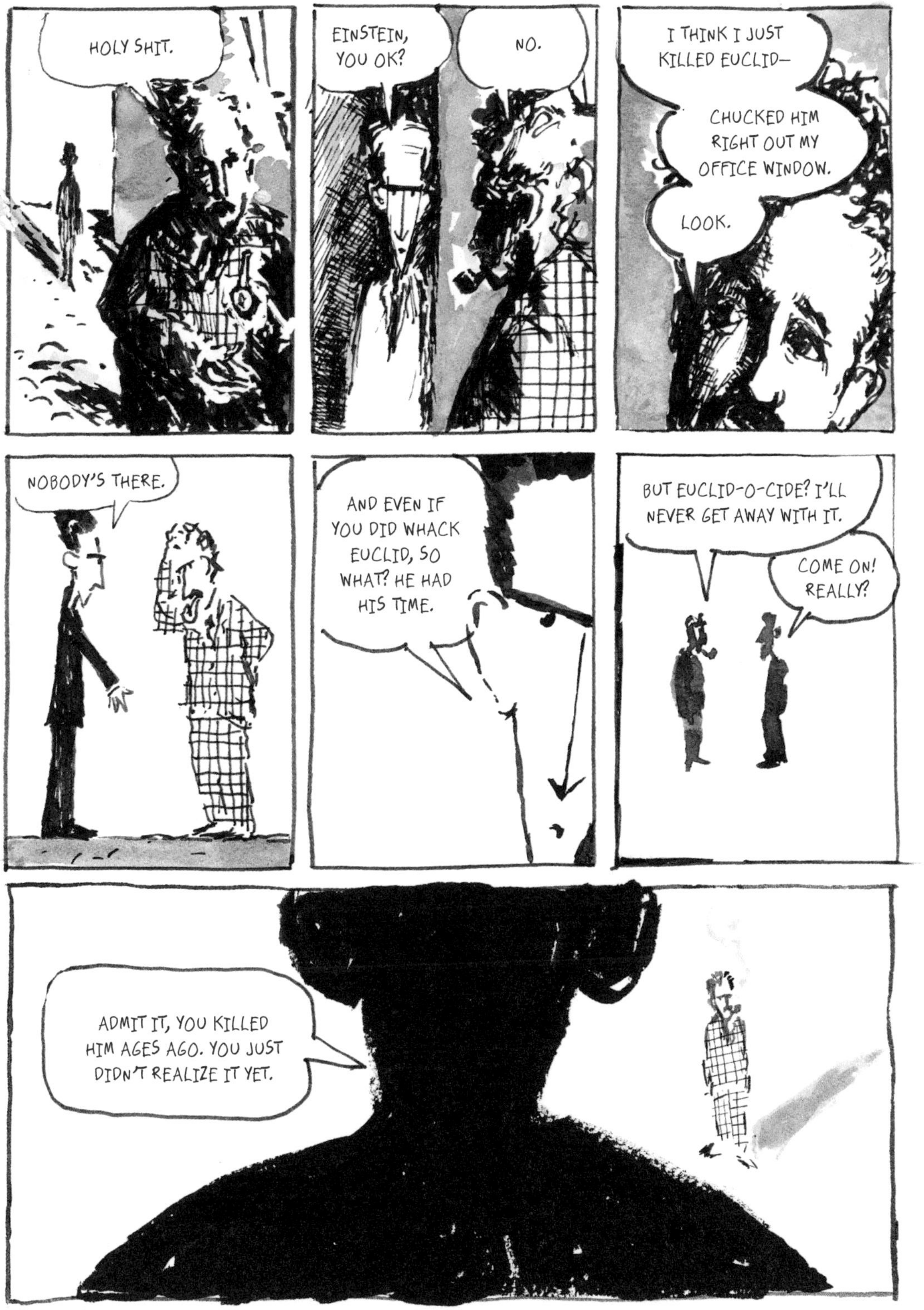
HOLY SHIT.
EINSTEIN, YOU OK?
NO.
I THINK I JUST KILLED EUCLID—
CHUCKED HIM RIGHT OUT MY OFFICE WINDOW.
LOOK.
NOBODY'S THERE.
AND EVEN IF YOU DID WHACK EUCLID, SO WHAT? HE HAD HIS TIME.
BUT EUCLID-O-CIDE? I'LL NEVER GET AWAY WITH IT.
COME ON! REALLY?
ADMIT IT, YOU KILLED HIM AGES AGO. YOU JUST DIDN'T REALIZE IT YET.

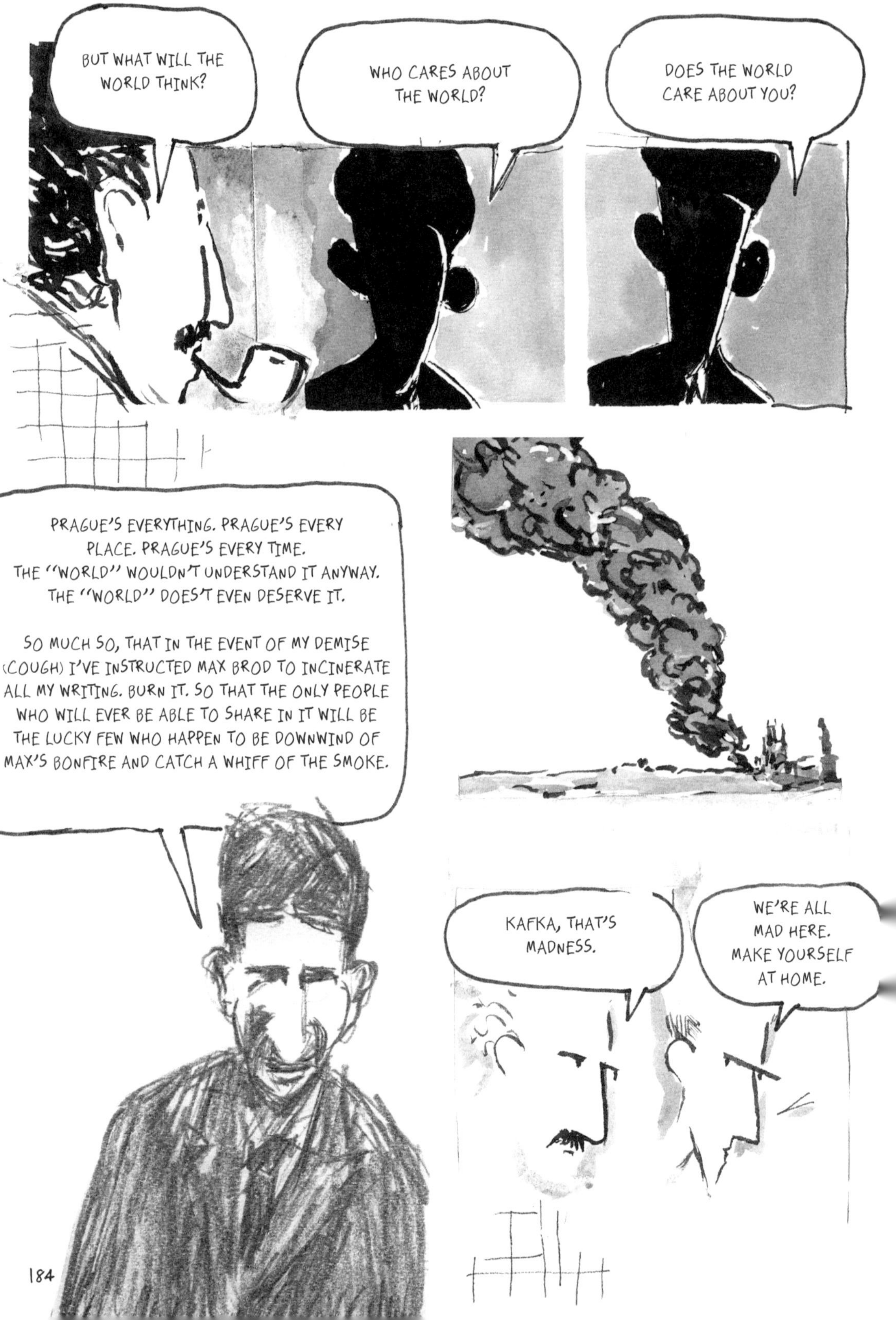
BUT WHAT WILL THE WORLD THINK?
WHO CARES ABOUT THE WORLD?
DOES THE WORLD CARE ABOUT YOU?
PRAGUE'S EVERYTHING. PRAGUE'S EVERY PLACE. PRAGUE'S EVERY TIME. THE "WORLD" WOULDN'T UNDERSTAND IT ANYWAY. THE "WORLD" DOES'T EVEN DESERVE IT.
SO MUCH SO, THAT IN THE EVENT OF MY DEMISE (COUGH) I'VE INSTRUCTED MAX BROD TO INCINERATE ALL MY WRITING. BURN IT. SO THAT THE ONLY PEOPLE WHO WILL EVER BE ABLE TO SHARE IN IT WILL BE THE LUCKY FEW WHO HAPPEN TO BE DOWNWIND OF MAX'S BONFIRE AND CATCH A WHIFF OF THE SMOKE.
KAFKA, THAT'S MADNESS.
WE'RE ALL MAD HERE. MAKE YOURSELF AT HOME.

Ceci n'est pas
Stay

Ceci n'est pas
Go

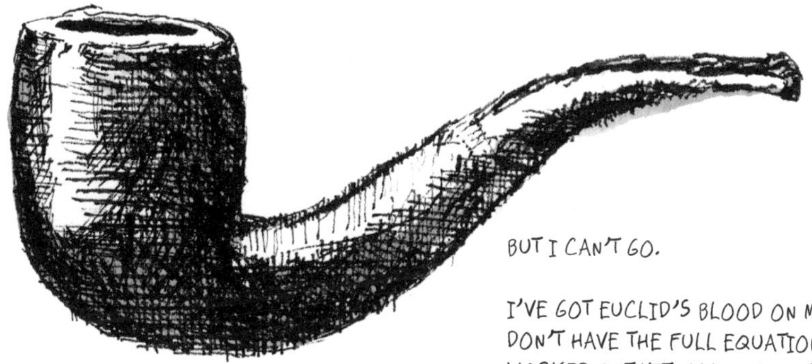

I CAN'T STAY IN PRAGUE.

I'VE ANSWERED QUESTIONS ABOUT GRAVITY THAT STUMPED NEWTON.

I'VE BEAT ABRAHAM.

I'VE FUCKING SOLVED GRAVITY, FOR CHRISSAKES!

AND IF I MAKE THEM STAY HERE, MY WIFE AND KIDS WILL NEVER SPEAK TO ME AGAIN.

BUT I CAN'T GO.

I'VE GOT EUCLID'S BLOOD ON MY HANDS, I DON'T HAVE THE FULL EQUATIONS FOR GRAVITY WORKED OUT YET, ALL I HAVE IS A PIECE OF THE TRUTH, AND I'M GOING TO HAVE TO EMBRACE THOSE ITALIANS WITH THEIR LUNATIC MATH.

WHATEVER I DO, I'M IRREPARABLY INFECTED WITH PRAGUE MADNESS, I'VE GLIMPSED A UNIVERSE WITHOUT TIME OR PLACE OR MEANING, AND I CAN'T UNSEE IT.

OH, AND I MIGHT BE WRONG.

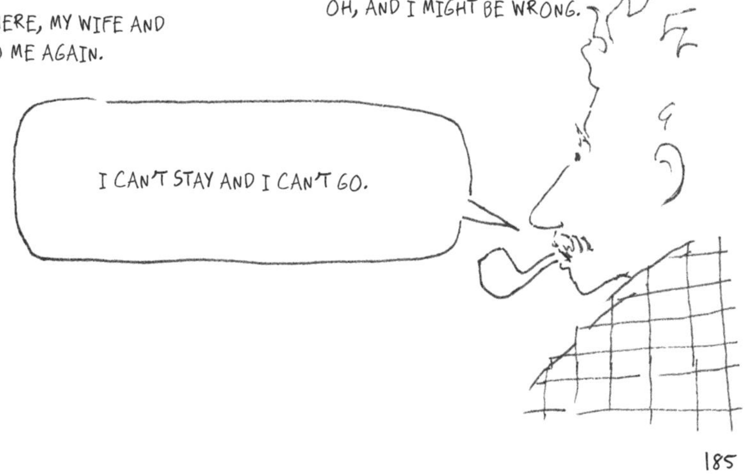
I CAN'T STAY AND I CAN'T GO.

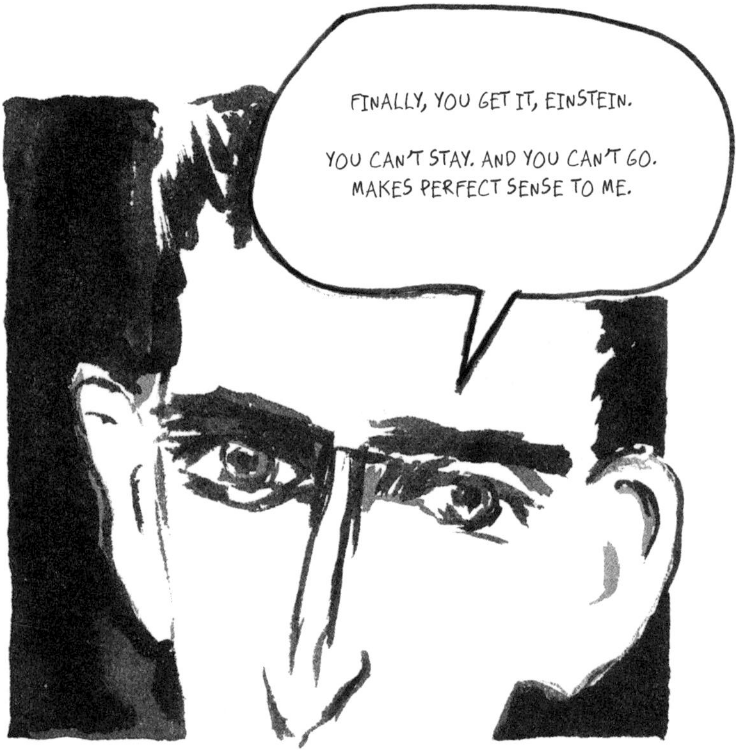
FINALLY, YOU GET IT, EINSTEIN.
YOU CAN'T STAY. AND YOU CAN'T GO. MAKES PERFECT SENSE TO ME.

WELCOME.

CHAPTER XIV

EINSTEIN'S
UNIFIED THEORY
of EVERYTHING
IT ALL

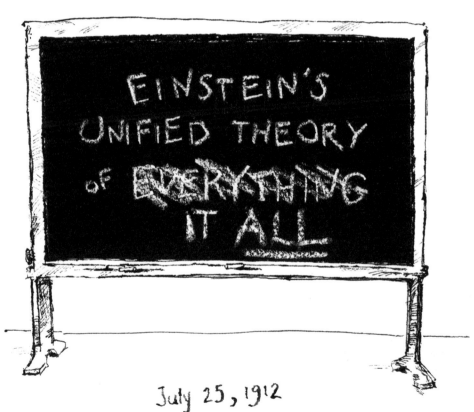

July 25, 1912

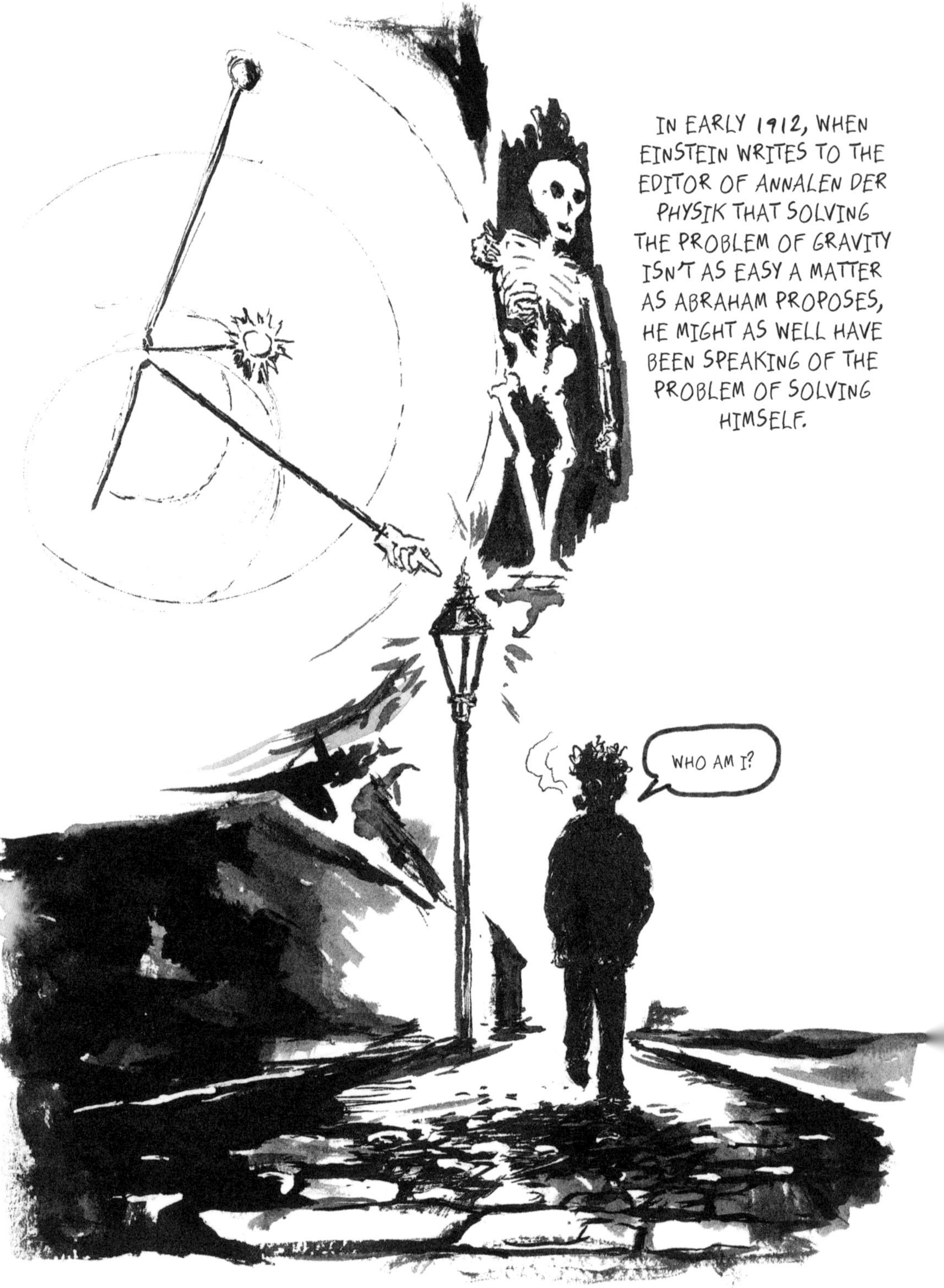
IN EARLY 1912, WHEN EINSTEIN WRITES TO THE EDITOR OF ANNALEN DER PHYSIK THAT SOLVING THE PROBLEM OF GRAVITY ISN'T AS EASY A MATTER AS ABRAHAM PROPOSES, HE MIGHT AS WELL HAVE BEEN SPEAKING OF THE PROBLEM OF SOLVING HIMSELF.
WHO AM I?

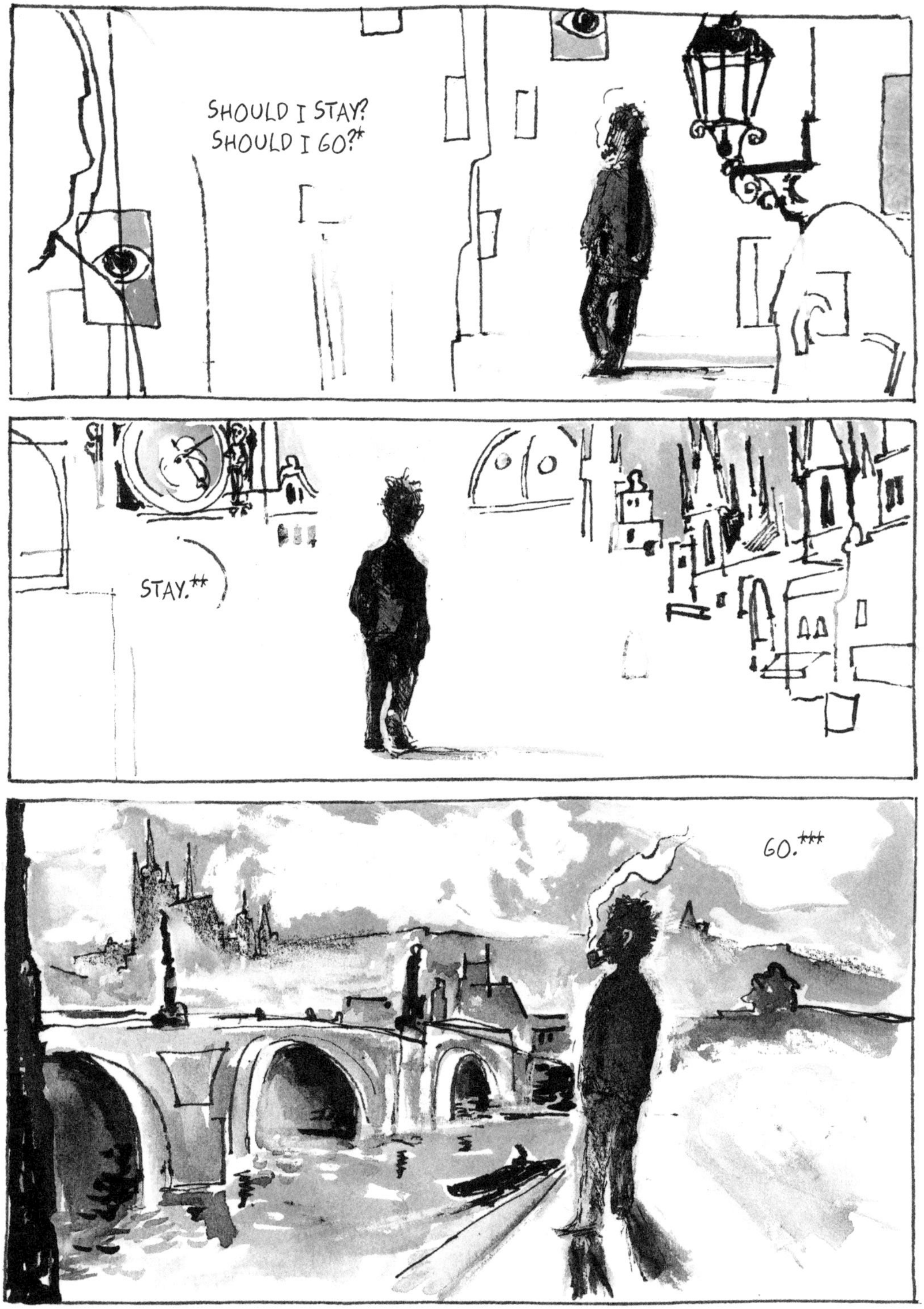
SHOULD I STAY?
SHOULD I GO?*
STAY.**
GO.***

* Even though by early 1912, Einstein has accepted a job back in Zurich, his bags are packed, and he's even found a suitable replacement in Prague, Philipp Frank, and Frank has even bought the entire Brazilian admiral's uniform from him (at a discount), complete with sword.

** If I stay, I'm just another world-denying recluse, luxuriating in my personal fantasies, the world be damned, a captive of the Prague madness.

*** If I go, I'm a Prague-stained refugee, smuggling dangerous new laws back to the meticulous, four-square streets of Switzerland.

WAIT A MINUTE . . .

HOLD ON . . .

WHOA! THAT'S IT!

stayOR go?

stay,
OR stay
go? OR

stAy
go?

IF TWO THINGS YOU THINK
ARE DIFFERENT ALWAYS DELIVER
THE SAME RESULT, THEN THEY'RE
NOT DIFFERENT AFTER ALL. THEY'RE
THE SAME. SO THE ANSWER
ISN'T STAY OR GO ...

stay
OR
stay go
OR
go?

THE ANSWER IS ...

GO
go ?
Ay, OR GO?
stay OR go ? ? GO?
go ? ?
stay ?

191

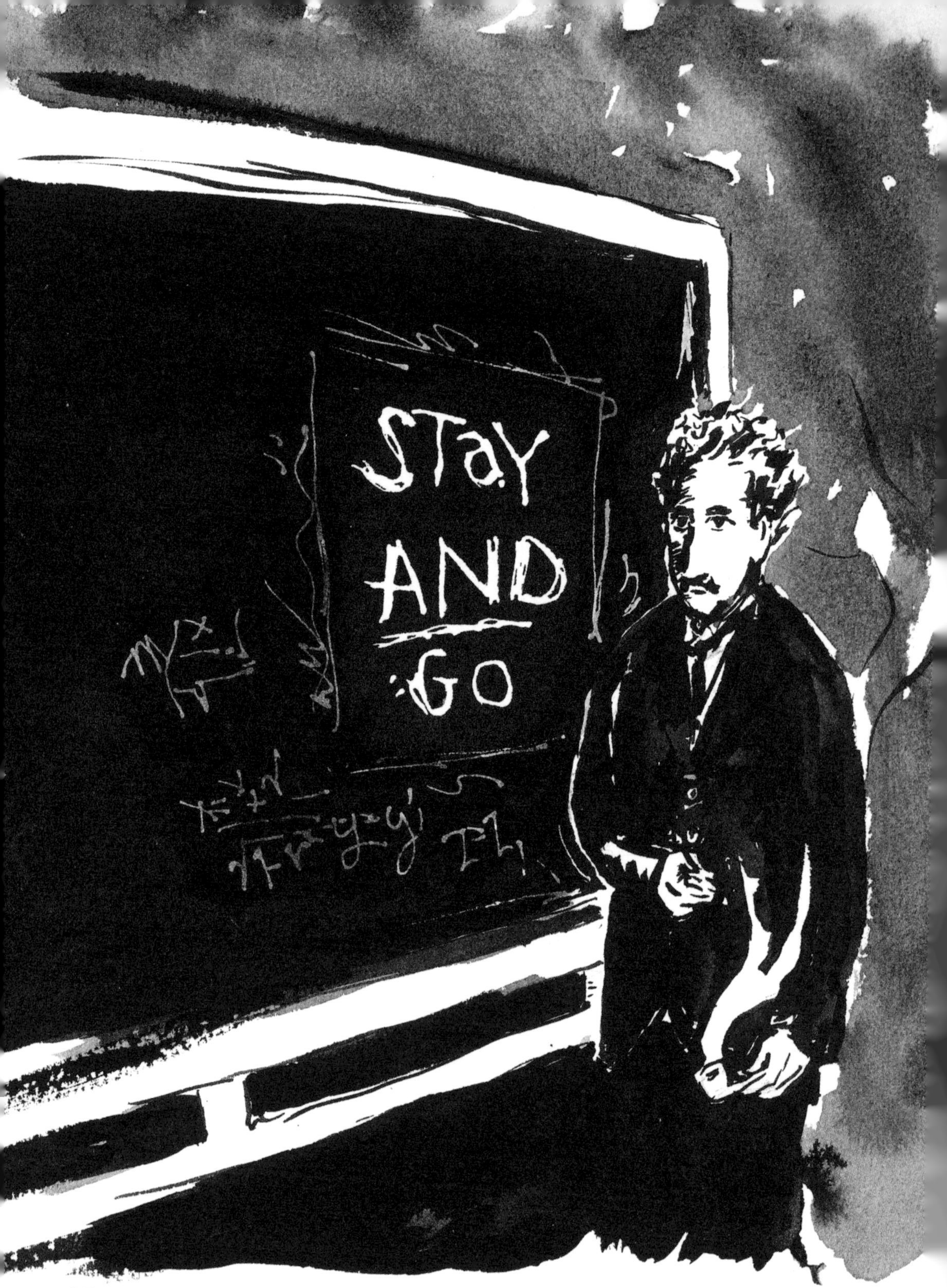
STAY
AND
GO

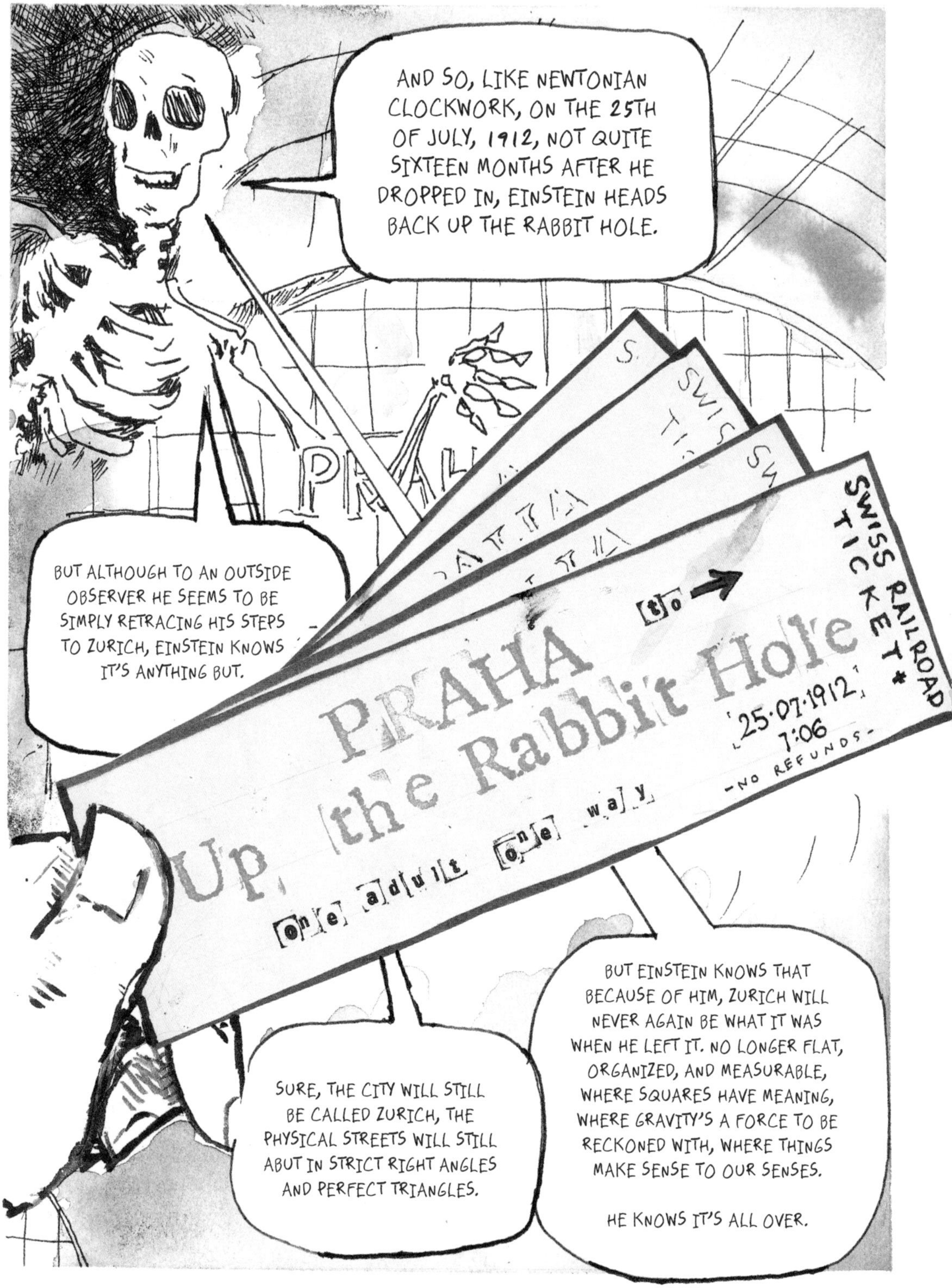
AND SO, LIKE NEWTONIAN CLOCKWORK, ON THE 25TH OF JULY, 1912, NOT QUITE SIXTEEN MONTHS AFTER HE DROPPED IN, EINSTEIN HEADS BACK UP THE RABBIT HOLE.
BUT ALTHOUGH TO AN OUTSIDE OBSERVER HE SEEMS TO BE SIMPLY RETRACING HIS STEPS TO ZURICH, EINSTEIN KNOWS IT'S ANYTHING BUT.
SWISS RAILROAD TICKET
PRAHA to
Up the Rabbit Hole
25·07·1912
7:06
-NO REFUNDS-
one adult one way
SURE, THE CITY WILL STILL BE CALLED ZURICH, THE PHYSICAL STREETS WILL STILL ABUT IN STRICT RIGHT ANGLES AND PERFECT TRIANGLES.
BUT EINSTEIN KNOWS THAT BECAUSE OF HIM, ZURICH WILL NEVER AGAIN BE WHAT IT WAS WHEN HE LEFT IT. NO LONGER FLAT, ORGANIZED, AND MEASURABLE, WHERE SQUARES HAVE MEANING, WHERE GRAVITY'S A FORCE TO BE RECKONED WITH, WHERE THINGS MAKE SENSE TO OUR SENSES.
HE KNOWS IT'S ALL OVER.

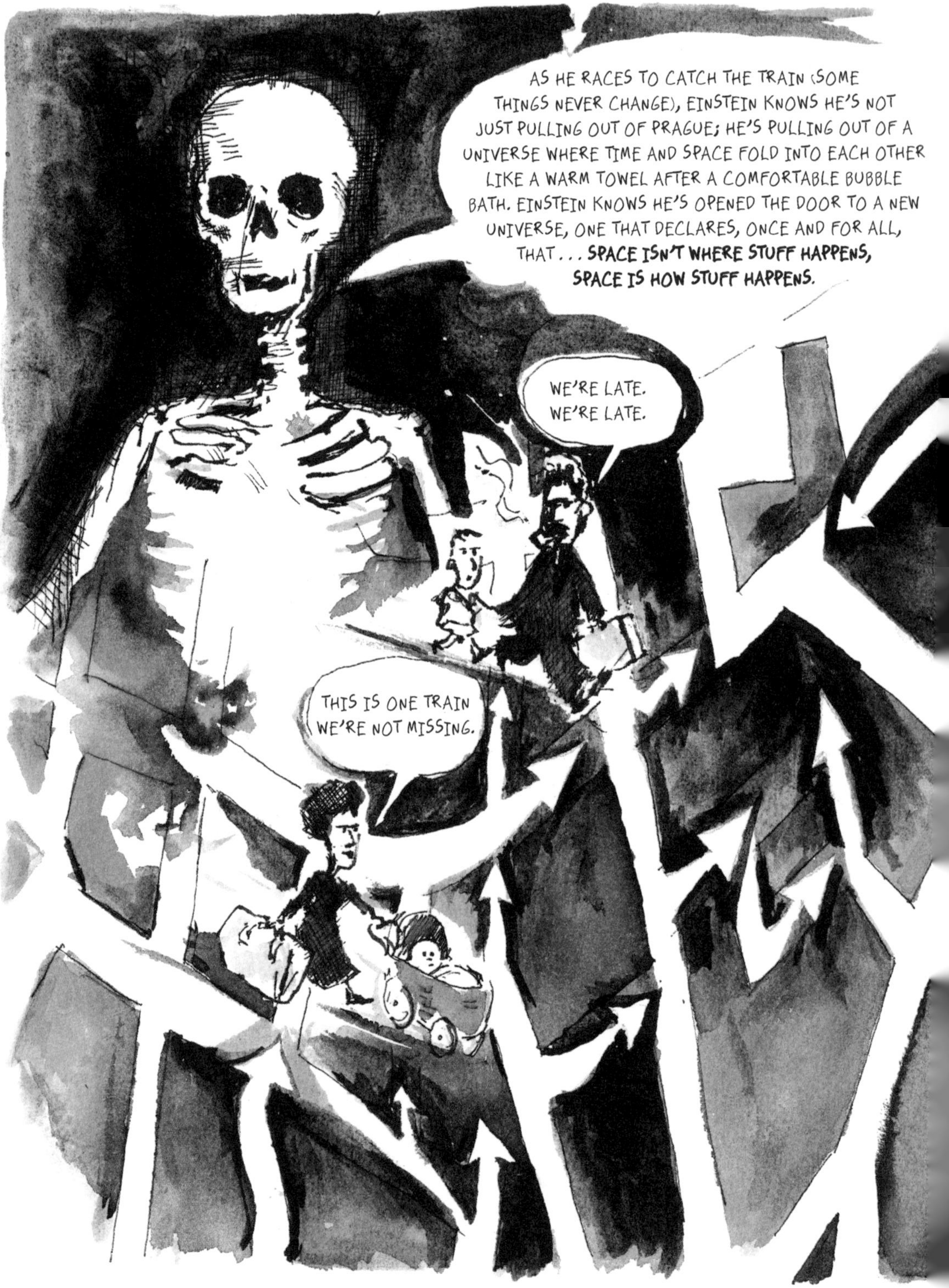
AS HE RACES TO CATCH THE TRAIN (SOME THINGS NEVER CHANGE), EINSTEIN KNOWS HE'S NOT JUST PULLING OUT OF PRAGUE; HE'S PULLING OUT OF A UNIVERSE WHERE TIME AND SPACE FOLD INTO EACH OTHER LIKE A WARM TOWEL AFTER A COMFORTABLE BUBBLE BATH. EINSTEIN KNOWS HE'S OPENED THE DOOR TO A NEW UNIVERSE, ONE THAT DECLARES, ONCE AND FOR ALL, THAT . . . **SPACE ISN'T WHERE STUFF HAPPENS, SPACE IS HOW STUFF HAPPENS.**
WE'RE LATE. WE'RE LATE.
THIS IS ONE TRAIN WE'RE NOT MISSING.

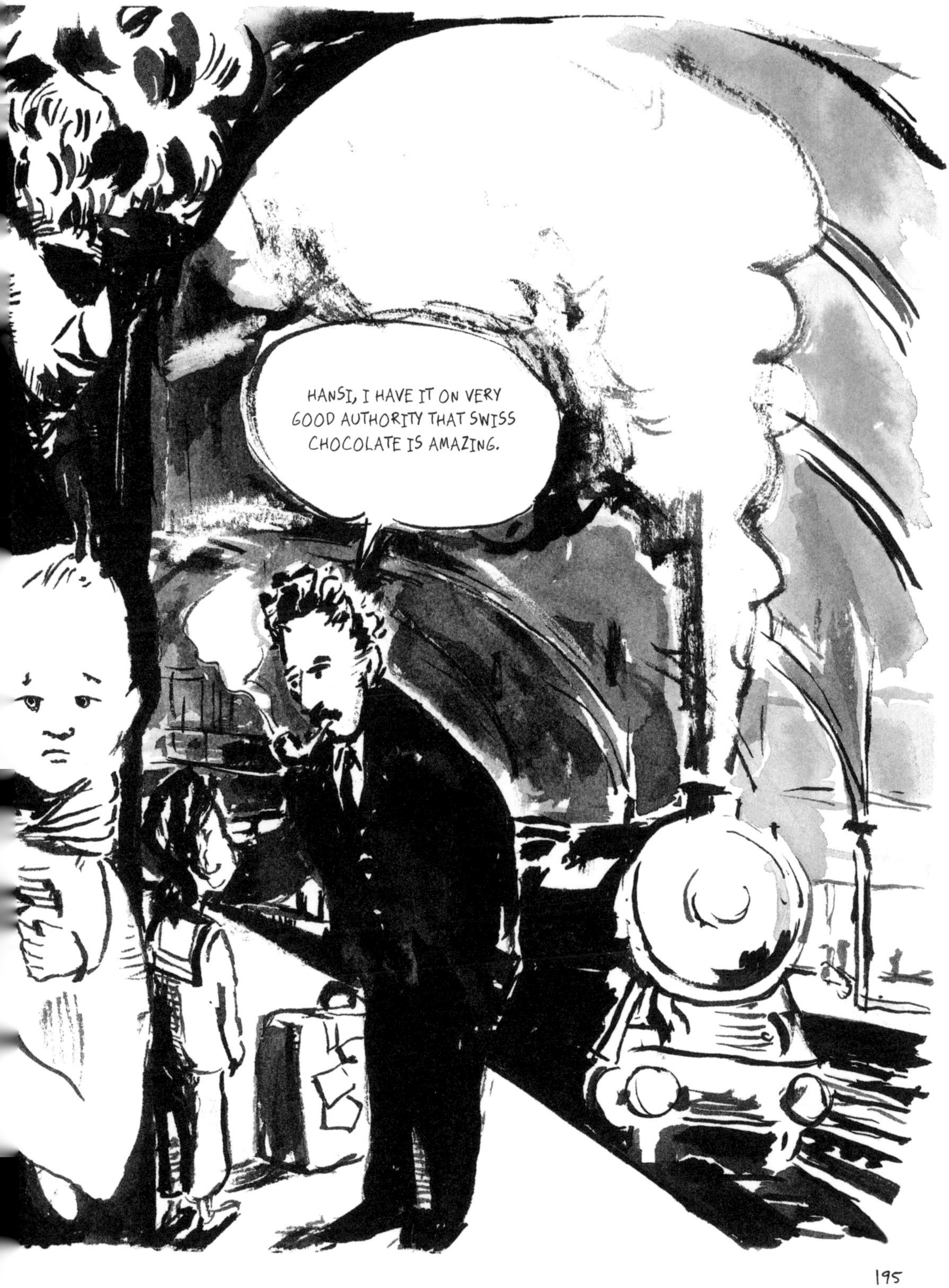
HANSI, I HAVE IT ON VERY GOOD AUTHORITY THAT SWISS CHOCOLATE IS AMAZING.

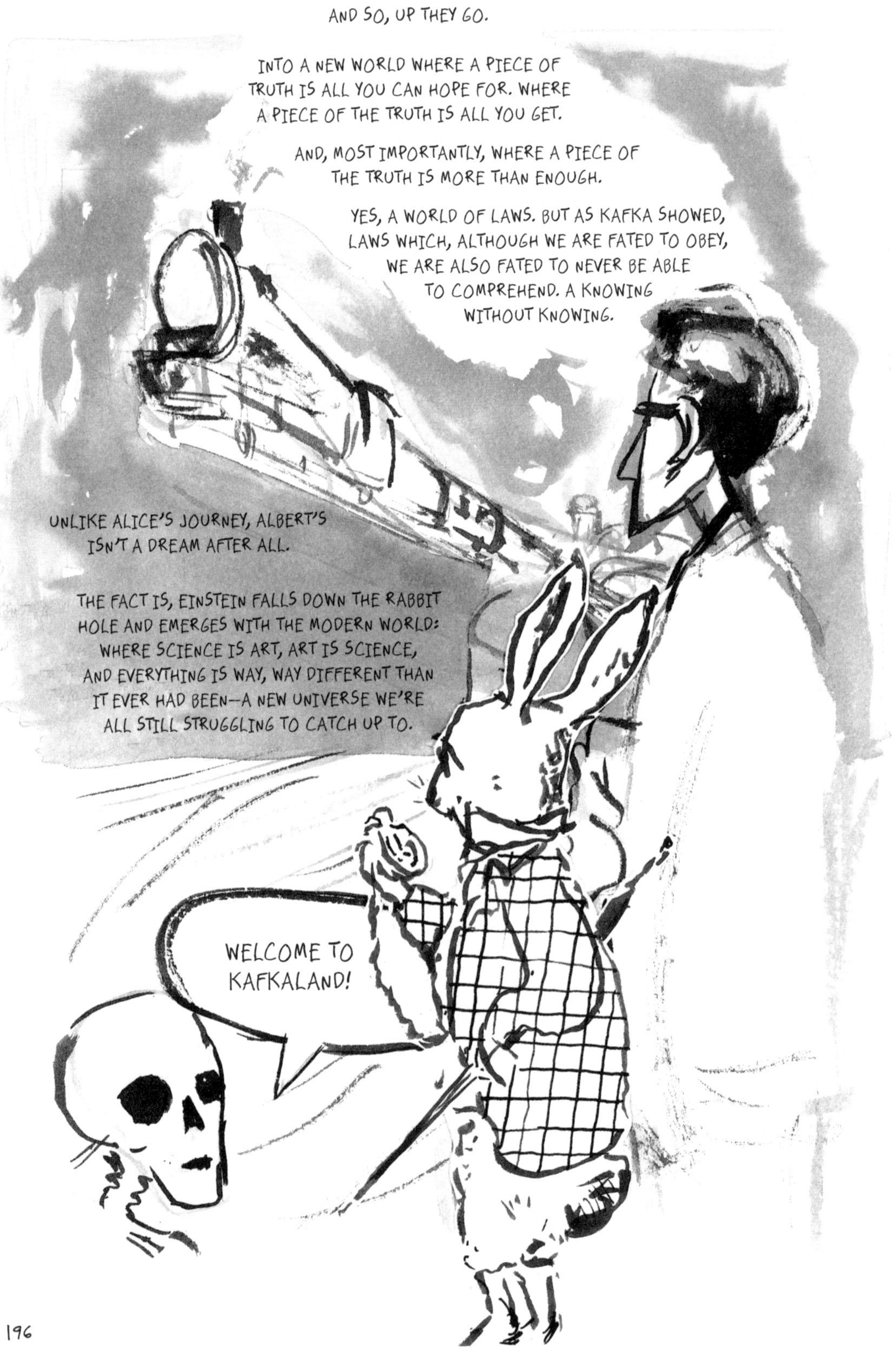
AND SO, UP THEY GO.
INTO A NEW WORLD WHERE A PIECE OF TRUTH IS ALL YOU CAN HOPE FOR. WHERE A PIECE OF THE TRUTH IS ALL YOU GET.
AND, MOST IMPORTANTLY, WHERE A PIECE OF THE TRUTH IS MORE THAN ENOUGH.
YES, A WORLD OF LAWS. BUT AS KAFKA SHOWED, LAWS WHICH, ALTHOUGH WE ARE FATED TO OBEY, WE ARE ALSO FATED TO NEVER BE ABLE TO COMPREHEND. A KNOWING WITHOUT KNOWING.
UNLIKE ALICE'S JOURNEY, ALBERT'S ISN'T A DREAM AFTER ALL.
THE FACT IS, EINSTEIN FALLS DOWN THE RABBIT HOLE AND EMERGES WITH THE MODERN WORLD: WHERE SCIENCE IS ART, ART IS SCIENCE, AND EVERYTHING IS WAY, WAY DIFFERENT THAN IT EVER HAD BEEN—A NEW UNIVERSE WE'RE ALL STILL STRUGGLING TO CATCH UP TO.
WELCOME TO KAFKALAND!

CODA

The Gravitating Mass of Kafkaland

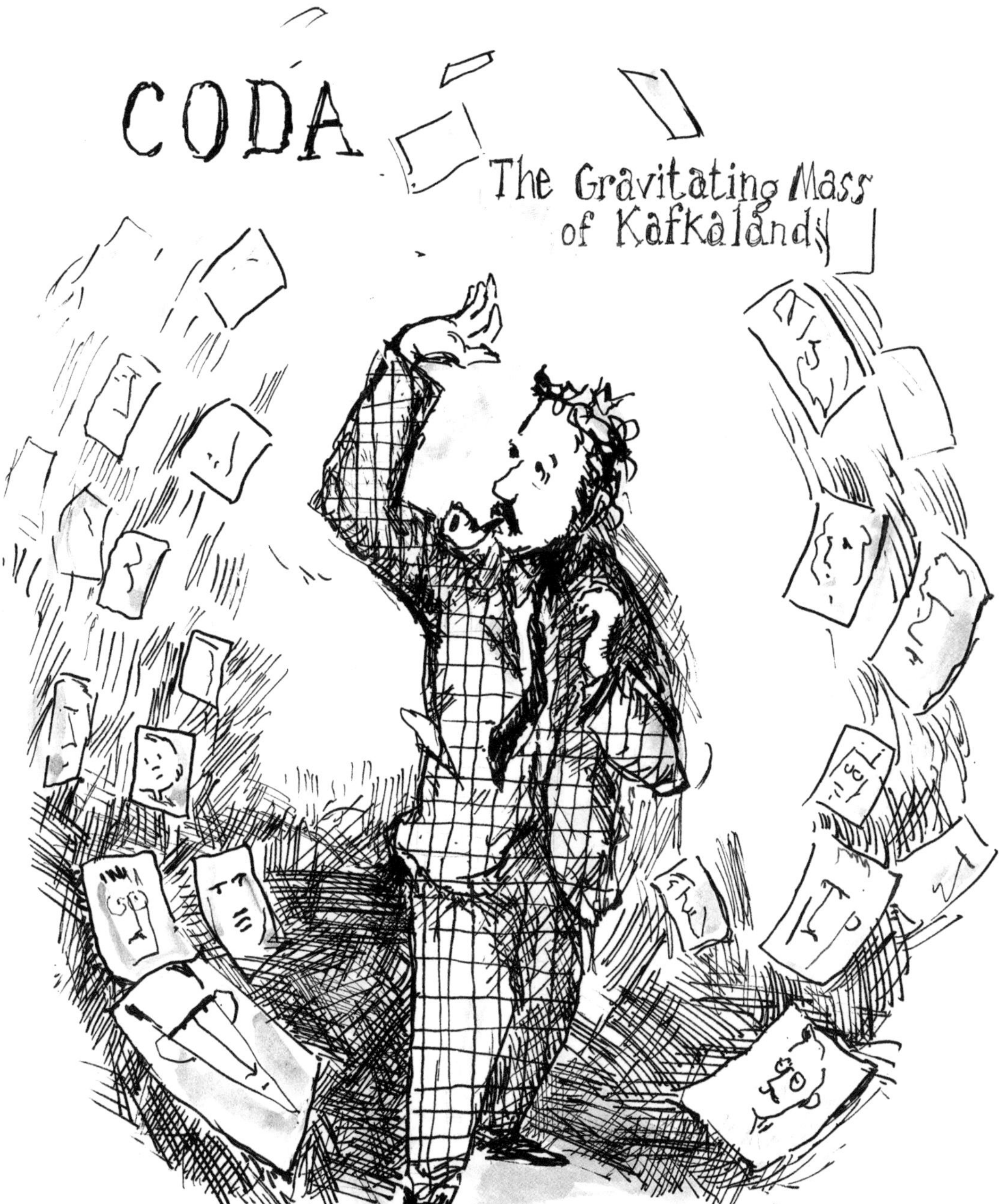

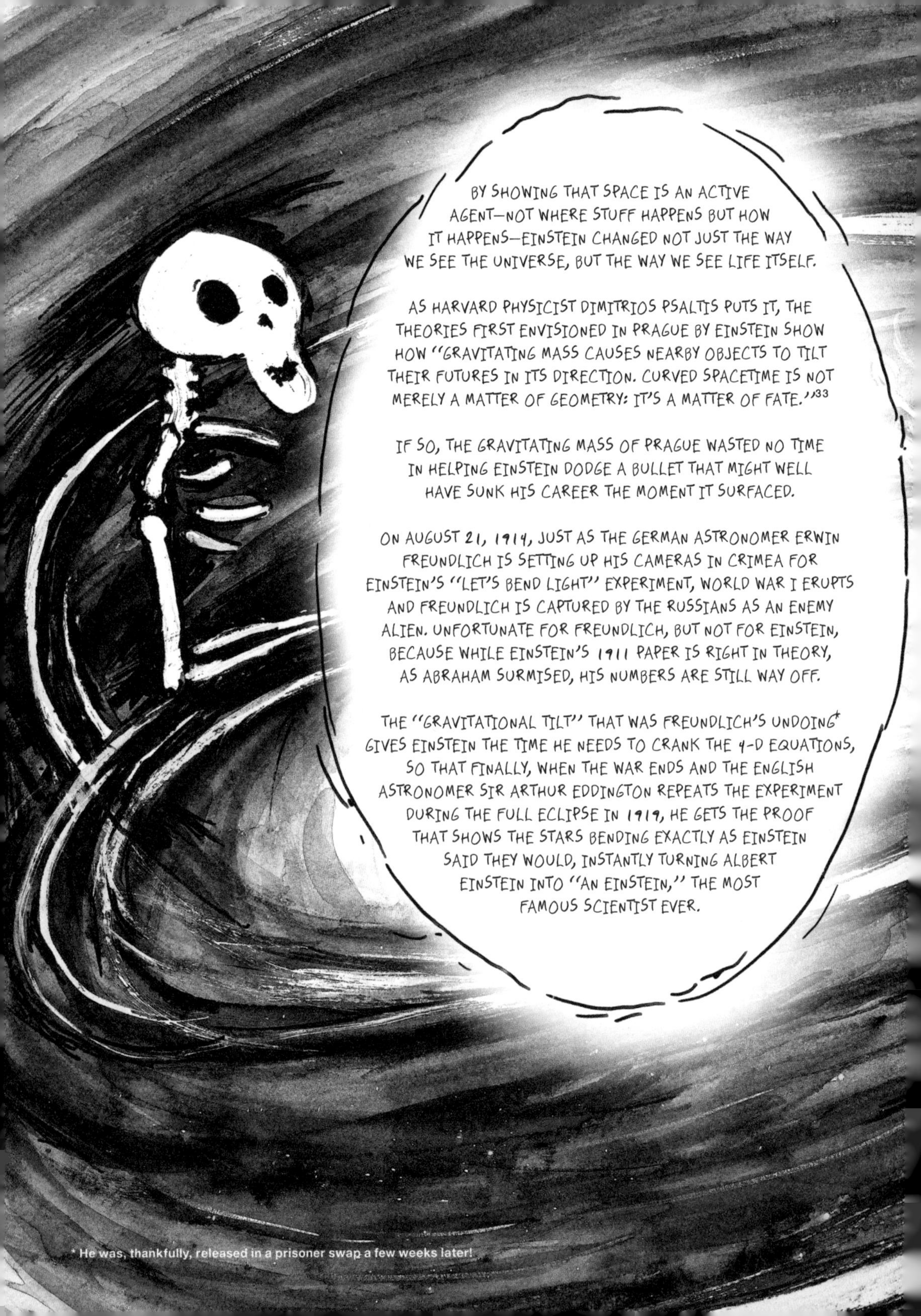
BY SHOWING THAT SPACE IS AN ACTIVE AGENT—NOT WHERE STUFF HAPPENS BUT HOW IT HAPPENS—EINSTEIN CHANGED NOT JUST THE WAY WE SEE THE UNIVERSE, BUT THE WAY WE SEE LIFE ITSELF.
AS HARVARD PHYSICIST DIMITRIOS PSALTIS PUTS IT, THE THEORIES FIRST ENVISIONED IN PRAGUE BY EINSTEIN SHOW HOW "GRAVITATING MASS CAUSES NEARBY OBJECTS TO TILT THEIR FUTURES IN ITS DIRECTION. CURVED SPACETIME IS NOT MERELY A MATTER OF GEOMETRY: IT'S A MATTER OF FATE."[33]
IF SO, THE GRAVITATING MASS OF PRAGUE WASTED NO TIME IN HELPING EINSTEIN DODGE A BULLET THAT MIGHT WELL HAVE SUNK HIS CAREER THE MOMENT IT SURFACED.
ON AUGUST 21, 1914, JUST AS THE GERMAN ASTRONOMER ERWIN FREUNDLICH IS SETTING UP HIS CAMERAS IN CRIMEA FOR EINSTEIN'S "LET'S BEND LIGHT" EXPERIMENT, WORLD WAR I ERUPTS AND FREUNDLICH IS CAPTURED BY THE RUSSIANS AS AN ENEMY ALIEN. UNFORTUNATE FOR FREUNDLICH, BUT NOT FOR EINSTEIN, BECAUSE WHILE EINSTEIN'S 1911 PAPER IS RIGHT IN THEORY, AS ABRAHAM SURMISED, HIS NUMBERS ARE STILL WAY OFF.
THE "GRAVITATIONAL TILT" THAT WAS FREUNDLICH'S UNDOING* GIVES EINSTEIN THE TIME HE NEEDS TO CRANK THE 4-D EQUATIONS, SO THAT FINALLY, WHEN THE WAR ENDS AND THE ENGLISH ASTRONOMER SIR ARTHUR EDDINGTON REPEATS THE EXPERIMENT DURING THE FULL ECLIPSE IN 1919, HE GETS THE PROOF THAT SHOWS THE STARS BENDING EXACTLY AS EINSTEIN SAID THEY WOULD, INSTANTLY TURNING ALBERT EINSTEIN INTO "AN EINSTEIN," THE MOST FAMOUS SCIENTIST EVER.

* He was, thankfully, released in a prisoner swap a few weeks later!

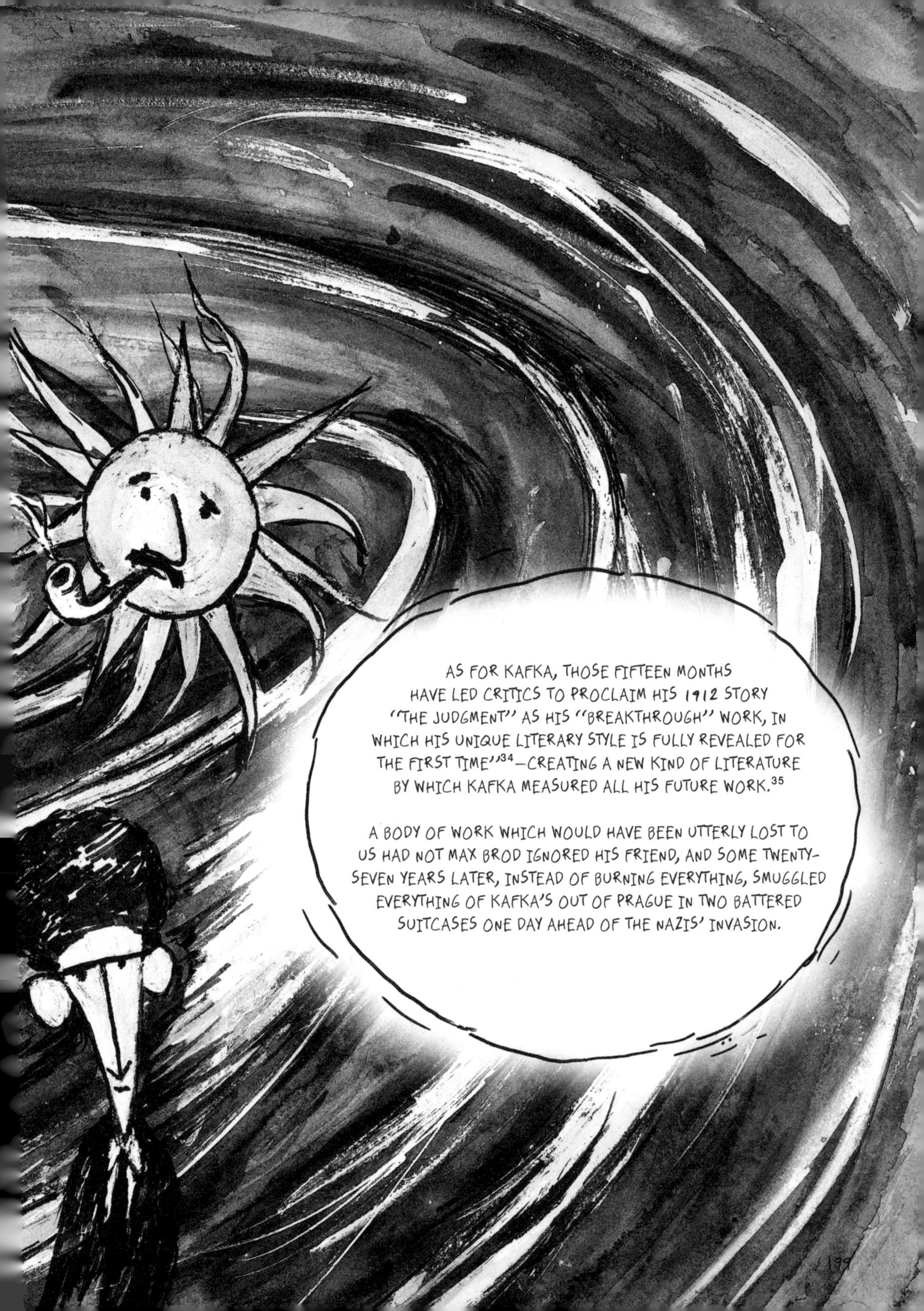
AS FOR KAFKA, THOSE FIFTEEN MONTHS HAVE LED CRITICS TO PROCLAIM HIS 1912 STORY "THE JUDGMENT" AS HIS "BREAKTHROUGH" WORK, IN WHICH HIS UNIQUE LITERARY STYLE IS FULLY REVEALED FOR THE FIRST TIME"[34]—CREATING A NEW KIND OF LITERATURE BY WHICH KAFKA MEASURED ALL HIS FUTURE WORK.[35]
A BODY OF WORK WHICH WOULD HAVE BEEN UTTERLY LOST TO US HAD NOT MAX BROD IGNORED HIS FRIEND, AND SOME TWENTY-SEVEN YEARS LATER, INSTEAD OF BURNING EVERYTHING, SMUGGLED EVERYTHING OF KAFKA'S OUT OF PRAGUE IN TWO BATTERED SUITCASES ONE DAY AHEAD OF THE NAZIS' INVASION.

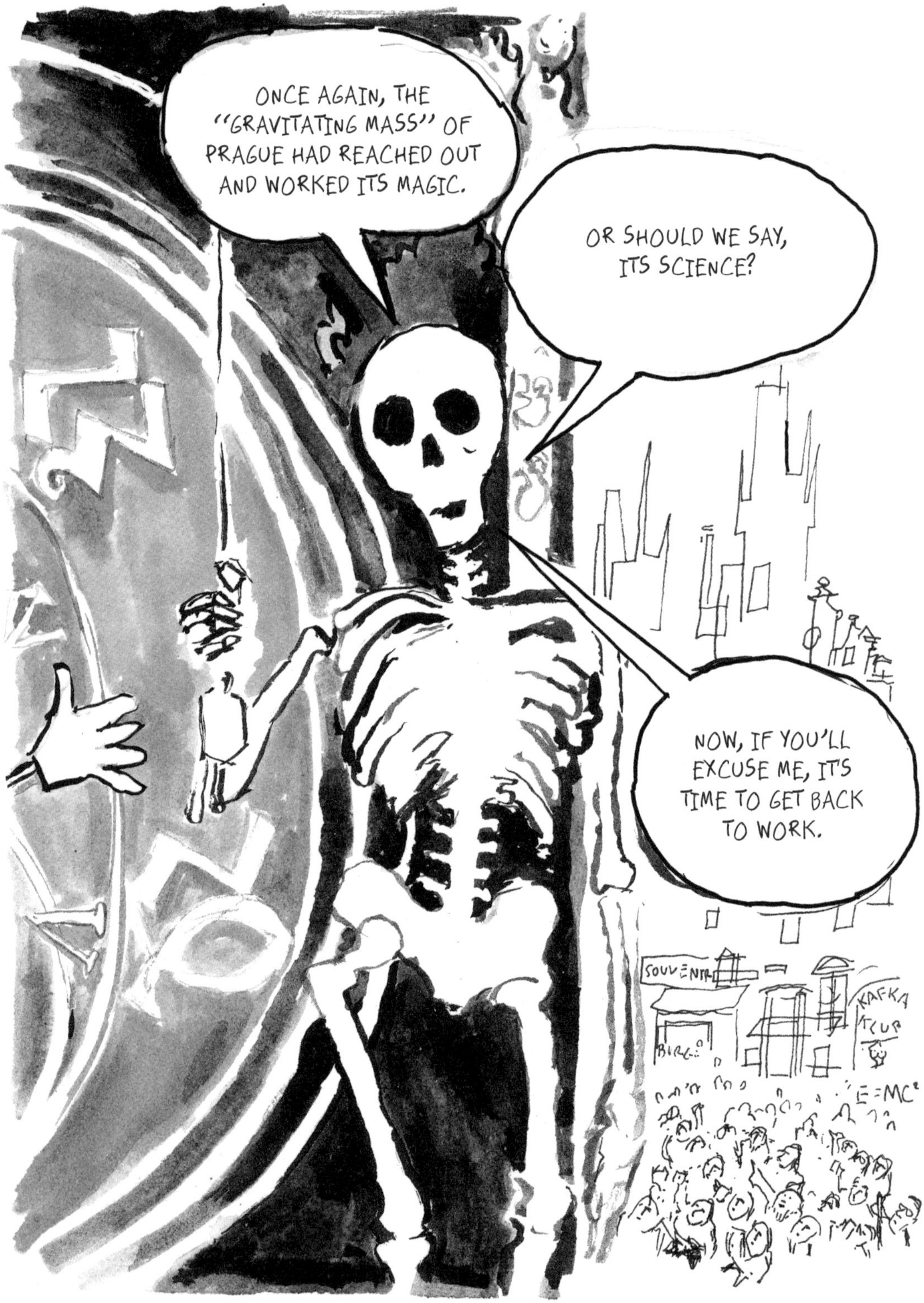
ONCE AGAIN, THE "GRAVITATING MASS" OF PRAGUE HAD REACHED OUT AND WORKED ITS MAGIC.
OR SHOULD WE SAY, ITS SCIENCE?
NOW, IF YOU'LL EXCUSE ME, IT'S TIME TO GET BACK TO WORK.
SOUVENIR
KAFKA
E=MC²

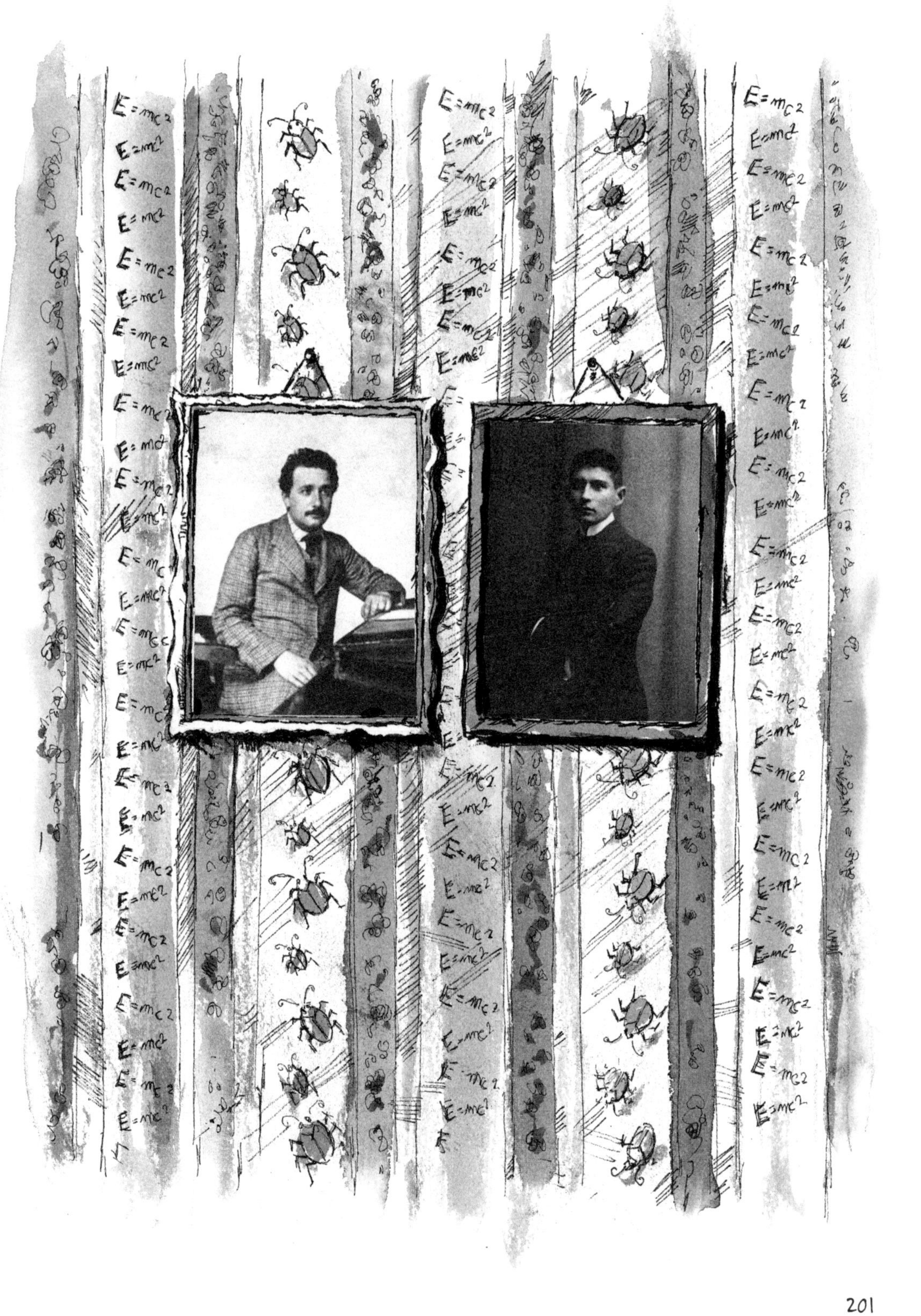
E=mc2

ACKNOWLEDGMENTS

To the scientists and scholars who offered their expertise and insight as I plunged into "Kafkaland"; Dan Hooper, Senior Member, KICP, Professor, Department of Astronomy and Astrophysics, University of Chicago, and Senior Scientist at Fermilab; Dr. Alex Gilman of Oxford and CERN; Peter Fenves, the Joan and Sarepta Harrison Professor of Literature at Northwestern University; Nicholas Sawicki, Department Chair, Art, Architecture, and Design at Lehigh University; in Prague, Tereza Matějčková, philosopher and member of the faculty at Karlova University; Kafkaist, translator, and photographer Věra Koubová; researcher and guide Marek Czerny, as well as philosopher and patient guide and translator, Marek Kettner; and from Italy, Marco Giovanelli, Associate Professor in the Department of Philosophy at Università di Torino. Also, a special thanks to my Hyde Park physics tutor who showed the patience of a secular saint, Bradford Boonstra. To Diana L. Kormos-Buchwald, General Editor and Director of the Einstein Papers Project and the Robert M. Abbey Professor of History at the California Institute of Technology; Dr. Stuart Gilman, who left no stone unturned until he put me in touch with Dr. Monica Kurzel-Runtscheiner and Dr. Susanne Hehenberger of the State Historical Museum in Vienna to find some evidence of Einstein's induction "costume."

For professional understanding, questioning, and inspiration: Kai Bird; Pat Byrnes; Sam Gross; my editor, Nancy Miller; my agent, Jennifer Lyons; and my consigliere and honorary cousin, Kenneth Gertz. To my dear friends Sylvie Seidman in Zurich and Kathy Roeder in NYC. And the Einstein Museum in Bern. To the Corporation of Yaddo for giving me the time and space to run the whole thing through the typewriter a couple more times, with a special shout-out to my fellow "fellow," River, for being my first audience. To the entire team at Bloomsbury, especially the welcome addition of art director Katya Mezhibovskaya. My mother, Joan, for asking tough questions since I was five; my kids, Noah, Milo, and Ruby, for pushing me; and, most of all, for my wife: first reader, last reader, and all the time visionary, Alex Sinclair.

FURTHER READING, LISTENING, VIEWING . . .

Even before I started my journey down the rabbit hole with Einstein, I was a fan who noticed that each book, article, podcast, and documentary tackling Einstein explained his genius in a completely different, completely un-understandable fashion. Regarding Kafka; as more than one critic has said, his work defies analysis. In response, all I could do was read, reread, keep turning over stones, move to Prague for a month, and trust that I'd find my own way in. (But I knew one thing: I knew that when it came time to "explain" relativity, I wasn't going to use any examples with flashlights on moving train cars.)

What follows is a heavily edited selection of the key sources that informed **Einstein in Kafkaland.**

THE EINSTEIN CANON
Kismet struck fast, as it usually does. Three years before embarking on this project I stumbled across Philipp Frank's **Einstein: His Life and Times** in a used bookstore in Wellfleet, MA. Frank succeeded Einstein in Prague, and his work,

besides being beautifully written, strikes me as the urtext on Einstein. Another primary resource by an Einstein colleague is **Subtle Is the Lord: The Science and the Life of Albert Einstein** by Abraham Pais. While it demanded a lot of rereading, with help from my tutor, it was particularly enlightening on the importance of the Prague interregnum. **Albert Einstein: Creator & Rebel** by Banesh Hoffmann with the collaboration of Einstein's secretary Helen Dukas widened the lens. (It seemed that almost anyone who shared a cup of coffee with Einstein was compelled to write a book about him.) Another fascinating overview from one of the players at the time is George Gamow's **Thirty Years that Shook Physics: The Story of Quantum Theory.**

As befits a figure as pivotal, fascinating, enigmatic, and universally recognized as Einstein, there are several "standard" biographies. While each reflects their times and author's individual perspective, triangulating them helped me discover what I call the "true zone." They include: **Einstein: The Life** and Times by Ronald W. Clark, Einstein: His Life and Universe by Walter Isaacson, The Private Lives of Albert Einstein by Roger Highfield and Paul Carter (rather controversial and written in the wake of the publication of his then unknown correspondence with Mileva), and the somewhat mistitled Einstein in Love: A Scientific Romance by Dennis Overbye, mistitled in that it is a tremendous overall biography and explanation of his thinking beyond just a love tale. Shelves of books attempt, in part or in whole, to explain this whole "theory of relativity" thing, from the charming illustrated Einstein for Beginners by Joseph Schwartz and Michael McGuinness to Harald Fritzsch's An Equation that Changed the World: Newton, Einstein, and the Theory of Relativity, as well as David Bodanis's E=mc2: A Biography of the World's Most Famous Equation and his lively take, Einstein's Greatest Mistake: The Life of a Flawed Genius, and Einstein: His Space and Times by Steven Gimbel. Then there's the considerably heavier Substance and Function & Einstein's Theory of Relativity by the philosopher Ernst Cassirer (phew) and John Rigden's Einstein 1905: The Standard of Greatness, which provides an excellent overview of Einstein's titanic achievements, from 1911 to his completion of general relativity in 1915. Notable in the "relativity made simple" category of books (most of which I honorably submit don't make it at all simple), is the useful Einstein's Universe by Nigel Calder, the somewhat more mind-bending (truth be told, they're all mind-

bending) Understanding Einstein's Theories of Relativity: Man's New Perspective on the Cosmos by Stan Gibilisco, and Hermann Bondi's Relativity and Common Sense: A New Approach to Einstein. See also the many excellent books on that big, huge, gnarly, frustratingly simple and demonically complicated concept we call "gravity" like Gravity: A Very Short Introduction by Timothy Clifton, Gravity: How the Weakest Force in the Universe Shaped Our Lives by Brian Clegg, Einstein's Masterwork: 1915 and the General Theory of Relativity by John Gribbin, The Perfect Theory: A Century of Geniuses and the Battle over General Relativity by Pedro Ferreira, The Ascent of Gravity: The Quest to Understand the Force That Explains Everything by Marcus Chown. And it would be a fool's errand to skip the master's own take, written in the thick of the moment and revised in 1952, Relativity: The Special and General Theory by Albert Einstein—who, not incidentally, at least in translation, is always an excellent writer, noted for his clear, direct, human style even in his academic papers and especially in his letters.

Zeroing in on the specific fifteen or so months of 1911's annus mirabilus, I owe a heavy debt to the wonderful Einstein in Bohemia by Michael D. Gordin, which came out just as I was diving into my own exploration of the events in question, as well as the copious and sage work done on Einstein's intellectual, creative, and scientific development spearheaded by Jürgen Renn in the collection he edited, The Genesis of General Relativity,

Sources and Interpretations, and all the wonderful and charmingly illustrated studies by Galina Weinstein, notably her *Genesis of General Relativity—Discovery of General Relativity*. Renn also provides some invaluable insights into Max Abraham in his paper "The Summit Almost Scaled: Max Abraham as a Pioneer of a Relativistic Theory of Gravitation."

I was especially smitten, and delighted, when I came across the work of Marco Giovanelli, especially his 2022 paper "Nothing but Coincidences: The Point-coincidence and Einstein's Struggle with the Meaning of Coordinates in Physics." As I mentioned above, I am grateful for his generous time as he walked me through his paper and answered my barrage of questions via a Zoom meeting from his office in Turin.

Another contemporary Italian scientist deserves my thanks: Carlo Rovelli, whose *Seven Brief Lessons on Physics* showed me how science writing can be mysterious and elegant at the same time.

I'm sorry to say that this laundry list is only a heavily edited selection which had to omit many videos, online lectures, and podcasts, films, etc. But before I leave the Einstein portion of this section I do want to make note of a few extraordinary resources I came across that really helped me get a picture of the thirty-two-year-old expat Einstein.

First, a pristine copy of volume one of Martin J. Klein's *Paul Ehrenfest: The Making of a Theoretical Physicist* practically jumped off the bookshelf and

smacked me in the face in a bunker-like used bookstore in Saratoga Springs (as far as I can tell there's no volume two!). This is a revealing, tragic, and triumphant tale of a truly fascinating genius.

And finally (for now), in a nod to the modern "library of Alexandria" that is the Internet, I was delighted to discover an amazing panel discussion, taped in glorious black and white video with some really cool midcentury modern set decorations, by Manchester University's Math Department in 1972: "Life and Work of Albert Einstein." But the real delight was hearing firsthand from two scientists, mathematician and physicist and Einstein's assistant in Berlin, Cornelius Lanczos, and, especially, the eminent astronomer Zdeněk Kopal, fresh from his groundbreaking work as part of the Apollo team that put people on the moon. Kopal, young, passionate, eloquent, makes no bones about the importance, indeed, the centrality of Einstein's "lost" year in Prague. I still get chills when I see him lean forward and pronounce: "His first full chair was in Prague, and he got it on the recommendation of Max Planck, Max Planck was his promotor . . . we believe that the time Einstein spent in Prague was the zenith of his creative activity," and then, with what I can only call a Cheshire Cat grin, Kopal adds, "Prague may have had nothing to do with it, it may have been coincidence, but, nevertheless his general relativity was formulated, and to a large extent, discovered there."

THE KAFKA CANON

Kafka must begin with Kafka's own words. But which of them? For me, the diaries. The language, the life, the nascent stories, it's like being present in the primordial goo, watching as the new life of modern prose comes to life. But which version of the diaries? Well, of course the classic Schocken one, first published in 1948. But, with a shock of the new, just as I was finishing writing this book, the copious, wonderful, magisterial Ross Benjamin translation came out, also published by Schocken, completely retranslated and reimagined. Dive into both. An essential companion is the incredible collection of Kafka's drawings edited by Andreas Kilcher and published by Yale Press in 2022, and benefiting from the fruits of the Kafkaesque battles over Max Brod's dispensation of the complete Kafka works (incidentally, thrillingly limned in Benjamin Balint's 2018 **Kafka's Last Trial**). Kafka saw himself as an artist (cartoonist?) almost as much as he saw himself as a writer, and these gnarly, honest explorations of Prague and his mind are essential companions to the prose.

Then, or perhaps, at the same time: Kafka's works. I am partial to the stories. The aphorisms. The 1952 Modern Library Collection, with an introduction by the **Partisan Review**'s editor Philip Rahv. This packs a punch and shines a light because it was probably the version read by most of the English-language writers who fell under Kafka's spell. You can read it forward, from his 1912 breakthrough "The

Judgment" on, or you can read it backward, from late, supposedly incinerated masterworks like "Josephine the Singer," or just fly in and out. And then, as you are getting your sea legs aboard the SS **Kafka**, dip in and out of the novels—and be aware there aren't many (notably **The Castle** and **The Trial**) and that they were mostly completed and patched together by Brod.

But even though it's fairly certain that Kafka's writing defies description in place of direct supraverbal encounter, that doesn't mean everyone from Borges to Updike haven't tried to describe and analyze the work. And this makes for some very entertaining and enlightening reading. By the way—I count along with these fascinating "analytic" encounters artistic encounters. It seems like the only way for writers (and artists and filmmakers and poets and . . .) to deal with the "word virus" of Kafka is to let works burst forth not unlike the creature that pops out of Sigourney Weaver in Ridley Scott's film **Alien**. My current favorite is Philip Roth's "**I Always Wanted You to Admire My Fasting**" or, Looking at **Kafka**, that concludes his 1975 collection **Reading Myself and Others** and, for me, predicts much of Roth's later blossoming. With that, here are some notable, and notably delicious, Kafka excursions I took along my way: **The City of K.: Franz Kafka & Prague**, a beautiful, oversize book that accompanied the 2002 exhibition at the Jewish Museum in NYC; Walter Benjamin's two essays on Kafka in **Illuminations**, the first of two collections of his work, this one edited and introduced by Hannah Arendt in 1968, and *He: Shorter Writings of Franz Kafka*, published by Riverrun in London in 2020, selected and with a fascinating, on-point introduction by Joshua Cohen. There is the sui generis memoir/impressionistic spiel by Kafka's contemporary, poet Johannes Urzidil, *There Goes Kafka*, published by Wayne State University Press in 1968, which the "gravitating mass of Kafkaland" practically thrust into my hands in Amaranth Books, a sleepy used book mecca around the corner from my house. *Franz Kafka and Prague: A Literary Guide* by Harald Salfellner is a rare example of a somewhat "touristic" tome that really delivers, beautifully designed and insightfully written; I picked up my copy of the 2019 Vitalis/Prague edition at the Jewish Museum in Prague. For two essential journeys into "Brod-land," there's the somewhat controversial biography (hagiography) Kafka's best friend published with Schocken in 1960 and Brod's totemic historical novel *Tycho Brahe's Path to God*, with an introduction by Stefan Zweig and an introduction that is a novel (and a good one) in itself by Peter Fenves (from Northwestern University Press in 2007), and which pulls many of the subrosa Kafka and Einstein and Prague strands together. From the shelves of Kafka biographies, I'll note two from opposite ends of the spectrum, the exhaustively researched 810 pages of *Franz Kafka: Representative Man: Prague, Germans, Jews, and the Crisis of Modernism* by Frederick R. Karl in 1991 and the, one can only say, Fellini-esque Kafka by the

Italian critic Pietro Citati (no subtitles for him!). And I'll round this snapshot of books with the you-can't-make-this-up *Franz Kafka: The Office Writings*, Kafka's actual insurance briefs, edited by Stanley Corngold, Jack Greenberg, and Benno Wagner and published by nothing less than Princeton University Press in 2008.

A startling and refreshing window into Kafka, the man and times, is *Kafka Goes to the Movies* by Hanns Zischler, who, full disclosure, translated my biography of Hannah Arendt into German but who, also, unbeknownst to me at the time, was a Kafka-freak of the first order and put together this thoroughly researched and wonderfully designed slim volume, published in America in 2002 thanks to the University of Chicago Press.

The final word on Kafka biographies must go to Reiner Stach, whose ten years of researching and writing led to the 1,557-page (excluding notes and index!) biography, published in three volumes, 2013's *Kafka: The Years of Insight* and *Kafka: The Decisive Years* and, finally, 2017's *Kafka: The Early Years*. What a testimony to the expansive power of Kafka's compact, almost minimal prose!

As for the academic papers, reviews, satires, responses: small forests worth of pulp have been devoted to them. I make special mention of just two, Professor Peter Pesic's "Before the Law: Einstein and Kafka" in *Literature and Theology 8* (1994): 174–192, and the truly Kafkaesque (Borgesian?) "Kafka and His Precursors" ("Kafka y sus precursors"), which originally appeared in the Argentine newspaper La

Nación on August 15, 1951, by Luis Borges.

And there are so many more. So many. All worthwhile.

And in the end, I would say just stop what you're doing and pick up a copy of **Alice's Adventures in Wonderland** by Lewis Carroll and read it right now! (I worked from a 192-page facsimile of the original 1865 first edition published by R. Clay, Son and Taylor, which came tucked into my four-LP Riverside Records box set (home of Thelonious Monk, no less), narrated by Cyril Ritchard with music by Alec Wilder.)

Much more than merely a fun children's story about a girl chasing a white rabbit, Carroll's book is a portal into a thought experiment that examines art, science, language, math, humor, horror, and joy. Beyond that, it's an uncanny precursor of so many of Einstein's innovations, as well as a kind of road map for the kind of mad courage it takes to make sense of an equally mad universe. I can only marvel at its "gravitating mass" for giving me the lens with which to envision this pivotal episode in Einstein's trajectory. And ours.

NOTES

1 According to Diana L. Kormos-Buchwald, PhD, Robert M. Abbey Professor of History at Caltech and the general editor and director of the Einstein Papers Project. vol. 7 and vol. 9–15, when I asked her on April 14, 2022, if the archive by chance had a copy of Einstein's diary from his time in Prague, she replied, "No, he did not have a diary, he was a modest academic at the time with limited social contacts." Sounds like a nobody to me.

2 Timo Hannu, "Did Franz Kafka Invent the Safety Helmet?" *Journal of Occupational Medicine* 69 (May 25, 2019): 188.

3 Freeman Dyson, (1923–2020).

4 Carlo Rovelli, born 1956.

5 Max Brod (1884–1968). German-speaking Prague author, composer, and journalist. Although prolific and talented, he is mostly remembered as Kafka's best friend.

6 Franz Kafka, "On the Examination of Forms by Trade Inspectors," June 22, 1911.

7 Legendary literary and philosophical society once frequented by Franz Brentano, father of phenomenology; it appears that due to demand, the lecture was moved to the auditorium at the physics institute.

8 Most influential physics journal, launched in Berlin in 1799, and still publishing robust, peer-reviewed, and Germanically uncompromising articles and letters to this day.

9 By no less a dignitary than Count Franz Thun und Hohenstein, the Governor of Bohemia acting as viceroy on behalf of His Imperial Apostolic Majesty Franz Josef.

10 Georg Pick, 1859–1942. Brilliant Austrian mathematician, murdered by the Nazis at Theresienstadt.

11 Max Planck, 1858–1947, dean of all physicists and head of the vaunted Prussian Academy, as well as co-editor of *Annalen der Physik*.

12 Ernst Mach, 1838–1916. Scientist and philosopher, the guru of "if you can't see it, or solid evidence of it in reality, it doesn't exist." Also renowned for what is known as the "Mach" number, the speed of sound, 761 mph. The term was bestowed on Mach long after his death by a Swiss physicist in honor of Mach's 1887 experiment in which he captured photographic evidence of an object moving faster than the speed of sound. Proof!

13 According to Reiner Stach, on p. 434 of volume one of his definitive three-volume biography of Kafka, *Kafka: The Early Years* (Princeton, Princeton University Press, 2017).

14 Einstein to Marić, February 2, 1902, https://einsteinpapers.press.princeton.edu/.

15 Egon Erwin Kisch, 1885–1948. Journalist, feuilletonist, Communist. Did a famous story on searching for the golem, the legendary homunculus made of clay, progenitor of everything from Frankenstein to Superman.

16 Rudolf Steiner, 1861–1925. Founder of theosophy, he very nearly took over the Madame Blavatsky
wing of occultism until a visionary Indian child was deemed to be even more visionary—Krishnamurti.

17 Creation of artist and cartoonist Josef Čapek (1887–1945), who coined the term *robot*.
Murdered by the Nazis in Bergen-Belsen for his cartoons.

18 Hugo Bergmann, 1883–1975. Philosopher, married to Fanta's daughter Else. Emigrated, worked with
philosopher Martin Buber to create a binational solution to the problems of the state of Israel.

19 Pioneering linguist, obsessed with Sanskrit and all it could possibly mean.

20 Ottoli Nagel, sister-in-law of Winternitz, pedantic pianist. According to Gordin (p. 95),
Einstein's most important musical connection at Fanta's.

21 Franz Kafka, 1883–1924.

22 Oskar Kraus, 1872–1942. Czech philosopher and lawyer, fierce critic of the theory of relativity.
Once got so incensed during a debate with Einstein that Einstein simply pulled out his violin
and started playing something mellow to soothe him.

23 Rudolf II, 1552–1612. Pretty much the maddest emperor ever, but in addition to hiring all the top alchemists
in Europe trying to turn lead into gold, he hired Tycho Brahe and Johannes Kepler, the two greatest astronomers
between Copernicus and Galileo.

24 His Serene Highness the Imperial Royal Viceroy Count Franz Thun-Hohenstein von Braun Councillor
of the Viceroyalty.

25 *The Bakersfield Californian*, September 2, 1911, p. 10.

26 From *Národní listy*: Edison visited Prague on September 15, 1911, and was received by the Mayor Karel Groš
at the Old Town Square. Edison signed himself in the official guest book and was mostly interested in the
mechanics of the astronomical clock.

27 Max Abraham, "Sulla teoria della gravitazione," *Atti Della Reale Accademia Dei Lincei, Rendiconti*, cl. sci. fis.,
mat. e nat. 20, no. 2 (1911): 678–682.

28 Highfield, p. 142.

29 In fact, the renowned physicist and student of Einstein John Archibald Wheeler, the person
who coined the term "black holes," called Ehrenfest Einstein's "closest in spirit of all his colleagues."

30 Einstein to S. Smolenski.

31 Impoverished med student who Einstein's parents hired to tutor their boy and wean him off
his kosher-obsessed, piss-off-my-agnostic-parents Judaism, Talmud changed his name to Talmey
when he moved to New York and wrote a biography about his years with Einstein.

32 "Wunsch, Indianer zu werden" is a prose sketch by Franz Kafka, written some time in 1911 or 1912.
Translated from the original German by Adam Tapper.

33 *Einstein's Shadow: A Black Hole, a Band of Astronomers, and the Quest to See the Unseeable* by Seth Fletcher
(New York City: Ecco, 2018), p. 11.

34 Gerhard Neumann, *Franz Kafka, "Das Urteil": Text, Materialien, Kommentar*
(München: Carl Hanser, 1981), pp. 189–190.

35 Reiner Stach, *Kafka: The Decisive Years*, trans. Shelley Frisch (Orlando: Harcourt, 2005), p. 114.,
cited in Yale.edu, Modernism Lab, Nathan Ernst.

WHAT THE HELL WAS GOING ON? A TIMELINE

[1911]

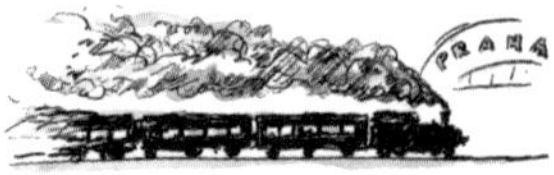
PRAHA

SATURDAY, APRIL 1 — The Einstein family pulls into Prague train station from Zurich.

WEDNESDAY, APRIL 5 — 100,000 people turn out in the rain in NYC to protest the March 25 loss of 146 workers in the Triangle Shirtwaist Factory fire.

SATURDAY, APRIL 8 — The premier of Winsor McCay's full-color animated film ***Little Nemo***.

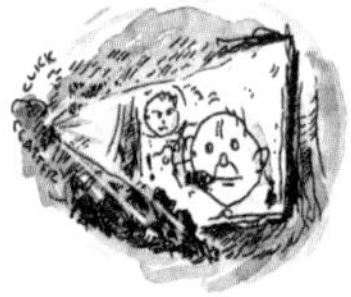

THURSDAY, APRIL 20 — Einstein teaches his first class at Prague's German university.

MONDAY, MAY 15 — The Standard Oil Trust is broken up by the U.S. Supreme Court.

WEDNESDAY, MAY 24 — Einstein gives his introductory lecture to the Prague public, Kafka in attendance.

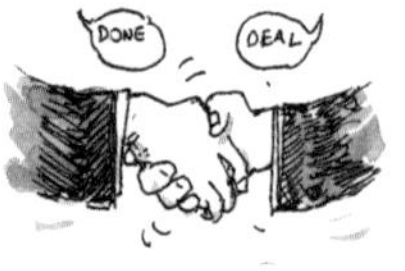
DONE
DEAL

THURSDAY, JUNE 1 — First ever group insurance policy written for a leather company in Passaic, NJ.

SUNDAY, JUNE 11 — Einstein's submits his first Prague paper to ***Annalen der Physik***, wherein he predicts that light can be bent by gravity.

TUESDAY, JUNE 13 — The ballet ***Pétrouchka*** premiers in Paris.

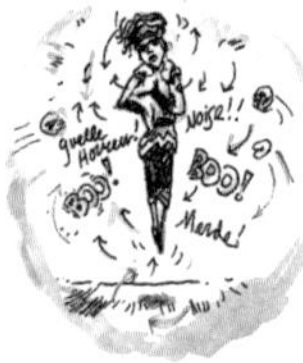

FRIDAY, JUNE 16 — IBM is incorporated.

SUNDAY, JULY 2 — First appearance of George Herriman's ***Krazy Kat*** cartoon in the funny pages.

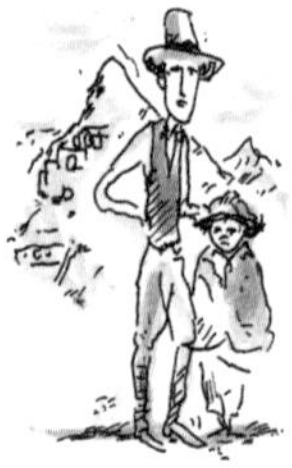

MONDAY, JULY 24 — Yale historian Hiram Bingham "discovers" the "lost city" of Machu Picchu after being led there by eight-year-old Melquiades Richarte.

MONDAY, AUGUST 21 — The ***Mona Lisa*** is stolen from the Louvre in broad daylight—well, the museum was closed, but still . . .

WEDNESDAY, AUGUST 23 — Einstein's official Imperial induction as a full professor at Karl-Ferdinand University.

FRIDAY, SEPTEMBER 1 — Einstein's first Prague paper, on bending light, is published in ***Annalen der Physik***.

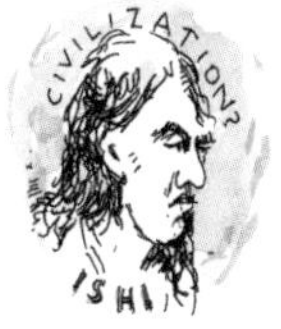
CIVILIZATION?
ISHI

SATURDAY, SEPTEMBER 2 — Last "wild" American "Indian" wanders out of the wilderness in Oroville, California.

FRIDAY, SEPTEMBER 15 — Thomas Edison visits Prague—all he's interested in is the workings of the 600-year-old astronomical clock.

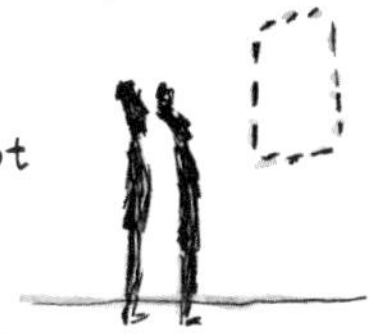

WEDNESDAY, SEPTEMBER 20 — Kafka and Brod visit the Louvre, check out the empty spot on the wall where the Mona Lisa once hung.

MONDAY, OCTOBER 30 — Einstein attends first Solvay conference of all the world's leading physicists; he calls it a "witches' sabbath."

WEDNESDAY, NOVEMBER 15 — Einstein writes to one of his best friends in Zurich, Heinrich Zangger, lamenting his role in Prague and impatient about hearing back from the Zurich university about a job ("they can kiss my . . ."), adding in a postscript, "I teach the foundations of poor, departed mechanics, which is so beautiful. What will its successor look like? I toil at that incessantly."

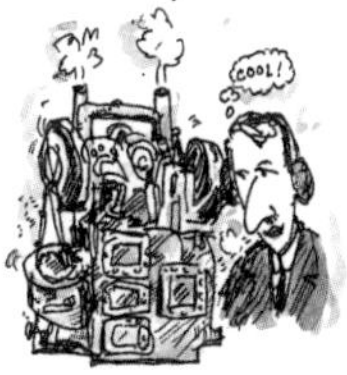
COOL!

MONDAY, DECEMBER 4 — American inventor Willis Carrier presents his designs for the first air conditioner.

THURSDAY, DECEMBER 14 — Roald Amundsen "discovers" the South Pole.

MONDAY, DECEMBER 18(ISH) — Max Abraham submits and quickly publishes his tidy four-page "takedown" of Einstein's "let's bend light" paper, his own "Sulla teoria della gravitazione."

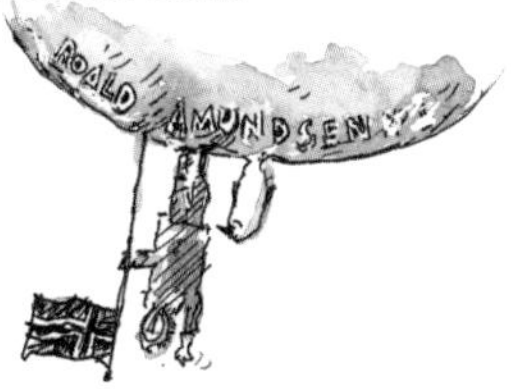
ROALD AMUNDSEN

SUNDAY, DECEMBER 31 — Marie Curie wins her second Nobel Prize.

[1912]

MONDAY, JANUARY 1 — Einstein invites Austrian physicist stranded in St. Petersburg Russia Paul Ehrhenfest to visit him in Prague on his round-Europe job search.

THURSDAY, JANUARY 4 — The shortest-ever distance between the Earth and the Moon in the 20th and 21st century (so far), a mere 356,375 km!

FRIDAY, JANUARY 9 — Einstein writes that he finds Abraham's thinking "atrocious." Abraham responds "I shall deal a death blow to relativity."

THURSDAY, JANUARY 18 — In the secret "Prague Conference," Vladimir Lenin forms the Bolshevik movement and outlines its goals; Joseph Stalin becomes the first editor of the newly created **Pravda** newspaper, official news organ of Bolshevism.

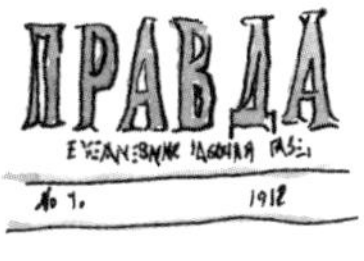
ПРАВДА
1912

WEDNESDAY, FEBRUARY 14 — Amateur British archeologist Charles Dawson writes to Sir Arthur Smith Woodward, Keeper of Geology at London's Natural History Museum, that he has discovered the "missing link" between ape and man, perpetrating one of the greatest hoaxes in scientific history. "'The Piltdown Man" wasn't definitively debunked until 1953.

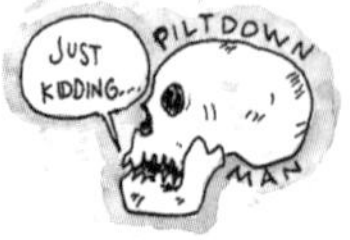
JUST KIDDING...
PILTDOWN MAN

FRIDAY, FEBRUARY 23 — Paul Ehrenfest arrives in Prague and is greeted at the train station by Albert and Mileva.

MONDAY, MARCH 11 — Einstein's "two letter" day.

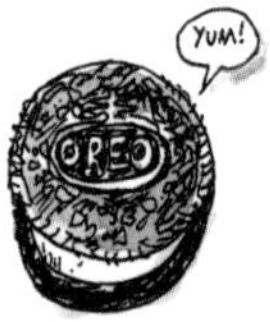
YUM!
OREO

THURSDAY, MARCH 14 — The OREO cookie debuts.

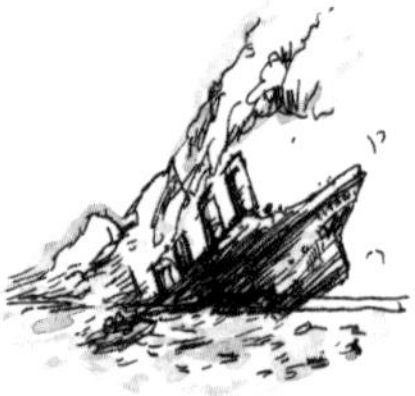

SUNDAY, APRIL 14 — The "unsinkable" RMS ***Titanic*** hits an iceberg and sinks on its maiden voyage.

FRIDAY, APRIL 26 — The first ever homer at Boston's newly opened Fenway Park is hit by Red Sox first baseman Hugh Bradley.

THURSDAY, MAY 23 — Einstein's second Prague paper, in which he modifies his first as a result of the findings of Abraham, Ehrenfest, and who knows what else, is published in **Annalen**.

WEDNESDAY, JUNE 26 — Mahler's Ninth (and final) symphony has its world premiere in Vienna.

MONDAY, JULY 1 — NYC's Woolworth Building tops out at 792 feet, becoming the tallest building in the world, a title it would hold until 1930.

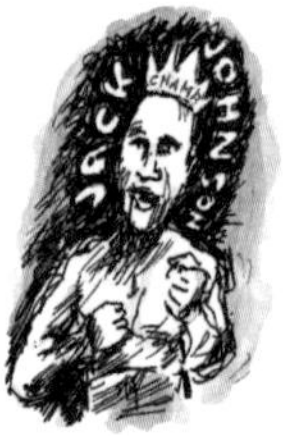
JACK JOHNSON
CHAMP

THURSDAY, JULY 4 — Jack Johnson beats "Fireman" Jim Flynn to become world heavyweight boxing champion.

THURSDAY, JULY 25 — Einstein and family depart Prague, taking the train back to Zurich.

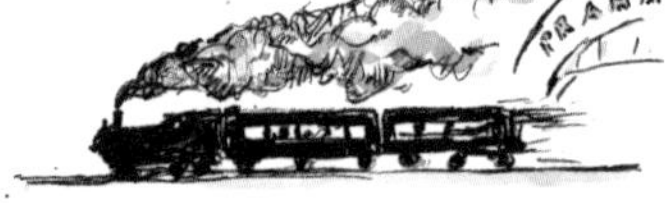
PRAHA